Lecture Notes in Mathematics

Edited by A. Dold and B. Eckmann

1044

Eckart Gekeler

Discretization Methods for Stable Initial Value Problems

Springer-Verlag
Berlin Heidelberg New York Tokyo 1984

Author

Eckart Gekeler
Mathematisches Institut A der Universität Stuttgart
Pfaffenwaldring 57, 7000 Stuttgart 80, Federal Republic of Germany

AMS Subject Classifications (1980): 65L07, 65L20, 65M05, 65M10, 65M15, 65M20

ISBN 3-540-12880-8 Springer-Verlag Berlin Heidelberg New York Tokyo
ISBN 0-387-12880-8 Springer-Verlag New York Heidelberg Berlin Tokyo

Printing and binding: Beltz Offsetdruck, Hemsbach/Bergstr.
2146/3140-543210

Introduction

In the past twenty years finite element analysis has reached a high standard and also great progress has been achieved in the development of numerical procedures for stiff, i.e., stable and ill-conditioned differential systems since the communication of Dahlquist 1963. Both fields together provide the ingredients for a method of lines solution for partial differential equations. In this method time and space discretization are carried out independently of each other, which has the advantage that often available subroutine packages can be applied in one or both directions. Finite element or finite difference methods are used for the discretization in space direction and finite difference methods as multistep or Runge-Kutta methods are used for the numerical solution of the resulting semi-discrete system in time, as a rule.

For example, if a hyperbolic initial boundary value problem with the differential equation

$$u_{tt} + u_t - u_{xx} = g(x,t)$$

and suitable initial and boundary conditions is discretized by a finite element method or more generally by a Galerkin procedure then the semi-discrete system of ordinary differential equations has the form

$$(*) \qquad My'' + Ny' + Ky = c(t)$$

where M, N, and K are real symmetric and positive definite matrices. M and N are well-conditioned but K is ill-conditioned in general, i.e, $\|K\|\|K^{-1}\| >> 0$. The finite element approximation of more general linear hyperbolic problems leads to similar systems. In engineering mechanics the basic partial differential equation is mostly not available because the body to be considered is too complex, instead the equations of motion are approximated by matrix structural analysis. The resulting 'equilibrium equations of dynamic finite element analysis' are then a large differential system for the displacements y being of the form (*) too.

If also a number of eigenvalues of the associated generalized eigenvalue problem is wanted then methods employing eigenvector expansions may be preferred in the solution of (*) (modal analysis). In the other case the numerical approximation leads immediately to the study of discretization methods for differential systems $y' = f(t,y)$ being stable in the sense that

$$(v - w)^T(f(t,v) - f(t,w)) \leq 0.$$

As the system (*) changes dimension and condition heavily with a refinement of the space discretization, methods are of particular interest here whose error propagation depends as little as possible on these data. Mathematically, the verification of this property or in other words of the uniformity of the error propagation with respect to a class of related problems can be established only by a-priori error estimations therefore particular emphasis is placed on them in this volume.

Three different classes of methods are at our disposal in the solution of initial value problems:

$$\begin{array}{l} \text{multistep methods} \\ \left.\begin{array}{l}\text{multiderivative methods} \\ \text{Runge-Kutta methods}\end{array}\right\} \text{multistage methods} \end{array} .$$

Multistep methods need a multiderivative or a Runge-Kutta method as start-procedure. Skilfully mixed procedures can have advantages over their components without inheriting the bad properties to the same degree. Multistep multiderivative methods are treated here from a rather general point of view. Runge-Kutta methods are intermediate-step methods actually, and they coincide with multiderivative methods for the linear differential system $y' = Ay$ with constant matrix A. Therefore these methods are both denoted as multistage methods and they have the same properties with respect to the test equation $y' = \lambda y$. Multistep Runge-Kutta methods are not fully investigated to date and besides there are many further combinations which are not treated here.

In the derivation of 'uniform' error bounds we are faced with two principal problems: the verification of 'uniform' stability in multistep methods and a suitable estimation of the discretization error in Runge-Kutta methods. The first difficulty is overcome by a uniform boundedness theorem being applied here in a version due to Crouzeix and Raviart. The second difficulty is overcome by the pioneering work in Crouzeix's thesis 1975. Furthermore, we should name Jeltsch and Nevanlinna whose contributions threw important light on the shape of the stability region.

In chapter I and II multistep multiderivative methods are considered for differential systems of first and second order. A-priori error bounds are derived for systems with constant coefficients and a survey is given on modern stability analysis. Over a long period the test equations $y' = \lambda y$ and $y'' = \lambda^2 y$ have been studied here only. Nevertheless, many important results have been produced in this way and a large variety of numerical schemes been widely used in the meantime.

In chapter III we leave the constant case and turn to linear systems with scalar time-dependence. Following a work of LeRoux [79a] and the dissertation of Hackmack [81] error bounds are established for linear multistep methods which show that a bad condition of the differential system does not affect seriously the error propagation here, too.

Chapter IV then deals with recent results on the error propagation in linear multistep methods and nonlinear differential systems of first order.

For a comparison with multistep multiderivative methods, Runge-Kutta methods are treated in chapter V but not to the same extent because we must refer here to a forthcoming book of Crouzeix and Raviart. These methods haven't lost anything of their fascination and today new variants are known in which the computational effort is reduced considerably.

In finite element analysis of elliptic boundary value problems a-priori error estimations play a large part and there are celebrated results among them. In Chapter VI some of these error bounds are combined with error bounds established in the first two chapters. Because of the special form of the latter results, error estimations are obtained for 'finite element multistep multiderivative' discretizations of parabolic and hyperbolic initial boundary value problems without further computations. The convergence order of the fully discrete schemes with respect to time and space discretization turns out to be the order of their components.

My thanks are due to Mrs. E. von Powitz for typing an early draft. I am also grateful to U. Hackmack for reading the manuscript and for some useful comments. Finally, I am indebted to S. Huber, K.-H. Hummel, and U. Ringler for computational examples and the plotting of the figures.

Table of Contents

I. Multistep Multiderivative Methods for Differential Systems of First Order

1.1. Consistence

Let us begin with an introduction of numerical approximation schemes for the general initial value problem

(1.1.1) $$y' = f(t,y),\ t > 0,\ y(0) = y_0.$$

We assume that $f\colon \mathbb{R}^+ \times \mathbb{R}^m \to \mathbb{R}^m$ is sufficiently smooth and denote by $\partial f/\partial y$ the Jacobi matrix of f. Let Δt be a small fixed time increment and recall that

(1.1.2) $$f^{(j)}(t,y) = (\partial f^{(j-1)}/\partial t)(t,y) + [(\partial f^{(j-1)}/\partial y)(t,y)]f(t,y), \quad j = 1,2,\ldots ,$$

where $f^{(0)} = f$. Then a general multistep multiderivative method - below briefly called multistep method - can be written as

(1.1.3) $$\sum_{i=0}^{k}\alpha_{0i}v_{n+i} + \sum_{i=0}^{k}\sum_{j=1}^{\ell}\Delta t^j\alpha_{ji}f_{n+i}^{(j-1)}(v_{n+i}) = 0, \qquad n = 0,1,\ldots .$$

By virtue of (1.1.2) the total derivatives $f^{(j)} = d^jf/dt^j$ are to be expressed here as far as possible by partial derivatives of f. v_n shall be an approximation to the solution $y_n = y(n\Delta t)$ of (1.1.1) at the time level $t = n\Delta t$ and we always assume in a multistep method that the initial values $v_0,\ldots,v_{k-1}$ are given in some way by an other method, e.g., by a Runge-Kutta method or a single-step multiderivative method.

A scheme (1.1.3) can be described in a twofold way by the polynomials

$$\rho_j(\zeta) = \sum_{i=0}^{k}\alpha_{ji}\zeta^i, \qquad j = 0,\ldots,\ell,$$

or by the polynomials

$$\sigma_i(\eta) = \sum_{j=0}^{\ell}\alpha_{ji}\eta^j, \qquad i = 0,\ldots,k.$$

We introduce the differential operator $\Theta = \partial/\partial t$ and the shift operator τ defined by $(\tau y)(t) = y(t+\Delta t)$. Furthermore, we use the notation

$$f^{(-1)}(t,v(t)) = v(t)$$

if thereby no confusion arises. Then we can write instead of (1.1.3)

(1.1.4) $$\sum_{j=0}^{\ell}\rho_j(T)\Delta t^j\Theta^j f_n^{(-1)}(v_n) \equiv \sum_{i=0}^{k}\sigma_i(\Delta t\Theta)T^i f_n^{(-1)}(v_n) = 0, \qquad n = 0,1,\ldots \quad .$$

In particular, we obtain for the differential equation $y' = \lambda y$

$$\pi(T,\Delta t\lambda)v_n \equiv \sum_{j=0}^{\ell}\rho_j(T)(\Delta t\lambda)^j v_n \equiv \sum_{i=0}^{k}\sigma_i(\Delta t\lambda)T^i v_n = 0$$

and $\pi(\zeta,\eta)$ is called the *characteristic polynomial* of the method (1.1.3).

Obviously not all polynomials $\rho_j(\zeta)$ as well as not all polynomials $\sigma_i(\eta)$ must have the same degree but we suppose that

(1.1.5) $$\alpha_{0k} \neq 0,\ \rho_\ell(\zeta) \not\equiv 0, \text{ and } \sigma_0(\eta) \not\equiv 0.$$

This condition guarantees that the method (1.1.3) is exactly a k-step method with derivatives up to order ℓ. That number is sometimes called the stage number and the method shortly a (k,ℓ)-method.

The *discretization error* or defect $d(\Delta t,y)$ of a method is obtained if the exact solution y is substituted into the approximation scheme:

(1.1.6) $$d(\Delta t,y)(t) = \sum_{j=0}^{\ell}\Delta t^j\rho_j(T)y^{(j)}(t).$$

<u>(1.1.7) Definition.</u> *The method* (1.1.3) *is consistent if there exists a positive integer* p *such that for all* $u \in C^{p_*}(\mathbb{R};\mathbb{R}^m)$, $p_* = \max\{p+1,\ell\}$,

$$\|d(\Delta t,u)(t)\| \leq \Gamma\Delta t^{p+1}$$

where Γ *does not depend on* Δt. *The maximum* p *is the order of the method.*

The following lemma generalizes a result due to Dahlquist [59, ch. 4]; see also Lambert [73 , § 3.3] and Jeltsch [76a]. It proves the important fact that consistent multistep methods allow an estimation of the discretization error which does not depend on the *data* of the differential equation.

<u>(1.1.8) Lemma.</u> *If the method* (1.1.3) *is consistent of order* p *then*

$$\|d(\Delta t,u)(t)\| \leq \Gamma\Delta t^p \int_t^{t+k\Delta t}\|u^{(p+1)}(\tau)\|d\tau + \sum_{j=p+1}^{\ell}\Delta t^j\rho_j(T)\|u^{(j)}(t)\| \quad \forall\, u \in C^{p_*}(\mathbb{R};\mathbb{R}^m)$$

where Γ *does not depend on* t, Δt, u, *and the dimension* m.

Proof. It suffices to prove the assertion for $p \geq \ell$. We write $0^0 = 1$ and substitute the Taylor expansions of u,

$$u^{(j)}(t+i\Delta t) = \sum_{\nu=0}^{p-j} \frac{(i\Delta t)^\nu}{\nu!} u^{(j+\nu)}(t) + \frac{1}{(p-j)!}\int_0^{i\Delta t} (i\Delta t-\tau)^{p-j} u^{(p+1)}(t+\tau)d\tau,$$

$j = 0,\dots,p$, into (1.1.6). By this way we obtain

$$\begin{aligned} d(\Delta t,u)(t) &= \sum_{j=0}^{\ell}\sum_{\nu=0}^{p-j}\Big(\sum_{i=0}^{k}\alpha_{ji}\frac{i^\nu}{\nu!}\Big)\Delta t^{j+\nu}u^{(j+\nu)}(t) \\ &+ \sum_{j=0}^{\ell}\sum_{i=0}^{k}\alpha_{ji}\frac{\Delta t^j}{(p-j)!}\int_0^{i\Delta t}(i\Delta t-\tau)^{p-j}u^{(p+1)}(t+\tau)d\tau. \end{aligned} \tag{1.1.9}$$

The assumption that the method is consistent of order p implies

$$\begin{aligned} &\sum_{j=0}^{\ell}\sum_{\nu=0}^{p-j}\Big(\sum_{i=0}^{k}\alpha_{ji}\frac{i^\nu}{\nu!}\Big)\Delta t^{j+\nu}u^{(j+\nu)}(t) \\ &= \sum_{\mu=0}^{p}\sum_{j=0}^{\min\{\mu,\ell\}}\Big(\sum_{i=0}^{k}\alpha_{ji}\frac{i^{\mu-j}}{(\mu-j)!}\Big)\Delta t^\mu u^{(\mu)}(t) = 0. \end{aligned} \tag{1.1.10}$$

Let

$$z_+ = \begin{cases} z \text{ if } z \geq 0, \\ 0 \text{ if } z < 0 \end{cases}$$

then we have by (1.1.9) and (1.1.10)

$$d(\Delta t,u)(t) = \int_0^{k\Delta t}\Big[\sum_{j=0}^{\ell}\sum_{i=0}^{k}\alpha_{ji}\frac{\Delta t^j}{(p-j)!}(i\Delta t-\tau)_+^{p-j}\Big]u^{(p+1)}(t+\tau)d\tau \tag{1.1.11}$$

or

$$\|d(\Delta t,u)(t)\| \leq \Big(\sum_{j=0}^{\ell}\sum_{i=0}^{k}|\alpha_{ji}|\frac{i^{p-j}}{(p-j)!}\Big)\Delta t^p\int_t^{t+k\Delta t}\|u^{(p+1)}(\tau)\|d\tau$$

which proves the assertion.

By (1.1.10) we immediately obtain the following result:

(1.1.12) Lemma. *The method* (1.1.3) *is consistent of order* p *iff*

$$\sum_{j=0}^{\min\{\mu,\ell\}}\sum_{i=0}^{k}\alpha_{ji}\frac{i^{\mu-j}}{(\mu-j)!} = 0, \qquad \mu = 0,1,\dots,p.$$

In particular, the method (1.1.3) is consistent iff

$$\rho_0(1) = \rho_0'(1) + \rho_1(1) = 0 \tag{1.1.13}$$

which are the well-known conditions for the consistence of *linear* multistep methods, i.e., methods with $\ell = 1$. A study of the trivial equation $y' = \kappa$ shows that we must always suppose that $\rho_1(1) \neq 0$ hence we may stipulate as customary that

(1.1.14) $$\rho_1(1) = -1.$$

Sometimes it is advantageous to use the consistence criteria in the subsequent form.

(1.1.15) Lemma. *The method* (1.1.3) *is consistent of order* p *iff the characteristic polynomial* $\pi(\zeta,\eta)$ *satisfies*

$$\pi(e^{\Delta t},\Delta t) = \chi_p \Delta t^{p+1} + \mathcal{O}(\Delta t^{p+2}),\ \Delta t \to 0,\ \chi_p \neq 0.$$

Proof. If the method (1.1.3) has order p then we obtain by a substitution of $u: t \to e^t$ into the discretization error $d(\Delta t,u)$

$$d(\Delta t,e^t) = \sum_{j=0}^{\ell} \Delta t^j \rho_j(T)e^t = \sum_{j=0}^{\ell} \Delta t^j \rho_j(e^{\Delta t})e^t$$

$$= \pi(e^{\Delta t},\Delta t)e^t = \chi_p \Delta t^{p+1} + \mathcal{O}(\Delta t^{p+2}),\ \Delta t \to 0,$$

where $\chi_p \neq 0$. For $t = 0$ this proves the necessity of the condition. On the other side, observe that

$$\pi(e^{\Delta t},\Delta t) = \sum_{j=0}^{\ell} \Delta t^j \sum_{i=0}^{k} \alpha_{ji} \sum_{\nu=0}^{\infty} \frac{(i\Delta t)^\nu}{\nu!} \doteq \sum_{\mu=0}^{\infty} \sum_{j=0}^{\min\{\mu,\ell\}} \Big(\sum_{i=0}^{k} \alpha_{ji} \frac{i^{\mu-j}}{(\mu-j)!}\Big) \Delta t^\mu .$$

If the condition of the lemma is fulfilled then we obtain from this the conditional equations of Lemma (1.1.12) which proves the sufficiency.

In other words, the lemma says that in a method of order p the value $\eta = 0$ is exactly a (p+1)-fold root of $\pi(e^\eta,\eta)$, or, $\zeta = 1$ is exactly a (p+1)-fold root of $\pi(\zeta,\log\zeta)$. The number $-\chi_p/\rho_1(1)$ is called the *error constant* of the method. In general, the comparison of the discretization error of two methods with the same order via (1.1.11), i.e., via their Peano kernels, is too involved. Hence under two otherwise comparable methods that with the smaller modulus of the error constant is usually preferred.

Recall now once more that

$$\pi(e^{\Delta t},\Delta t) = \sum_{j=0}^{\ell} \Delta t^j \rho_j(e^{\Delta t}) = \sum_{i=0}^{k} \sigma_i(\Delta t)e^{i\Delta t}$$

and assume that the polynomials $\rho_j(\zeta)$ have a common divisor,

$$\rho_j(\zeta) = \phi(\zeta)\tilde{\rho}_j(\zeta),\ \phi(1) \neq 0, \qquad j = 0,\dots,\ell.$$

If the method with the characteristic polynomial $\pi(\zeta,\eta)$ has order p and $\rho_1(1) \neq 0$ then we obtain by Lemma (1.1.15)

$$\tilde{\pi}(e^{\Delta t},\Delta t) = \sum_{j=0}^{\ell}\Delta t^j\tilde{\rho}_j(e^{\Delta t}) = \frac{\chi_p}{\phi(1)}\,\Delta t^{p+1} + \mathcal{O}(\Delta t^{p+2}), \qquad \Delta t \to 0.$$

Consequently, the method with the characteristic polynomial $\tilde{\pi}(\zeta,\eta)$ has the same order and the same error constant as the method with the characteristic polynomial $\pi(\zeta,\eta)$. If the polynomials $\sigma_i(\eta)$ have a common divisor,

$$\sigma_i(\eta) = \psi(\eta)\hat{\sigma}_i(\eta),\ \psi(0) \neq 0, \qquad i = 0,\dots,k,$$

then

$$\hat{\pi}(e^{\Delta t},\Delta t) = \sum_{j=0}^{\ell}\Delta t^j\hat{\rho}_j(e^{\Delta t}) = \sum_{i=0}^{k}\hat{\sigma}_i(\Delta t)e^{i\Delta t} = \frac{\chi_p}{\psi(0)}\,\Delta t^{p+1} + \mathcal{O}(\Delta t^{p+2}) \qquad \Delta t \to 0,$$

and the method with the characteristic polynomial $\hat{\pi}(\zeta,\eta)$ has again the same order as that with the polynomial $\pi(\zeta,\eta)$. Moreover, let

$$\psi(\eta) = \sum_{\nu=0}^{r}\gamma_\nu\eta^\nu,\ \gamma_0 = \psi(0) \neq 0, \qquad 1 \leq r < \ell,$$

then

$$\rho_0(\zeta) = \gamma_0\hat{\rho}_0(\zeta)$$

and

$$\rho_1(\zeta) = \gamma_1\hat{\rho}_0(\zeta) + \gamma_0\hat{\rho}_1(\zeta).$$

Because $\rho_0(1) = \gamma_0\hat{\rho}_0(1) = 0$ by (1.1.13) we obtain

$$\rho_1(1) = \psi(0)\hat{\rho}_1(1)$$

and the error constants of both methods coincide, too. Thus we assume henceforth without loss of generality that the polynomials $\rho_j(\zeta)$, $j = 0,\dots,\ell$, have no common roots and that the polynomials $\sigma_i(\eta)$, $i = 0,\dots,k$, have no common roots.

1.2. Uniform Stability

Instead of the general problem (1.1.1) we consider in the remaining part of this chapter the linear initial value problem

$$y' = Ay + c(t),\ t > 0,\ y(0) = y_0, \tag{1.2.1}$$

with constant leading matrix A. In Section 1.4 however this system arises from a transformation of a linear differential system of second order. For the problem (1.2.1) the multistep method (1.1.3) has the form

$$\sum_{i=0}^{k}\sigma_i(\Delta tA)T^i v_n = -\sum_{j=1}^{\ell}\sum_{\nu=0}^{j-1}\Delta t^j A^\nu \rho_j(T)c_n^{(j-1-\nu)}, \qquad n = 0,1,\ldots,$$

or

$$\sum_{i=0}^{k}\sigma_i(\Delta tA)T^i v_n = -\sum_{i=0}^{k}T^i\sum_{j=1}^{\ell}\sigma_{ij}(\Delta tA)\Delta t^j c_n^{(j-1)}, \qquad n = 0,1,\ldots, \tag{1.2.2}$$

where

$$\sigma_{ij}(\eta) = \sum_{m=j}^{\ell}\alpha_{mi}\eta^{m-j}. \tag{1.2.3}$$

In order to estimate the global error we have to write every method as a single-step procedure. For this we introduce the following notations,

$$V_n = (v_{n-k+1},\ldots,v_n)^T,$$

$$C_n = (0,\ldots,0,\sigma_k(\Delta tA)^{-1}\sum_{i=0}^{k}T^i\sum_{j=1}^{\ell}\sigma_{ij}(\Delta tA)\Delta t^j c_{n-k}^{(j-1)})^T, \tag{1.2.4}$$

and the Frobenius matrix associated with the characteristic polynomial $\pi(\zeta,\eta)$ of the method (1.1.3),

$$F_\pi(\eta) = \begin{bmatrix} 0 & 1 & & & \\ & \ddots & \ddots & & \\ & & \ddots & \ddots & \\ & & & 0 & 1 \\ -\sigma_0(\eta)/\sigma_k(\eta) & \cdots & \cdots & \cdots & -\sigma_{k-1}(\eta)/\sigma_k(\eta) \end{bmatrix}. \tag{1.2.5}$$

Let the Euclid norm and the associated matrix lub norm (spectral norm) be denoted by $|\cdot|$, let $|||h|||_n = \max_{0\le t\le n\Delta t}|h(t)|$, and let Sp(A) be the set of the eigenvalues, i.e., the spectrum of the matrix A. κ and Γ denote always positive constants which are not necessarily the same in two different contexts.

Instead of (1.2.2) we now can write

(1.2.6) $$V_n = F_\pi(\Delta tA)V_{n-1} - C_n, \qquad n = k,k+1,\dots .$$

For $C_n = 0$, $n = k,k+1,\dots$, the increasing or decreasing of the sequence $\{\|V_n\|\}_{n=k}^{\infty}$ depends on the eigenvalues of $F_\pi(\Delta tA)$ which are the roots of the polynomial $\pi(\zeta,\eta)$. For simplicity we *define* the region of absolute stability - below briefly called stability region - in the following way:

<u>(1.2.7) Definition.</u> *Let* $\overline{\mathbb{C}} = \mathbb{C} \cup \{\infty\}$ *and let* $\pi(\zeta,\infty) = \rho_\ell(\zeta)$. *The stability region* S *of a multistep method* (1.1.3) *consists of the* $\eta \in \overline{\mathbb{C}}$ *with the following properties:*
(i) $\sigma_k(\eta) \neq 0$, $\eta \in S \cap \mathbb{C}$,
(ii) all roots $\zeta_i(\eta)$ *of* $\pi(\zeta,\eta)$ *satisfy* $|\zeta_i(\eta)| \leq 1$,
(iii) all roots $\zeta_i(\eta)$ *of* $\pi(\zeta,\eta)$ *with* $|\zeta_i(\eta)| = 1$ *- i.e. the unimodular roots - are simple roots of* $\pi(\zeta,\eta)$.

By (i) we exclude from S isolated points where $\pi(\zeta,\eta)$ degenerates. E.g., by Stoer and Bulirsch [80 , Theorem 6.3.4] the Frobenius matrix $F_\pi(\eta)$ is undiagonable iff its characteristic polynomial $\pi(\zeta,\eta)$ has a multiple root. Hence the definition says that η lies in the stability region S iff the sequence $\{\|F_\pi(\eta)^n\|\}_{n=1}^{\infty}$ remains bounded. The proof of the *uniform* boundedness with respect to $\eta \in S$ is the crucial step of the otherwise simple error estimation.

We postpone a study of some properties of S to the next section. But it should be remarked here that the time step Δt is usually a small number therefore only methods are of practical interest in this context whose stability region contains the negative real line including zero in a neighborhood of zero. The property $0 \in S$ is the classical D-stability introduced by Dahlquist [59] for linear multistep methods. Together with consistence it implies the convergence of multistep methods in well-conditioned stable or unstable differential systems if the considered time interval is fixed. See e.g. Stoer and Bulirsch [80 , Theorem 7.2.10.3] and Brown [74].

Let now y be the exact solution of the slightly modified initial value problem (1.2.1),

(1.2.8) $$y' = Ay + c(t) + h(t),\ t > 0,\ y(0) = y_0.$$

The function h plays here only the role of a perturbation but it is needed in the last chapter for the error estimation of full-discrete approximation schemes to initial *boundary* value problems. If y is substituted into the discretization error $d(\Delta t,u)$ then we obtain after a repeated application of the differential equation (1.2.8)

(1.2.9) $$\textstyle\sum_{i=0}^{k}\sigma_i(\Delta tA)T^i y_n = -\sum_{i=0}^{k}\sum_{j=1}^{\ell}\sigma_{ij}(\Delta tA)\Delta t^j T^i(c_n^{(j-1)} + h_n^{(j-1)}) + d(\Delta t,y)_n .$$

On the other side, let v_n, $n = k,k+1,\dots$, be defined by (1.2.2) as above. Subtracting

(1.2.2) from (1.2.9) and writing the result in single-step form we obtain

(1.2.10) $$E_n = F_\pi(\Delta tA)E_{n-1} + D(\Delta t,y)_n - H_n, \qquad n = k,k+1,\ldots,$$

where

(1.2.11) $$D(\Delta t,y)_n = (0,\ldots,0,\sigma_k(\Delta tA)^{-1}d(\Delta t,y)_{n-k})^T.$$

The following theorem gives an error estimation which holds uniformly for all normal matrices A satisfying the spectral condition that $Sp(\Delta tA)$ is contained in a fixed closed subset R of the stability region S.

(1.2.12) Theorem. *(i) Let the (m,m)-matrix* A *in* (1.2.8) *be diagonable,* $A = X\Lambda X^{-1}$*; let the solution* y *of* (1.2.8) *be (p+1)-times continuously differentiable.*
(ii) Let the method (1.2.2) *be consistent of order* $p \geq \ell \geq 1$ *with the stability region* S.
(iii) Let $Sp(\Delta tA) \subset R \subseteq S$ *where* R *is closed in* $\overline{\mathbb{C}}$.
Then for $n = k,k+1,\ldots,$

$$|X^{-1}(y_n - v_n)| \leq \kappa_R|X^{-1}|\Big[|Y_{k-1} - V_{k-1}| + \Delta t^p\int_0^{n\Delta t}|y^{(p+1)}(\tau)|d\tau + n\Delta t \max_{0\leq i\leq \ell-1}|||h^{(i)}|||_n\Big].$$

Proof. $\Lambda = (\lambda_1,\ldots,\lambda_m)$ is the diagonal matrix of the eigenvalues of A. (1.2.10) yields

(1.2.13) $$|X^{-1}E_n| \leq |F_\pi(\Delta t\Lambda)^{n-k+1}||X^{-1}E_{k-1}| + \textstyle\sum_{\nu=k}^n|F_\pi(\Delta t\Lambda)^{n-\nu}||X^{-1}(D(\Delta t,y)_\nu - H_\nu)|,$$

$n = k,k+1,\ldots,$ and we have

$$|F_\pi(\Delta t\Lambda)^n| = \max_{1\leq\mu\leq m}|F_\pi(\Delta t\lambda_\mu)^n|.$$

The Uniform Boundedness Theorem which is proved in the Appendix yields immediately

$$\sup_{\eta\in R}\sup_{n\in\mathbb{N}}|F_\pi(\eta)^n| \leq \kappa_R.$$

Poles of $\sigma_k(\eta)^{-1}$ cannot lie in $\overline{S}$ hence

$$|\sigma_k(\Delta t\Lambda)^{-1}| \leq \sup_{\eta\in S}|\sigma_k(\eta)^{-1}| \leq \kappa.$$

Therefore we obtain from (1.2.13) by Lemma (1.1.8)

(1.2.14) $$|X^{-1}E_n| \leq \kappa_R\Big[|X^{-1}E_{k-1}| + |X^{-1}|\Delta t^p\int_0^{n\Delta t}|y^{(p+1)}(\tau)|d\tau + \textstyle\sum_{\nu=k}^n|X^{-1}H_\nu|\Big].$$

H_n has the same form as C_n in (1.2.4) with c_n replaced by h_n so we have

$$\sum_{\nu=k}^{n}|X^{-1}H_\nu| = \sum_{\nu=k}^{n}|X^{-1}\sigma_k(\Delta tA)^{-1}\sum_{i=0}^{k}T^i\sum_{j=1}^{\ell}\sigma_{ij}(\Delta tA)\Delta t^j h_{\nu-k}^{(j-1)}|$$

$$\le \kappa n\Delta t\sum_{i=0}^{k}\sum_{j=1}^{\ell}|\sigma_k(\Delta t\Lambda)^{-1}\sigma_{ij}(\Delta t\Lambda)\Delta t^{j-1}||X^{-1}|\,|||h^{(j-1)}|||_n.$$

If $R \subset S$ is bounded then $\Delta t\Lambda$ is bounded by assumption and we obtain

$$(1.2.15) \qquad |\sigma_k(\Delta t\Lambda)^{-1}\sigma_{ij}(\Delta t\Lambda)| \le \kappa_R, \quad i = 0,\ldots,k,\ j = 1,\ldots,\ell.$$

If $R \subset S$ is unbounded then $\sigma_k(\eta)$ is necessarily a polynomial of exact degree ℓ by Lemma (A.1.3) whereas the polynomials $\sigma_{ij}(\eta)$ have degree less than ℓ by (1.2.3). Accordingly, (1.2.15) is valid in this case, too. Therefore we have

$$\sum_{\nu=k}^{n}|X^{-1}H_\nu| \le \kappa_R n\Delta t|X^{-1}|\max_{0\le i\le \ell-1}|||h^{(i)}|||_n$$

and a substitution of this bound into (1.2.14) proves the theorem.

The stability region S has not necessarily a non-empty interior and it is not necessarily a closed set in $\overline{\mathbb{C}}$. For instance the *Milne-Simpson* method is the linear 2-step method of order 4 with the characteristic polynomial

$$(1.2.16) \qquad \pi(\zeta,\eta) = \zeta^2 - 1 - \frac{\eta}{3}(\zeta^2 + 4\zeta + 1).$$

This method has the stability region S with

$$\overline{S} = \{\eta = i\tilde{\eta},\ -\sqrt{3} \le \tilde{\eta} \le \sqrt{3}\}$$

but the polynomial (1.2.16) has double unimodular roots for $\eta = \pm i\sqrt{3}$ hence we obtain $S = \overline{S} \setminus \{\pm i\sqrt{3}\}$. The conjecture suggested by this example reveals to be generally true: By Theorem (A.1.4) the roots $\zeta_i(\eta)$ of $\pi(\zeta,\eta)$ are continuous functions in simply connected open domains with exception of the points where $\sigma_k(\eta) = 0$. Therefore $\overline{S}$ consists of the $\eta \in \overline{\mathbb{C}}$ with $\mathrm{spr}(F_\pi(\eta)) \le 1$ up to isolated points. For $\eta \in \overline{S}\setminus S$ we thus must have at least one unimodular root of multiplicity greater than one.

As is proved in the next section, the interior $\mathring{S}$ of S consists of the $\eta \in \overline{\mathbb{C}}$ with $\mathrm{spr}(F_\pi(\eta)) < 1$ where $\mathrm{spr}(F_\pi(\eta))$ denotes the spectral radius of the Frobenius matrix. We now introduce the regions where $\mathrm{spr}(F_\pi(\eta))$ is uniformly less than one in a slight modification of Stetter [73 , Definition 2.3.15 and § 4.6].

(1.2.17) Definition. *The region of μ-exponential stability* $S_\mu \subset S$ *of a method* (1.1.3) *consists of the* $\eta \in \overline{\mathbb{C}}$ *with the property that all roots* $\zeta_i(\eta)$ *of* $\pi(\zeta,\eta)$ *satisfy* $|\zeta_i(\eta)| \le 1 - 2\mu$, $0 < \mu < 1/2$.

By a continuity argument, too, S_μ is closed in $\bar{\mathbb{C}}$ for $\mu > 0$. Using this concept and the Uniform Boundedness Theorem we easily obtain exponential stability. The exponential decreasing factor $\mu/\Delta t$ is however somewhat vague at present.

<u>(1.2.18) Theorem</u>. *Let the assumptions of Theorem* (1.2.12) *be fulfilled and let* $Sp(\Delta tA) \subset S_\mu$ *for a fixed* $\mu > 0$. *Then for* $n = k, k+1, \dots,$

$$|X^{-1}(y_n - v_n)|$$

$$\leq \kappa_\mu |X^{-1}| \Big[e^{-(\mu/\Delta t)(n-k)\Delta t} |Y_{k-1} - V_{k-1}| + \Delta t^p \int_0^{n\Delta t} e^{-(\mu/\Delta t)((n-k)\Delta t - \tau)} |y^{(p+1)}(\tau)| d\tau$$

$$+ \Delta t \sum_{\nu=k}^n e^{-(\mu/\Delta t)(n-\nu)\Delta t} \max_{(\nu-k)\Delta t \leq \tau \leq \nu\Delta t} \max_{0 \leq i \leq \ell-1} |h^{(i)}(\tau)| \Big].$$

Proof. The result follows in the same way as in Theorem (1.2.12) if we substitute the bound

$$(1.2.19) \qquad |F_\pi(\Delta t\Lambda)^n| \leq \kappa_\mu (1 - \mu)^n \leq \kappa_\mu e^{-\mu n}$$

into (1.2.13). Hence we have to prove (1.2.19). But if $Sp(\Delta tA) \subset S_\mu$ then

$$(1.2.20) \qquad \sup_{n \in \mathbb{N}} |[(1 - \mu)^{-1} F_\pi(\Delta t\Lambda)]^n| \leq \sup_{n \in \mathbb{N}} \sup_{\eta \in S_\mu} |[(1 - \mu)^{-1} F_\pi(\eta)]^n|$$

and we have

$$spr([(1 - \mu)^{-1} F_\pi(\eta)]) \leq (1 - 2\mu)/(1 - \mu) < 1 - \mu < 1$$

for all $\eta \in S_\mu$. Consequently, the Uniform Boundedness Theorem yields

$$(1.2.21) \qquad \sup_{n \in \mathbb{N}} \sup_{\eta \in S_\mu} |[(1 - \mu)^{-1} F_\pi(\eta)]^n| \leq \kappa_\mu .$$

A substitution of this bound into (1.2.20) proves (1.2.19).

If the stability region S is closed in $\bar{\mathbb{C}}$ then we can obviously choose R = S in Theorem (1.2.12) and the constant κ_R becomes a fixed constant depending only on the data of the multistep method. This remains also true if for instance $Sp(\Delta tA) \subset \mathbb{R}$ and $\mathbb{R} \cap S$ is closed in $\bar{\mathbb{C}}$. A similar remark holds for the constant κ_μ in Theorem (1.2.18) and for all subsequent error estimations which contain a spectral condition of the form (1.2.12)(iii).

1.3. General Properties of the Region S of Absolute Stability

The closed hull $\bar{S}$ of the stability region S contains no singularity points of $\pi(\zeta,\eta) = \sum_{i=0}^{k}\sigma_i(\eta)\zeta^i$, i.e., no points η^* with $\sigma_k(\eta^*) = 0$. Let $\eta^* \in \bar{S}$ and let ξ^* be a root of $\pi(\zeta,\eta^*)$ of multiplicity r. In a neighborhood of ξ^* every root $\zeta_i(\eta)$ of $\pi(\zeta,\eta)$ then can be written as a *Puiseux series*; cf. Theorem (A.1.4). In a somewhat simplified form we write instead of (A.1.5)

$$(1.3.1) \quad \zeta_i(\eta) = \xi^* + \chi(\eta - \eta^*)^{p/q} + \mathcal{O}((\eta - \eta^*)^s), \; \chi \neq 0, \; \eta \to \eta^*,$$

with $p, q \in \mathbb{N}$ having no common factor and $s > p/q$. Naturally, (1.3.1) represents q different roots for $\eta \neq \eta^*$ in dependence of the chosen branch of $(\eta - \eta^*)^{1/q}$. The general connection between p, q, r, and χ is ruled by the *Puiseux diagram* in Appendix A.1. If $|\xi^*| = 1$ and $\eta^* \in \bar{S}$ is not an isolated stability point then Corollary (A.1.21) says that $p \in \{1, \ldots, \ell\}$ and $q \in \{1, 2\}$. In general, q can be less that r and there can be several different growth parameters χ to the r-fold root ξ^*. By Corollary (A.1.24) however this case does not occur if the method (1.1.3) is *linear*. Then we have p = 1 and q = r, i.e.,

$$(1.3.2) \quad \zeta_i(\eta) = \xi^* + \chi(\eta - \eta^*)^{1/r} + \mathcal{O}((\eta - \eta^*)^s), \; \chi \neq 0, \; s > 1/r, \; \eta \to \eta^*.$$

The expansion (1.3.1) shows that the roots of $\pi(\zeta,\eta)$ depend continuously on η. If $|\xi^*| = 1$ then there exist in every neighborhood values of η such that $|\zeta_i(\eta)| > 1$ for some index i. These both statements lead to the following result:

(1.3.3) Lemma. $\eta \in \mathring{S}$ *iff* $\mathrm{spr}(F_\pi(\eta)) < 1$.

Let now $\mathcal{L} = \{\eta \in \mathbb{C}, \; |\zeta_i(\eta)| = 1 \text{ for some } i\}$ be the *root locus curve* then the lemma says that the boundary ∂S of S is a subset of $\mathcal{L}$. $\mathcal{L}$ consists of a finite number of analytic curves which intersect themselves in finitely many points. ∂S is *not analytic* in the points where $\pi(\zeta,\eta)$ has unimodular roots of multiplicity *greater than one*, i.e., in $\bar{S} \setminus S$. The complement of $\mathcal{L}$ in $\mathbb{C}$ consists of a finite number of connected components $\mathcal{C}_1, \ldots, \mathcal{C}_r$. If $\eta_\rho \in \mathcal{C}_\rho$ then $\mathcal{C}_\rho \subset \mathring{S}$ iff $\eta_\rho \in \mathring{S}$ and if $\eta_\rho \notin \mathring{S}$ then $\mathcal{C}_\rho \cap \mathring{S} = \emptyset$. Therefore, if $\mathcal{L}$ is known, one can test whether a component $\mathcal{C}_\rho$ belongs to $\mathring{S}$ by testing just one sample point of $\mathcal{C}_\rho$. See e.g. Jeltsch [76a, § 4.2.3, 78c]. $\mathcal{L}$ is *symmetric to the real line* and can be written geometrically as

$$\mathcal{L} = \{\eta \in \mathbb{C}, \; \pi(e^{i\phi},\eta) = 0, \; 0 \leq \phi < 2\pi\}.$$

Using a root solving algorithm we obtain by this way a method for the computation of

$\mathcal{B}$ where ϕ is naturally to be discretized.

A method with $0 \in S$ is sometimes also called *zero-stable*. A consistent and zero-stable method is *convergent* in the classical sense as already mentioned above. For these methods we now study the behavior of $\mathrm{spr}(F_\pi(\eta))$ in a neighborhood of $\eta = 0$ more exactly.

(1.3.4) Definition. *Let the method defined by* $\pi(\zeta,\eta)$ *be convergent, i.e., consistent and zero-stable. Then*
(i) the root $\zeta_1(\eta)$ *with* $\lim_{\eta\to 0}\zeta_1(\eta) = 1$ *is the principal root of* $\pi(\zeta,\eta)$,
(ii) the roots $\zeta_i(\eta)$ *with* $\lim_{\eta\to 0}|\zeta_i(\eta)| = 1$ *are the essential roots of* $\pi(\zeta,\eta)$ *numbered by* $\zeta_1(\eta),\ldots,\zeta_{k_*}(\eta)$, $1 \le k_* \le k$.
(iii)

$$\chi_i^* = \zeta_i'(0)/\zeta_i(0) = -\rho_1(\zeta_i(0))/[\rho_0'(\zeta_i(0))\zeta_i(0)], \qquad i = 1,\ldots,k_*,$$

are the growth parameters of the essential roots.

(i) and (ii) are well-defined only in a neighborhood of zero. Note that $\chi_1^* = 1$ and $0 \in \partial S$ holds for a convergent method. χ_i^*, $i = 2,\ldots,k_*$, are nonzero at least if the method is linear by the irreducibility of $\pi(\zeta,\eta)$. The following result shows that in multistep methods the principal root $\zeta_1(\eta)$ retains the approximation properties of single-step methods.

(1.3.5) Lemma. *Let* $0 \in S$ *then the method defined by* $\pi(\zeta,\eta)$ *is consistent of order* p *iff*

$$\zeta_1(\eta) = e^\eta - \chi_p\eta^{p+1} + \mathcal{O}(\eta^{p+2}),\ \chi_p \ne 0, \qquad \eta \to 0, \tag{1.3.6}$$

where χ_p *is the error constant.*

Proof. In a sufficiently small neighborhood of $\eta = 0$ we define

$$\pi^*(\zeta,\eta) = \pi(\zeta,\eta)/(\zeta - \zeta_1(\eta)) \tag{1.3.7}$$

and then obtain by (1.1.13)

$$\pi^*(1,0) = \rho_0'(1) = -\rho_1(1) \ne 0. \tag{1.3.8}$$

If the method is consistent of order p then Lemma (1.1.15) yields by a substitution of e^η for ζ in $\pi(\zeta,\eta)$

$$\pi(e^\eta,\eta) = (e^\eta - \zeta_1(\eta))\pi^*(e^\eta,\eta) = \chi_p\eta^{p+1} + \mathcal{O}(\eta^{p+2}), \qquad \eta \to 0. \tag{1.3.9}$$

This implies (1.3.6) because of (1.3.8). On the other side, if (1.3.6) holds then a substitution into the first equation of (1.3.9) yields

$$\pi(e^\eta,\eta) = \pi^*(e^\eta,\eta)(\chi_p\eta^{p+1} + \mathcal{O}(\eta^{p+2})), \qquad \eta \to 0.$$

Therefore the method has order p by Lemma (1.1.15) and χ_p is the error constant by (1.3.8).

The essential roots satisfy $\zeta_i(\eta) = \zeta_i(0)(1 + \chi_i^*\eta + \mathcal{O}(\eta^2))$ or

(1.3.10) $$|\zeta_i(\eta)| = 1 + \mathrm{Re}(\chi_i^*)\mathrm{Re}(\eta) - \mathrm{Im}(\chi_i^*)\mathrm{Im}(\eta) + \mathcal{O}(|\eta|^2), \ \eta \to 0, \ i = 1,\dots,k_*.$$

This equation has some interesting consequences because there exists a $\tilde{\delta} > 0$ such that

(1.3.11) $$\mathrm{spr}(F_\pi(\eta)) = \max_{1\le i\le k_*}|\zeta_i(\eta)|, \qquad |\eta| \le \tilde{\delta}:$$

(1.3.12) Lemma. *Let* $\mathcal{A}(\alpha,\rho) = \{\eta \in \overline{\mathbb{C}}, \ |\arg\eta - \pi| \le \alpha, \ |\eta| \le \rho\}$ *be an angular domain. Let the method defined by* $\pi(\zeta,\eta)$ *be convergent and let the growth parameters* χ_i^*, $i = 1,\dots,k_*$, *be nonzero. Then*

(i) $\mathcal{A}(\alpha,\rho) \subset S$ *for some* $\alpha > 0$ *and* $\rho > 0$ *iff* $\mathrm{Re}(\chi_i^*) > 0$, $i = 1,\dots,k_*$,

(ii) $\mathcal{A}(\alpha,\rho) \subset \overline{\mathbb{C}} \setminus S$ *for some* $\alpha > 0$ *and* $\rho > 0$ *iff* $\mathrm{Re}(\chi_i^*) < 0$ *for some* i, $1 \le i \le k_*$,

(iii) there exists a $\rho > 0$ *such that* $\overline{\mathbb{C}} \setminus S$ *contains the ball* $\{\eta \in \mathbb{C}, \ |\eta - \rho| < \rho\}$.

Proof. (Cf. also Crouzeix and Raviart [80].) The first two assertions follow from (1.3.10) and (1.3.11). For the third assertion observe that

$$|\zeta_1(\eta)| = 1 + \mathrm{Re}(\eta) + \mathcal{O}(|\eta|^2), \qquad \eta \to 0.$$

Substituting $\eta = \rho + \rho(\cos\phi + i\sin\phi)$, $0 \le \phi < 2\pi$, we obtain

$$|\zeta_1(\eta)| = 1 + \rho(1 + \cos\phi) + \mathcal{O}(\rho^2(1 + \cos\phi)), \qquad \rho \to 0, \ 0 \le \phi < 2\pi,$$

which proves the result.

The Milne-Simpson method defined by (1.2.16) has the growth parameters $\chi_1^* = 1$ and $\chi_2^* = -1/3$. This example shows that S is not necessarily bounded away from zero if $\mathrm{Re}(\chi_i^*) < 0$ for some i, $1 \le i \le k_*$.

By (1.3.10), there exists a δ_1, $0 < \delta_1 < \tilde{\delta}$, such that

$$|\zeta_i(\eta)| \le 1 + (\mathrm{Re}(\chi_i^*)/2)\eta, \qquad -\delta_1 \le \eta \le 0, \ i = 1,\dots,k_*.$$

If $\chi = \min_{1\le i\le k_*}\{\mathrm{Re}(\chi_i^*)/2\} > 0$ then a substitution of these bounds into (1.3.11) yields

$$\mathrm{spr}(F_\pi(\eta)) \le 1 + \chi\eta, \qquad -\delta_1 \le \eta \le 0.$$

Observing now that there is a $0 < \delta \le \delta_1$ such that all essential roots of $\pi(\zeta,\eta)$ are simple for $-\delta \le \eta \le 0$, the Uniform Boundedness Theorem yields for $F_\pi(\eta)/\mathrm{spr}(F_\pi(\eta))$:

<u>(1.3.13) Corollary.</u> *Let the method defined by* $\pi(\zeta,\eta)$ *be convergent and let* $\mathrm{Re}(\chi_i^*) > 0$, $i = 1,\dots,k_*$. *Then there are positive constants* δ, κ_1, *and* κ_1^* *such that*

$$|F_\pi(\eta)^n| \le \kappa_1 e^{\kappa_1^* \eta}, \qquad -\delta \le \eta \le 0.$$

If $\overset{\circ}{S} \ne \emptyset$ then there exists by Lemma (1.3.3) and the continuity of $\mathrm{spr}(F_\pi(\eta))$ a $\mu_0 > 0$ such that $S_\mu \subset \overset{\circ}{S}$ for $0 < \mu \le \mu_0$. The boundary ∂S_μ of S_μ is a subset of the root locus curve

$$\mathfrak{L}_\mu = \{\mu \in \bar{\mathbb{C}},\ \pi((1 - 2\mu)e^{i\phi},\eta) = 0,\ 0 \le \phi < 2\pi\}$$

as again follows from Jeltsch [76a, § 4.2.3] and $\mathfrak{L}_\mu$ depends continuously on μ.

<u>(1.3.14) Lemma.</u> *Let the method defined by* $\pi(\zeta,\eta)$ *be convergent, let* $\mathrm{Re}(\chi_i^*) > 0$, $i = 1, \dots, k_*$, *and let* $[-s, 0) \subset \overset{\circ}{S}$, $0 < s \le \infty$. *For* $\eta_1 \le \eta_0 \le 0$ *then there exist positive constants* κ_s *and* κ_s^* *depending only on* η_1 *such that*

$$|F_\pi(\eta)^n| \le \kappa_s e^{\kappa_s^* n\eta_0}, \qquad -s \le \eta \le \eta_0.$$

Proof. Let $\delta > 0$ be the constant defined in Corollary (1.3.13) then we have $[-s, -\delta] \subset S_{\mu_1}$ for some $\mu_1 > 0$. Hence we obtain in the same way as in Theorem (1.2.18)

$$(1.3.15)\quad |F_\pi(\eta)^n| \le \kappa_2 e^{-\mu_1 n} \le \kappa_2 e^{|\mu_1/\eta_1| n\eta_0}, \qquad -s \le \eta \le -\delta.$$

This proves the assertion for $\eta_0 \le -\delta$. If $-\delta < \eta_0 \le 0$ then we obtain by (1.3.15) and Corollary (1.3.13)

$$|F_\pi(\eta)^n| \le \max\{\kappa_2 e^{|\mu_1/\eta_1| n\eta_0}, \kappa_1 e^{\kappa_1^* n\eta_0}\} \le \kappa_s e^{\kappa_s^* n\eta_0}, \qquad -s \le \eta \le \eta_0.$$

Let us now return once more to the initial value problem (1.2.8). If the matrix A is hermitean and negative definite, $A \le -\gamma I$, $\gamma > 0$, then the spectral condition (iii) of Theorem (1.2.12) reads

$$Sp(\Delta tA) \subset [-\Delta t|A|, -\Delta t\gamma] \subset [-s, -\Delta t\gamma] \subset S.$$

Setting $\eta_0 = \Delta t\gamma$ in Lemma (1.3.14) and substituting the result into (1.2.13) we obtain uniform error bounds with exponential decreasing multiplication factor for the class of initial value problems (1.2.8) with hermitean and uniformly negative definite matrix A:

(1.3.16) Corollary. *Let the assumption of Theorem* (1.2.12) *be fulfilled but let the matrix* A *of the initial value problem* (1.2.8) *be hermitean and negative definite,* $A \leq -\gamma I$, $0 < \gamma \leq \gamma_0$.
(ii) Let $0 < \Delta t \leq \Delta t_0$, $0 \in S$, $[-s, 0) \subset \mathring{S}$, $0 < s \leq \infty$, *and* $\mathrm{Re}(\chi_i^*) > 0$, $i = 1,\dots,k_*$. *Then the assertion of Theorem* (1.2.18) *holds with* ($|X| = |X^{-1}| = 1$,) $\kappa_\mu = \kappa_s$, *and* $\mu/\Delta t = \kappa_s^*\gamma$ *where* κ_s^* *depends only on* Δt_0 *and* γ_0.

Examples of linear multistep methods with $[-\infty, 0) \subset \mathring{S}$ are given in Appendix A.4.

1.4. Indirect Methods for Differential Systems of Second Order

For an approximate solution of the initial value problem

(1.4.1) $$y'' = Ay + By' + c(t) + h(t),\ t > 0,\ y(0) = y_0,\ y'(0) = y_0^*,$$

by a multistep method of the type (1.1.3) we have to write this problem as first order problem,

(1.4.2) $$z' = A^*z + c^*(t) + h^*(t),\ t > 0,\ z(0) = z_0,$$

where $z = (y, y')^T$ and

$$A^* = \begin{bmatrix} 0 & I \\ A & B \end{bmatrix}.$$

However, A was always supposed to be a diagonable matrix in the above error estimations hence A* must be also a diagonable matrix in order that the results of Section 1.1 to 1.3 apply. In matrix structural analysis it is frequently supposed that the damping matrix B has the same system of eigenvectors as the leading matrix A and that both are diagonable. If

$$\Lambda = (\lambda_1,\dots,\lambda_m) \text{ and } \Phi = (\phi_1,\dots,\phi_m)$$

denote the diagonal matrices of the eigenvalues of A and B respectively then the system (1.4.1) can be decoupled after a suitable transformation into m scalar differential equations,

$$\tilde{y}'' = \lambda_i \tilde{y} + \phi_i \tilde{y}' + \tilde{c}_i(t), \qquad i = 1,\ldots,m, \tag{1.4.3}$$

which can be solved independently of each other. This *modal analysis* is advantageous if the decomposition of the matrix A is available at a reasonable price or if the eigenvalues of A must be computed anyhow by other reasons. Every solution of the scalar differential equation (1.4.3) oscillates for $\lambda_i < 0$ and $\phi_i \leq 0$ only if $|\phi_i| < 2\sqrt{|\lambda_i|}$ therefore the value $2\sqrt{|\lambda_i|}$ is called the *critical value* for the eigenvalue λ_i.

The following lemma provides the technical tools for an error estimation in the present case of *'orthogonal' damping*.

<u>(1.4.4) Lemma</u>. *Let the real symmetric (m,m)-matrices* A *and* B *have the same system of eigenvectors,* $A = X\Lambda X^T$, $B = X\Phi X^T$, $X^TX = I$, *let* $0 < \gamma I \leq -\Lambda$, $0 \leq -\Phi \leq \sqrt{\delta}2(-\Lambda)^{1/2}$, $0 \leq \delta < 1$, *and let* $P = (I,(-A)^{-1/2})$ *be a block diagonal matrix. Then the matrix* A^* *is diagonable,*

$$A^* = X^*\Lambda^*X^{*-1},$$

with

$$\mathrm{Re}(\Lambda^*) \leq 0, \quad |X^{*-1}| \leq 2[(1+\gamma^{-1})/(1-\delta)]^{1/2}, \quad |PX^*| \leq 2.$$

Proof. We have

$$A^* = X\begin{bmatrix} 0 & I \\ \Lambda & \Phi \end{bmatrix}X^T$$

and the eigenvalues of A^* are therefore

$$\lambda_\mu^* = [\phi_\mu \pm (\phi_\mu^2 + 4\lambda_\mu)^{1/2}]/2, \qquad \mu = 1,\ldots,m. \tag{1.4.5}$$

These are by assumption 2m different numbers with $\mathrm{Re}(\lambda_\mu^*) \leq 0$ and nonzero imaginary part therefore A^* is diagonable with $\mathrm{Re}(\Lambda^*) \leq 0$. Let now $\Lambda^* = (\Xi_1, \Xi_2)$ be the block diagonal matrix consisting of the diagonal matrices of the eigenvalues (1.4.5) with positive and negative sign respectively. Then we obtain

$$X^* = X\begin{bmatrix} I & I \\ \Xi_1 & \Xi_2 \end{bmatrix}, \quad X^{*-1} = (\Xi_2 - \Xi_1)^{-1}\begin{bmatrix} \Xi_2 & -I \\ -\Xi_1 & I \end{bmatrix}X^T,$$

and

$$PX^* = X\begin{bmatrix} I & 0 \\ 0 & (-\Lambda)^{-1/2} \end{bmatrix}X^TX\begin{bmatrix} I & I \\ \Xi_1 & \Xi_2 \end{bmatrix} = X\begin{bmatrix} I & I \\ (-\Lambda)^{-1/2}\Xi_1 & (-\Lambda)^{-1/2}\Xi_2 \end{bmatrix}.$$

By direct computation we find that

$$|(-\Lambda)^{-1/2}\Xi_i| = 1, \qquad i = 1,2,$$

hence we obtain

$$|PX^*| \le (\|PX^*\|_\infty \|PX^*\|_1)^{1/2} = (2\cdot 2)^{1/2} = 2.$$

For the estimation of $|X^{*-1}|$ we observe that

$$X^{*-H}X^{*-1} = X\begin{bmatrix} \Xi_1\bar{\Xi}_1 + \Xi_2\bar{\Xi}_2 & -(\Xi_1^H + \Xi_2^H) \\ -(\Xi_1 + \Xi_2) & 2I \end{bmatrix}(\Xi_2 - \Xi_1)^{-2}X^T = -X(4\Lambda + \Phi^2)^{-1}\begin{bmatrix} -2\Lambda & -\Phi \\ -\Phi & 2I \end{bmatrix}X^T.$$

Because of $0 \le -\Phi \le \sqrt{\delta}2(-\Lambda)^{1/2}$, $0 \le \delta < 1$, we thus derive

$$|X^{*-1}|^2 \le 2\text{spr}\left[(4\Lambda + \Phi^2)^{-1}\begin{bmatrix} -\Lambda & (-\Lambda)^{1/2} \\ (-\Lambda)^{1/2} & I \end{bmatrix}\right] = 2|(4\Lambda + \Phi^2)^{-1}(I - \Lambda)|$$

$$\le 2|(-4\Lambda + 4\delta\Lambda)^{-1}(I - \Lambda)| = 2|\Lambda^{-1}(I - \Lambda)|/(1 - \delta) \le 2(1 + \gamma^{-1})/(1 - \delta).$$

Let now $(y, y')^T = z$ be the solution of the initial value problem (1.4.2) and let $v_n^* = (v_n, w_n)^T$, $n = k,k+1,\ldots$, be the numerical approximation obtained by the scheme (1.2.2), i.e.,

$$\sum_{i=0}^k \sigma_i(\Delta t A^*)T^i v_n^* = -\sum_{i=0}^k T^i \sum_{j=1}^{\ell} \sigma_{ij}(\Delta t A^*)\Delta t^j c_n^{*(j-1)}, \quad n = 0,1,\ldots, \tag{1.4.6}$$

then we have the following corollary to Theorem (1.2.12).

<u>(1.4.7) Theorem</u>. *Let the initial value problem* (1.4.2) *and the numerical approximation* (1.4.6) *fulfil the assumptions of Theorem* (1.2.12) *and Lemma* (1.4.4). *Then for* $n = k$, $k+1,\ldots$,

$$|y_n - v_n| + |A|^{-1/2}|y_n' - w_n| \le \kappa_R[(1+\gamma^{-1})/(1-\delta)]^{1/2}\Big[|Y_{k-1} - V_{k-1}| + |Y_{k-1}' - W_{k-1}|$$
$$+ \Delta t^p \int_0^{n\Delta t} (|y^{(p+1)}(\tau)| + |y^{(p+2)}(\tau)|)d\tau + n\Delta t \max_{0\le i\le \ell-1} |||h^{(i)}|||_n\Big].$$

Proof. Let $E_n^* = Z_n - V_n^*$ where $Z_n = (z_{n-k+1},\ldots,z_n)^T$ and $V_n^* = (v_{n-k+1}^*,\ldots,v_n^*)^T$ then we obtain by Theorem (1.2.12)

$$|X^{*-1}E_n^*| \le \kappa_R|X^{*-1}|\Big[|Z_{k-1} - V_{k-1}^*| + \Delta t^p \int_0^{n\Delta t} |z^{(p+1)}(\tau)|d\tau + n\Delta t \max_{0\le i\le \ell-1} |||h^{(i)}|||_n\Big] \tag{1.4.8}$$

which up to the pre-factor yields the right side of the error bound and the pre-factor on the right side of the assertion is the bound of $|X^{*-1}|$ from Lemma (1.4.4). For an arbitrary regular matrix P we have

$$|PE_n^*| \le |PX^*||(PX^*)^{-1}PE_n^*| = |PX^*||X^{*-1}E_n^*|.$$

Choosing for P the block diagonal matrix $P = (I, (-A)^{-1/2})$ and substituting the last bound of Lemma (1.4.4) we find

$$|Y_n - V_n| + |A|^{-1/2}|Y_n' - W_n| \le |Y_n - V_n| + |A^{-1/2}(Y_n' - W_n)| \le \sqrt{2}\cdot 2|X^{*-1}E_n^*|.$$

This estimation and (1.4.8) prove the assertion.

It should be emphasized that in the error bound of this theorem the initial error on the right side is multiplied neither by $|A|^{1/2}$ nor by Δt^{-1} unlike the subsequent error estimations of this and the following chapter. On the other side, if the second or the first term on the left side is omitted then we obtain the following bounds respectively:

$$\left.\begin{array}{l}|y_n - v_n| \le \\ |y_n' - w_n| \le |A|^{1/2}\times\end{array}\right\}\text{ right side of the inequality in Theorem (1.4.7).}$$

This result reflects a fact which is well-known in elliptic finite element analysis namely that in ill-conditioned problems the approximation of the derivative of the solution is always worse than that of the solution itself.

The next result concerns an other special form of the damping matrix B in (1.4.1) which appears e.g. in Dietrich [81].

(1.4.8) Theorem. *Let the initial value problem* (1.4.2) *and the numerical approximation* (1.4.6) *fulfil the assumptions of Theorem* (1.2.12). *Let the real* (m,m)*-matrices* A *and* B *be symmetric and negative definite and skew-symmetric respectively,* $A \le -\gamma I < 0$, $B^T = -B$. *Then for* $n = k,k+1,\ldots,$

$$|y_n - v_n| + |y_n' - w_n| \le \kappa_R(1+\gamma^{-1})^{1/2}\Big[|(-A)^{1/2}(Y_{k-1} - V_{k-1})| + |Y_{k-1}' - W_{k-1}|$$
$$+ \Delta t^p \int_0^{n\Delta t} (|(-A)^{1/2}y^{(p+1)}(\tau)| + |y^{(p+2)}(\tau)|)d\tau + n\Delta t \max_{0\le i\le \ell-1} |||h^{(i)}|||_n\Big].$$

Proof. We consider instead of (1.2.10) the modified error equation

$$(1.4.9)\qquad QE_n^* = F_\pi(\Delta t Q A^* Q^{-1})QE_{n-1}^* + QD^*(\Delta t,z)_n - H_n^*,\qquad n = k,k+1,\ldots,$$

where $Q = ((-A)^{1/2}, I)$ is a block diagonal matrix and hence

$$\text{(1.4.10)}\qquad A^{**} \equiv QA^*Q^{-1} = \begin{bmatrix} 0 & (-A)^{1/2} \\ -(-A)^{1/2} & B \end{bmatrix}.$$

By assumption, this matrix is real and skew-symmetric hence normal, $A^{**H}A^{**} = A^{**}A^{**H}$, hence unitary diagonable. Therefore we obtain in the same way as in Theorem (1.2.12) by the Uniform Boundedness Theorem (cf. Appendix)

$$\text{(1.4.11)}\qquad |QE_n^*| \le \kappa_R[|QE_{k-1}^*| + \textstyle\sum_{\nu=k}^n |QD^*(\Delta t,z)_\nu - QH_\nu^*|], \qquad n = k,k+1,\ldots .$$

But

$$QD^*(\Delta t,z)_n = Q(0,\ldots,0,\sigma_k(\Delta tA^*)^{-1}d(\Delta t,z)_{n-k})^T = (0,\ldots,0,\sigma_k(\Delta tA^{**})^{-1}d(\Delta t,Qz)_{n-k})^T,$$

$$QH_n^* = Q(0,\ldots,0,\sigma_k(\Delta tA^*)^{-1}\textstyle\sum_{i=0}^k\sum_{j=1}^{\ell}\sigma_{ij}(\Delta tA^*)\Delta t^j h_{n-k}^{*(j-1)})^T$$

$$= (0,\ldots,0,\sigma_k(\Delta tA^{**})^{-1}\textstyle\sum_{i=0}^k\sum_{j=1}^{\ell}\sigma_{ij}(\Delta tA^{**})\Delta t^j Qh_{n-k}^{*(j-1)})^T,$$

and $Qh^*(t) = Q(0,\, h(t))^T = h(t)$. Consequently, as in the proof of Theorem (1.2.12),

$$\textstyle\sum_{\nu=k}^n |QD^*(\Delta t,z)_\nu| \le \Gamma\Delta t^p \int_0^{n\Delta t} |Qz^{(p+1)}(\tau)|d\tau \le \Delta t^p \int_0^{n\Delta t} (|(-A)^{1/2}y^{(p+1)}(\tau)| + |y^{(p+2)}(\tau)|)d\tau$$

and

$$\textstyle\sum_{\nu=k}^n |QH_\nu^*| \le \kappa_R n\Delta t \max_{0\le i\le \ell-1} |||h^{(i)}|||_n .$$

Finally, $|QE_{k-1}^*| \le |(-A)^{1/2}(Y_{k-1} - V_{k-1})| + |Y'_{k-1} - W_{k-1}|$ and

$$|QE_n^*| = |((-A)^{1/2}(Y_n - V_n),\, (Y'_n - W_n))| \ge 2^{-1/2}(\gamma^{1/2}|Y_n - V_n| + |Y'_n - W_n|)$$

which proves the desired result.

Under the assumption of Theorem (1.4.7) and (1.4.8) the spectral condition (1.2.12)(iii) claims for the leading matrices A^* and A^{**} respectively that

$$Sp(\Delta tA^*) \subseteq \{\eta \in \mathbb{C},\ -2\sqrt{\delta}\Delta t|A|^{1/2} \le Re(\eta) \le 0,\ |Im(\eta)| \le \Delta t|A|^{1/2}\} \subset R \subseteq S$$

and

$$Sp(\Delta tA^{**}) \subseteq \{\eta \in \mathbb{C},\ Re(\eta) = 0,\ |Im(\eta)| \le \Delta t(|B| + |A|^{1/2})\} \subset R \subseteq S.$$

This leads to the demand for multistep methods whose stability region S contains a large interval $\{\eta = i\tilde{\eta},\ |\tilde{\eta}| \le s\}$ on the imaginary axis. The special class of single-step multiderivative methods studied in the next section has even the property that S contains the entire imaginary axis. For multistep methods the problem of stability on the imaginary axis is discussed more detailled in Section 1.7.

1.5. Diagonal Padé Approximants of the Exponential Function

In view of Theorem (1.4.7) and (1.4.8) it suggests itself to ask whether there are numerical schemes which *do not need the diagonability* of the matrix A for uniform error bounds. In order to answer this question we consider single-step multiderivative procedures of order 2ℓ defined by the characteristic polynomials

$$(1.5.1)\qquad \pi(\zeta,\eta) = \pi_\ell(\zeta,\eta) = \sigma^{(\ell)}(-\eta)\zeta - \sigma^{(\ell)}(\eta)$$

where

$$(1.5.2)\qquad \frac{\sigma^{(\ell)}(\eta)}{\sigma^{(\ell)}(-\eta)} = e^\eta + \mathcal{O}(\eta^{2\ell+1}), \qquad \eta \to 0,$$

are the well-known diagonal Padé approximants of the exponential function:

$$(1.5.3)\qquad \sigma^{(\ell)}(\eta) = \sum_{j=0}^{\ell} \frac{\ell!}{(\ell-j)!}\,\frac{(2\ell-j)!}{(2\ell)!}\,\frac{\eta^j}{j!}, \qquad \ell = 1,2,\dots\quad ;$$

see e.g. Grigorieff [72]. For instance we obtain

$$\sigma^{(1)}(\eta) = 1 + (\eta/2)$$
$$\sigma^{(2)}(\eta) = 1 + (\eta/2) + (\eta^2/12)$$
$$\sigma^{(3)}(\eta) = 1 + (\eta/2) + (\eta^2/10) + (\eta^3/120)$$
$$\sigma^{(4)}(\eta) = 1 + (\eta/2) + (\eta^2/28) + (\eta^3/84) + (\eta^4/180).$$

Below these schemes are briefly called (ℓ,ℓ)-schemes and the (1,1)-scheme is the trapezoidal rule.

(1.5.4) Lemma. *For* $\ell = 1,2,\dots$, *all roots of the polynomials* $\sigma^{(\ell)}(\eta)$ *defined by* (1.5.3) *have negative real parts.*

Proof. Birkhoff and Varga [65]; see also Grigorieff [72].

In the next lemma, $\mathrm{Re}(A) \geq 0$ means for an arbitrary real or complex (m,m)-matrix A that $\mathrm{Re}(x^H A x) \geq 0\ \forall\, x \in \mathbb{C}^m$, i.e., that the hermitean part $\mathrm{Re}(A) = (A + A^H)/2$ is positive semidefinite.

(1.5.5) Lemma. *Let* $\sigma(\eta)$ *be an arbitrary real polynomial of exact degree* ℓ *such that* $\mathrm{Re}(\eta) < 0$ *holds for all roots* η *of* $\sigma(\eta)$. *Then, for every matrix* A *with* $\mathrm{Re}(A) \leq 0$

$$|\sigma(-A)^{-1}| \leq \kappa \text{ and } |\sigma(-A)^{-1}\sigma(A)| \leq 1.$$

Proof. We follow Gekeler and Johnsen [77] and assume without loss of generality that $\sigma(\eta)$ is a normed polynomial. Then we have

$$\sigma(-A) = \prod_{i=1}^{\ell}(-A - \eta_i I)$$

and the components $(-A - \eta_i I)$ are regular matrices with $\mathrm{Re}(-A - \eta_i I) > 0$. If a matrix Q satisfies $\mathrm{Re}(Q) \geq \omega I > 0$ then $|Q^{-1}| \leq \omega^{-1}$ therefore we obtain

$$|\sigma(-A)^{-1}| \leq \prod_{i=1}^{\ell}|(-A - \eta_i)^{-1}| \leq \prod_{i=1}^{\ell}(-\mathrm{Re}(\eta_i))^{-1} \leq \kappa$$

where the constant κ does not depend on A.

In order to prove the second assertion we observe that

$$\sigma(-A)^{-1}\sigma(A) = \prod_{i=1}^{\ell}(-A - \eta_i I)^{-1}(A - \eta_i I).$$

Therefore we have to show that

$$G(A,\eta) \equiv (-A - \eta I)^{-1}(A - \eta I)$$

satisfies

(1.5.6) $$|G(A,\eta)| \leq 1, \qquad \eta < 0,$$

and

(1.5.7) $$|G(A,\eta)G(A,\bar{\eta})| \leq 1 \qquad \mathrm{Re}(\eta) < 0.$$

Let first $\eta < 0$ and let $y = G(A,\eta)x$ then $(-A - \eta I)y = (A - \eta I)x$ or

$$-A(y + x) = \eta(y - x).$$

Because $\mathrm{Re}(A) \leq 0$ we obtain

$$0 \leq \mathrm{Re}((y + x)^H(-A)(y + x)) = \mathrm{Re}(\eta(y + x)^H(y - x)) = \eta(|y|^2 - |x|^2), \qquad \eta < 0.$$

Hence we have $|y| < |x|$ for arbitrary $x \in \mathbb{C}^m$ which proves (1.5.6). In order to prove (1.5.6) let $\mathrm{Re}(\eta) < 0$ and let $y = G(A,\eta)G(A,\bar{\eta})x$ then

$$(A^2 + 2\mathrm{Re}(\eta)A + |\eta|^2 I)y = (A^2 - 2\mathrm{Re}(\eta)A + |\eta|^2 I)x$$

or

$$(A^2 + |\eta|^2 I)(y - x) = -2\,\mathrm{Re}(\eta)A(y + x).$$

If A is regular then we multiply this equation by $(y - x)^H(2\mathrm{Re}(\eta)A)^{-1}$ and obtain

$$\mathrm{Re}((y - x)^H(|\eta|^2 A^{-1} + A)(y - x)/2\mathrm{Re}(\eta)) = -(|y|^2 - |x|^2).$$

The left side is nonnegative because $\mathrm{Re}(A) \le 0$, $\mathrm{Re}(A^{-1}) \le 0$, and $\mathrm{Re}(\eta) < 0$ hence we have again $|y| \le |x|$ for arbitrary $x \in \mathbb{C}^m$. This proves (1.5.7) in the case where A is regular. If A is singular then we find in the same way that (1.5.7) holds for $A - \varepsilon I$ and $\varepsilon > 0$. As the right side of (1.5.7) does not depend on ε the inequality must hold for $\varepsilon = 0$, too.

Now we return to the initial value problem (1.4.1) and the associated first order problem (1.4.2). An (ℓ,ℓ)-scheme is consistent of oder 2ℓ by (1.5.2) and Lemma (1.3.5), and it yields for (1.4.2) the computational device

(1.5.8)
$$\begin{aligned}&\sigma^{(\ell)}(-\Delta tA^*)v^*_{n+1} - \sigma^{(\ell)}(\Delta tA^*)v^*_n \\ &= - \textstyle\sum_{j=1}^{\ell}\sigma_j^{(\ell)}(-\Delta tA^*)c^{*(j-1)}_{n+1} + \sum_{j=1}^{\ell}\sigma_j^{(\ell)}(\Delta tA^*)c^{*(j-1)}_{n}, \quad n = 0,1,\ldots \quad .\end{aligned}$$

The associated Frobenius matrix has obviously the form

$$F_\ell(\eta) = (\sigma^{(\ell)}(-\eta))^{-1}\sigma^{(\ell)}(\eta).$$

If the (m,m)-matrices A and B in the initial value problem (1.4.1) are arbitrary real symmetric and negative semidefinite matrices then the matrix A^{**} defined by (1.4.10) is no longer unitary diagonable because it is no longer normal. But A^{**} satisfies still the assumption of Lemma (1.5.5),

$$\mathrm{Re}(A^{**}) = \begin{bmatrix} 0 & 0 \\ 0 & B \end{bmatrix} \le 0.$$

Accordingly, we deduce from Lemma (1.5.4) and (1.5.5) that

$$|\sigma^{(\ell)}(-\Delta tA^{**})^{-1}| \le \kappa \text{ and } |F_\pi(\Delta tA^{**})| \le 1.$$

Substituting these bounds into the modified error equation (1.4.9) we obtain again (1.4.10) without using this time the Uniform Boundedness Theorem. The rest of the error estimation is then the same as in Theorem (1.4.8).

We assemble the result of this section in the following theorem.

<u>(1.5.9) Theorem.</u> *Let* $\ell \in \mathbb{N}$ *be fixed. In the initial value problem* (1.4.1) *let* A, B *be real symmetric,* $A \le -\gamma I < 0$, $B \le 0$, *and let the solution* y *be* $(2\ell+2)$*-times continuously differentiable. Then the numerical approximation* $v^*_n = (v_n, w_n)^T$ *defined by* (1.5.8) *satisfies for* $n = 1,2,\ldots,$

$$\begin{aligned}|y_n - v_n| + |y'_n - w_n| &\le \kappa(1+\gamma^{-1})^{1/2}\Big[|(-A)^{1/2}(y_0 - v_0)| + |y'_0 - w_0| \\ &+ \Delta t^{2\ell}\int_0^{n\Delta t}(|(-A)^{1/2}y^{(2\ell+1)}(\tau)| + |y^{(2\ell+2)}(\tau)|)d\tau + n\Delta t \max_{0\le i\le\ell-1}|||h^{(i)}|||_n\Big].\end{aligned}$$

1.6. Stability in the Left Half-Plane

We have seen in Section 1.3 that the stability region S can be found geometrically by a plot of the root locus curve $\mathfrak{Z}$. However, if it is demanded that S has a certain shape suggested from practical aspects then general statements on appropriate methods are to be derived in an analytic way using tools from algebra and analytic function theory. The first result in this direction has been provided by Dahlquist [63] who has defined a method to be *A-stable* if $\{\eta \in \mathbb{C},\ \mathrm{Re}\eta < 0\} \subset \mathring{S}$ and has shown that linear multistep methods of order greater than two cannot have this property. With this somewhat depressing result a very interesting development of numerical analysis has begun which is by no means finished up today. A complete description of the present state of knowledge would go beyound the scope of this volume therefore we restrict ourselves to a survey of the most important results of the recent years and omit the proofs of some classical results.

In many applications A-stability is a too severe requirement therefore Widlund [67] and Cryer [73] have introduced the following concept.

(1.6.1) Definition. *A method* (1.1.3) *is*
(i) $A(\alpha)$*-stable if* $\{\eta \in \mathbb{C},\ \eta \neq 0,\ |\pi - \arg\eta| < \alpha\} \subset \mathring{S}$, $\quad 0 < \alpha \leq \pi/2$,
(ii) $A(0)$*-stable if it is* $A(\alpha)$*-stable for some* $\alpha \in (0,\pi/2]$,
(iii) A_0*-stable if* $(-\infty,\ 0) \subset \mathring{S}$.

By Lemma (A.1.17) a convergent method cannot be $A(\alpha)$-stable with $\alpha > \pi/2$ in a neighborhood of $\eta = 0$. Obviously, an $A(0)$-stable method is A_0-stable but the converse is not true: The linear method with the polynomial

$$\pi(\zeta,\eta) = \zeta^2 - \zeta - \frac{\eta}{4}(\zeta^2 + 2\zeta + 1)$$

is convergent and A_0-stable but not $A(0)$-stable (Cryer [73]) because $\rho_1(\zeta) = -\frac{1}{4}(\zeta^2 + 2\zeta + 1)$ has a double root $\zeta = 1$; cf. Corollary (A.1.21)(ii).

(1.6.2) Theorem. *There is only one* A_0*-stable linear k-step method of order* $p \geq k + 1$*: the trapezoidal rule of order* $p = 2$ *with the polynomial* $\pi(\zeta,\eta) = \zeta - 1 - (\eta/2)(\zeta + 1)$.

Proof. See Cryer [73 , Theorem 3.3].

(1.6.3) Theorem. *For each* $\alpha \in [0,\ \pi/2)$ *there exists an* $A(\alpha)$*-stable linear k-step method of order* $p = k = 3$ *and* $p = k = 4$.

Proof. See Widlund [67].

Some well-known $A(\alpha)$-stable methods are represented in Appendix A.4.

In order to deduce necessary and sufficient conditions for A_0-stability we now recall the *Routh-Hurwitz criterion*. Let Δ_p denote the *Hurwitz matrix* of the polynomial $p(z) = \sum_{i=0}^{k} s_i z^i$,

$$(1.6.4)\quad \Delta_p = \begin{bmatrix} s_{k-1} & s_{k-3} & \cdots & & s_{-k+1} \\ s_k & s_{k-2} & \cdots & & s_{-k+2} \\ & s_{k-1} & s_{k-3} & \cdots & \\ & s_k & s_{k-2} & \cdots & \\ & 0 & s_{k-1} & \cdots & \\ & 0 & & & \\ \cdots & \cdots & \cdots & \cdots & \cdots \\ 0 & & & & s_\omega \ldots s_0 \end{bmatrix}, \qquad \omega = 2[k/2],\ s_i = 0 \text{ for } i < 0,$$

then this well-known result reads as follows; cf. e.g. Lambert [73]:

<u>(1.6.5) Lemma</u>. *All roots* z^* *of* $p(z) = \sum_{i=0}^{k} s_i z^i$ *satisfy* $\mathrm{Re}z^* < 0$ *iff all leading principal minors of the Hurwitz matrix are positive, i.e., the determinants of all matrices arising from* Δ_p *by cancelling the* r *last rows and columns for* $r = k-1,\ldots,0$.

It can be shown that the Routh-Hurwitz criterion implies

<u>(1.6.6) Lemma</u>. *If all roots* z^* *of* $p(z) = \sum_{i=0}^{k} s_i z^i$ *satisfy* $\mathrm{Re}z^* < 0$ *then* $s_i > 0$ *for* $i = 0,\ldots,k$.

Both results cannot be applied directly to the polynomial $\pi(\zeta,\eta)$ therefore the *Möbius transformation* is generally introduced in this context,

$$z = (\zeta + 1)/(\zeta - 1),\ \zeta = (z + 1)/(z - 1),$$

which maps the unit disk of the ζ-plane onto the left half-plane. In particular, $\zeta = 0$ is mapped into $z = -1$ and $\zeta = 1$ is mapped into $z = \infty$. Let

$$(1.6.7)\quad \begin{aligned} \Psi(z,\eta) &= (z - 1)^k \pi\left(\frac{z + 1}{z - 1},\eta\right) = \sum_{i=0}^{k}\sum_{j=0}^{\ell} \alpha_{ji}\eta^j (z + 1)^i (z - 1)^{k-i} \\ &= \sum_{i=0}^{k}\sum_{j=0}^{\ell} a_{ji}\eta^j z^i = \sum_{i=0}^{k} s_i(\eta) z^i = \sum_{j=0}^{\ell} r_j(z)\eta^j \end{aligned}$$

then

$$s_k(\eta) = \sum_{j=0}^{\ell}\sum_{i=0}^{k} \alpha_{ji}\eta^j \not\equiv 0$$

if the method is convergent and the method is A_0-stable iff all roots of $\Psi(z,\eta)$ lie in the left half-plane, $\mathrm{Re}z < 0$, for $\eta \in (-\infty, 0)$. The following lemma shows roughly speaking that it is unnecessary for the verification of A_0-stability to check the intermediate leading principal minors of the Hurwitz matrix $\Delta_\Psi(\eta)$.

(1.6.8). Lemma. *A method* (1.1.3) *is A_0-stable iff the following three conditions are fulfilled:*
(i) For some $\eta^* \in (-\infty, 0)$ *all roots of* $\Psi(z,\eta^*)$ *lie in the left half-plane.*
(ii) $s_k(\eta) \neq 0 \;\forall\; \eta \in (-\infty, 0)$.
(iii) $\det(\Delta_\Psi(\eta)) \neq 0 \;\forall\; \eta \in (-\infty, 0)$.

Proof. See Friedli and Jeltsch [78].

This communication contains also an algorithm to determine the Hurwitz determinant $\det(\Delta_\Psi(\eta))$ with polynomial entries by means of Sturmian sequences. As this is somewhat cumbersome we now give some further necessary conditions for A_0-stability at least which are more easily to check.

In linear methods one writes customarily

$$\pi(\zeta,\eta) = \rho(\zeta) - \eta\sigma(\zeta),\; \rho(\zeta) = \sum_{i=0}^k \alpha_i\zeta^i,\; \sigma(\zeta) = \sum_{i=0}^k \beta_i\zeta^i,$$

and

$$\Psi(z,\eta) = r(z) - \eta s(z),\; r(z) = \sum_{i=0}^k a_i z^i,\; s(z) = \sum_{i=0}^k b_i z^i.$$

Then we obtain

$$\rho(1) = \lim_{z\to\infty}\rho\left(\frac{z+1}{z-1}\right) = \lim_{z\to\infty}(z-1)^{-k} r(z) = a_k,\; \sigma(1) = b_k,$$

hence, if the method is convergent and (1.1.14) is stipulated,

$$a_k = 0,\; b_k = \sigma(1) = 1.$$

The conditional equations for consistence order $p \geq 1$ read here (cf. Lemma (1.1.12))

$$a_i = 2\sum_{j\geq 0}[b_{i+1+2j}/(1+2j)], \qquad k - p \leq i \leq k,$$

with the convention that $a_i = 0$ for $i < 0$ and $b_j = 0$ for $j > k$; cf. Widlund [67]. Accordingly, Lemma (A.1.3) and Lemma (1.6.6) for $z \to 0$ and $z \to \infty$ yield:

(1.6.9) Lemma. (Cryer [73].) *If the linear method with* $\sigma(1) = 1$ *is A_0-stable then* $a_i \geq 0$ *and* $b_j \geq 0$ *for* $j = 0,\dots,k$. *Furthermore,* $\sum_{i=0}^k a_i > 0$, $\sum_{i=0}^k b_i > 0$, *and* $\beta_k \neq 0$.

Returning to the general case and the notation (1.6.7) we prove an auxilary result which is due to Jeltsch [77].

(1.6.10) Lemma. *Let the polynomial* $\pi(\zeta,\eta)$ *be irreducibel.*
(i) If $\eta \in S \cap \mathbb{C}$ *then* $s_i(\eta) \neq 0$, $i = 0,\ldots,k$, *and*

$$\mathrm{Re}(s_{i+1}(\eta)/s_i(\eta)) > 0, \qquad i = 0,\ldots,k-1.$$

(ii) If $\infty \in S$ *then the degree of* $s_i(\eta)$ *is* ℓ, $i = 0,\ldots,k$, *and*

$$\mathrm{Re}(a_{\ell,i+1}/a_{\ell i}) > 0, \qquad i = 0,\ldots,k-1.$$

Proof. If $\eta \in S \cap \mathbb{C}$ then all roots of $\Psi(z,\eta)$ lie in the left half-plane, $\mathrm{Re}z < 0$. Let $(a)_i = a(a+1)\cdots(a+i-1)$, $i \in \mathbb{N}$, then by a repeated application of Theorem (A.1.46) we find that all roots z^* of

$$\frac{\partial^i \Psi}{\partial z^i}(z,\eta) = (1)_i s_i(\eta) + (2)_i s_{i+1}(\eta) z + \ldots + (k-i+1)_i s_k(\eta) z^{k-i}$$

satisfy $\mathrm{Re}z^* < 0$, too. Therefore $s_i(\eta) \neq 0$, $i = 0,\ldots,k$, and the sum of the reciprocals of all roots z^* has negative real part. Accordingly, Vieta's root criterion yields

$$-\mathrm{Re}\left[\frac{(2)_i s_{i+1}(\eta)}{(1)_i s_i(\eta)}\right] < 0, \qquad i = 0,\ldots,k-1.$$

This proves the first assertion. The second assertion follows in the same way by considering $\eta^\ell \Psi(z,\eta^{-1})$.

It is now convenient to introduce the following notation:

(1.6.11) Definition. *A method* (1.1.3) *is asymptotically* $A(\alpha)$*-stable if for all* $\theta \in (\pi-\alpha, \pi+\alpha)$ *there exists a* $\rho_\theta > 0$ *such that* $\{\eta = \rho e^{i\theta}, \rho > \rho_\theta\} \subset \mathring{S}$.

Asymptotic $A(0)$-stability and asymptotic A_0-stability are defined in an analogous way but observe that in these definitions the point ∞ itself is always excluded. Then the behavior of a method at the point ∞ is entirely ruled by the following simple résult:

(1.6.12) Lemma. *A method with the polynomial* $\pi(\zeta,\eta)$ *is asymptotically* $A(\alpha)$-, $A(0)$-, *or* A_0*-stable iff the method with the polynomial* $\eta^\ell \pi(\zeta,\eta^{-1}) \equiv \pi_*(\zeta,\eta)$ *is* $A(\alpha)$-, $A(0)$-, *or* A_0*-stable in a neighborhood of zero.*

Proof. It suffices to prove the assertion for $A(\alpha)$-stability. We first observe that $\arg(re^{i\theta}) \in (\pi-\alpha, \pi+\alpha)$ iff $\arg(\overline{re^{i\theta}}) = \arg(re^{-i\theta}) \in (\pi-\alpha, \pi+\alpha)$ hence we may consider the polynomial $\pi(\zeta, \overline{\eta})$. Substituting $\eta = \rho^{-1}e^{i\theta}$, $\rho \to 0$, we obtain with $\tilde{\eta} = \rho e^{i\theta}$

$$\tilde{\eta}^{\ell}\pi(\zeta,\overline{\eta}) = \tilde{\eta}^{\ell}\sum_{j=0}^{\ell}\rho_j(\zeta)\overline{\eta}^{j} = \tilde{\eta}^{\ell}\sum_{j=0}^{\ell}\rho_j(\zeta)\tilde{\eta}^{-j} = \sum_{j=0}^{\ell}\rho_{\ell-j}(\zeta)\tilde{\eta}^{j} = \pi_*(\zeta,\tilde{\eta})$$

which proves the assertion.

Naturally, $\pi_*(\zeta,\eta)$ is not necessarily the characteristic polynomial of a *consistent* method. But for every algebraic polynomial $\pi(\zeta,\eta)$ the shape of the stability region S near $\eta = 0$ is determined by the behavior of those roots $\zeta_i(\eta)$ which become unimodular in $\eta = 0$. If a 'method' with the general polynomial $\pi(\zeta,\eta)$ is A_0-stable near $\eta = 0$ with a possible exception of the point $\eta = 0$ itself then Corollary (A.1.21) implies that all roots $\zeta_i(\eta)$ with $|\zeta_i(0)| = 1$ have near $\eta = 0$ the form (A.1.18) with $q \in \{1, 2\}$ and $p \in \{1, \ldots, \ell\}$. The roots of $\pi_*(\zeta,\eta)$ must have this property if the method is asymptotically A_0-stable. Together with Lemma (1.6.10) for $\eta \in (-\infty, 0)$ we thus can state:

(1.6.13) Corollary. (Jeltsch [77].) *Let the method* (1.1.3) *be convergent. Then the following conditions are necessary for A_0-stability:*
(i) $\alpha_{\ell k} \neq 0$, $s_i(\eta) \neq 0$, $\eta \in (-\infty, 0)$, $i = 0,\ldots,k$, *and*

$$\mathrm{Re}(s_{i+1}(\eta)/s_i(\eta)) > 0 \qquad \eta \in (-\infty, 0),\ i = 0,\ldots,k-1.$$

(ii) $\infty \in \overline{S}$ *and all roots* $\zeta_i(\eta)$ *of* $\pi_*(\zeta,\eta) = \eta^{\ell}\pi(\zeta,\eta^{-1})$ *with* $|\zeta_i(0)| = 1$ *have near* $\eta = 0$ *the form*

$$\zeta_i(\eta) = \zeta_i(0) + \chi\eta^{p/q} + \mathcal{O}(\eta^s),\ \chi \neq 0,\ p \in \{1, \ldots, \ell\},\ q \in \{1, 2\},\ s > p/q.$$

$\alpha_{\ell k} \neq 0$ follows also directly from Lemma (A.1.3).

Obviously, an A_0-stable method is A_0-stable near $\eta = 0$, and a convergent method has no multiple unimodular roots in $\eta = 0$. Hence, as in linear methods the growth parameters χ_i^* are nonzero, Lemma (1.3.12) yields a necessary and sufficient algebraic condition for a *linear* convergent method to be A_0-stable near $\eta = 0$. On the other side, also the growth parameters $\hat{\chi}_i^*$ defined in Section 2.1 are nonzero in linear methods being not necessarily consistent (η^2 replaced by η). Hence, by (1.6.13)(ii), Lemma (2.1.25) with respect to $\pi_*(\zeta,\eta)$ yields a necessary and sufficient algebraic condition for a *linear* method to be asymptotically A_0-stable.

The next result is also due to Jeltsch [77].

(1.6.14) Lemma. *A method* (1.1.3) *is A(0)-stable iff it is A_0-stable, A(0)-stable near $\eta = 0$, and asymptotically A(0)-stable.*

Proof. The necessity of the three conditions is obvious. For the sufficiency we observe that the algebraic function $\zeta(\eta)$ defined by $\pi(\zeta(\eta),\eta) = 0$ satisfies $\zeta(\overline{\eta}) = \overline{\zeta}(\eta)$ because all coefficients of the algebraic equation $\pi(\zeta,\eta) = 0$ are real. Therefore we can restrict ourselves to the upper half-plane, $\mathrm{Im}\eta > 0$. By assumption, there exist two pairs of positive numbers, (α_0, ρ_0) and $(\alpha_\infty, \rho_\infty)$, such that

$$\mathcal{A}^*(\alpha_0,\rho_0) = \{\eta \in \mathbb{C},\ 0 < |\eta| \le \rho_0,\ \pi - \alpha_0 < \arg\eta < \pi\} \subset \mathring{S},$$

$$\mathcal{A}^*(\alpha_\infty,\rho_\infty) = \{\eta \in \mathbb{C},\ |\eta| \ge \rho_\infty,\ \pi - \alpha_\infty < \arg\eta < \pi\} \subset \mathring{S}.$$

As $\pi(\zeta,\eta)$ is irreducible there are only a finite number of branching points and singularities hence all roots of $\pi(\zeta,\eta)$ are continuous in a set $\{\eta \in \mathbb{C},\ -\infty < \mathrm{Re}\eta < 0,\ 0 < \mathrm{Im}\eta < \gamma\}$, $\gamma > 0$. Accordingly, as $(-\infty, 0) \subset \mathring{S}$ by assumption, i.e., as all roots are less than one in absolute value on the negative real line there exists an $\alpha_1 > 0$ such that

$$\{\eta \in \mathbb{C},\ \rho_0 \le |\eta| \le \rho_\infty,\ \pi - \alpha_1 < \arg\eta < \pi\} \subset \mathring{S}$$

and the method is $A(\alpha)$-stable with $\alpha = \min\{\alpha_0,\alpha_1,\alpha_\infty\}$.

Lemma (A.1.40) yields necessary and sufficient algebraic conditions for $A(0)$-stability near $\eta = 0$ and, by Lemma (1.6.12), for asymptotic $A(0)$-stability, too. In particular, the irreducibility of $\pi(\zeta,\eta)$ implies for *linear* methods (1.1.3) that $p_\nu = 1$ in (A.1.38). Therefore we can state e.g. the following corollary to Lemma (1.6.14).

<u>(1.6.15) Corollary</u>. (Jeltsch [76b].) *Let the linear method* (1.1.3) *with the polynomial* $\pi(\zeta,\eta) = \rho(\zeta) - \eta\sigma(\zeta)$ *be convergent. Then the following conditions (i) - (iv) are necessary and sufficient for $A(0)$-stability.*
(i) The method is A_0-stable.
(ii) The unimodular roots of $\sigma(\zeta)$ *are simple.*
(iii) If ζ is a unimodular root of $\rho(\zeta)$ *then* $\mathrm{Re}[\sigma(\zeta)/(\zeta\rho'(\zeta))] > 0$.
(iv) If ζ is a unimodular root of $\sigma(\zeta)$ *then* $\mathrm{Re}[\rho(\zeta)/(\zeta\sigma'(\zeta))] > 0$.

Further useful stability concepts are those of relative stability and of stiff stability:

<u>(1.6.16) Definition.</u> *Let Ω be the largest star into which the principal root $\zeta_1(\eta)$ of the consistent method* (1.1.3) *has an analytic continuation. Then*

$$\mathcal{R} = \{\eta \in \Omega,\ |\zeta_i(\eta)| < |\zeta_1(\eta)|,\ i = 2,\dots,k\}$$

is the region of relative stability.

Notice that $|\zeta_1(\eta)|$ is not necessarily bounded by one in $\mathfrak{R}$ hence relative stability deals also with unstable differential equations. Obviously, a necessary and sufficient condition for a consistent method (1.1.3) to be relatively stable in a (full) neighborhood of $\eta = 0$ is that it is 'strongly D-stable' in $\eta = 0$ which means that all roots of $\pi(\zeta,0)/(\zeta - 1) = \rho_0(\zeta)/(\zeta - 1)$ are less than one in absolute value.

(1.6.17) Definition. (Gear [69], Jeltsch [76b, 77].) *Let*

$$R_1 = \{\eta \in \overline{\mathbb{C}},\ \mathrm{Re}\eta < -a\},\ R_2 = \{\eta \in \overline{\mathbb{C}},\ \mathrm{Re}\eta \le -b,\ |\mathrm{Im}\eta| < c\},$$

$$R_3 = \{\eta \in \mathbb{C},\ |\mathrm{Re}\eta| < b,\ |\mathrm{Im}\eta| < c\}.$$

Then a convergent method is stiffly stable iff there exist positive numbers a, b, c *such that*
(i) $R_1 \cup R_2 \subset \mathring{S}$ *and* $R_3 \subset \mathfrak{R}$,
(ii) the method is A_0*-stable.*

Condition (ii) is introduced here in order to deal with the demand of stiff stability in the original meaning of Gear [69].

(1.6.18) Lemma. (Jeltsch [76b].) *If a convergent linear method* (1.1.3) *satisfies* (1.6.17)(i) *then it is* A_0*-stable hence stiffly stable.*

Proof. We have to show that $(-b, 0) \subset \mathring{S}$ and reconsider the polynomial $\Psi(z,\eta) = r(z) - \eta s(z)$ introduced above. As the method is convergent and stiffly stable $r(z)$ and $s(z)$ have only roots with $\mathrm{Re} z \le 0$, and $a_{k-1} = 2b_k$. For an $\eta^* \in (-b, 0)$, $\Psi(z,\eta^*)$ is a polynomial with positive coefficients and we have to show that its roots lie in the left half-plane, $\mathrm{Re} z < 0$. Clearly $\Psi(z,\eta^*)$ has no positive roots. Moreover, $\Psi(0,\eta^*) = a_0 - \eta^* b_0 \ne 0$ since otherwise ρ and σ in $\pi(\zeta,\eta) = \rho(\zeta) - \eta\sigma(\zeta)$ would have a common factor. Finally, let z be a root of $\Psi(z,\eta^*)$ with $\mathrm{Re}\, z \ge 0$ and $\mathrm{Im}\, z > 0$. Then $\overline{z}$ is a root, too. But then $\zeta = (z + 1)/(z - 1)$ and $\overline{\zeta}$ are two roots of $\pi(\zeta,\eta^*)$ of the same modulus. Hence η^* does not belong to the region of relative stability which is a contradiction. Consequently, the roots of $\Psi(z,\eta^*)$ lie in the left half-plane, $\mathrm{Re}\, z < 0$ and the roots of $\pi(\zeta,\eta^*)$ are less than one in absolute value for $\eta^* \in (-b, 0)$.

(1.6.19) Lemma. (Jeltsch [77].) *A convergent method* (1.1.3) *is stiffly stable iff the following three conditions are fulfilled:*
(i) $\rho_0(\zeta)$ *has the single unimodular root* $\zeta = 1$.
(ii) The method is A_0*-stable.*
(iii) There exists a $\rho > 0$ *such that* $\{\eta \in \mathbb{C},\ |\eta + \rho| < \rho\} \subset \mathring{S}$ *for the method with the polynomial* $\pi_*(\zeta,\eta) = \eta^\ell \pi(\zeta,\eta^{-1})$.

Proof. As already mentioned, the first condition is equivalent to the existence of a rectangle $R_3 \subset \mathfrak{R}$. The necessity of condition (ii) is trivial and on the other side this condition implies the existence of a set $R_2 \subset \mathring{S}$ following the pattern of Lemma (1.6.14). Condition (iii) finally is equivalent to the existence of a set $R_1 \subset \mathring{S}$ because

$$\frac{1}{\rho(e^{i\theta}-1)} = -\frac{1}{2\rho} - i\,\frac{\sin\theta}{2\rho(1-\cos\theta)},\ \lim_{\theta\to 0}\frac{\sin\theta}{1-\cos\theta} = \infty,$$

and thus the straight lines $\mathrm{Re}\,\eta = -1/2\rho < 0$ are mapped onto the circles $\eta = \rho(e^{i\theta}-1)$ by $\eta \to 1/\eta$.

Notice that the first two conditions of Lemma (A.1.53) - spoken out for $\pi_*(\zeta,\eta)$ - are equivalent to asymptotic $A(\pi/2)$-stability by Corollary (A.1.21) and that the polynomial (A.1.52) appearing in the third condition is linear if the method (1.1.3) is linear. Hence this latter condition is empty in linear methods and, accordingly, the disk condition (1.6.19)(iii) and asymptotic $A(\pi/2)$-stability are equivalent in this case. Thus we can state:

(1.6.20) Corollary. *Let the method* (1.1.3) *be linear, convergent, strongly D-stable in* $\eta = 0$, *and* A_0*-stable. Then it is stiffly stable iff it is asymptotically* $A(\pi/2)$*-stable.*

1.7. Stability on the Imaginary Axis

Methods with a large stability interval on the imaginary axis are of particular interest in the solution of differential systems of second order by the class of indirect methods studied in Section 1.4. The following notation has become customary in the meanwhile here.

(1.7.1) Definition. *A multistep multiderivative method is* I_r*-stable if* $\{i\eta, -r < \eta < r\} \subseteq S$, $0 < r \leq \infty$.

Recently, Jeltsch and Nevanlinna [81 , 82a, 82b] have developed an algebraic comparison theory for numerical methods with respect to their stability regions which allows the treatment of I_r-stable methods from a rather general point of view. This technique uses as a fundamental tool results on the shape of the 'order star' having been found by Wanner, Hairer, and Norsett [78a] in the necessary global form. Locally, i.e., in a neighborhood of $\eta = 0$, the order star is described by Lemma (A.1.17).

In this section we give a survey on the present state of knowledge in I_r-stability. For the proofs however the reader is referred to the original contributions.

As the class of single-step multiderivative methods coincides with the class of Runge-Kutta methods for the test equation $y' = \lambda y$, the below presented results hold also literally for these latter methods.

Let us first recall that a method is A-stable iff $\{\eta \in \mathbb{C}, \operatorname{Re}\eta < 0\} \subseteq \mathring{S}$. The characteristic polynomial $\pi(\zeta,\eta)$ of a single-step multiderivative method is linear with respect to ζ hence we have $S = \overline{S}$ here. Thus a method of this class is I_∞-stable if it is A-stable. The same is true for *linear* multistep methods, too, by remark (i) after Corollary (A.1.24). In the case of general multistep multiderivative methods the implication of I_∞-stability by A-stability is ruled by Lemma (A.1.53).

As concerns the implication of A-stability by I_∞-stability, the following results are due to Wanner, Hairer, and Norsett [78b]:

<u>(1.7.2) Theorem.</u> *A k-step ℓ-derivative method* (1.1.3) *of order* p *is A-stable if it is* I_∞*-stable and* $p \geq 2\ell - 1$.

<u>(1.7.3) Theorem.</u> *A k-step ℓ-derivative method* (1.1.3) *of order* p *is A-stable if it is* I_∞*-stable,* $p \geq 2\ell - 3$, *and the coefficients of the leading polynomial* $\sigma_k(\eta)$ *have alternating signs.*

In particular, an I_∞-stable consistent *linear* multistep method is A-stable which has been proved in an independent way by Jeltsch [78a].

Now, recalling the result of Dahlquist [63] namely that an A-stable linear multistep method has order $p \leq 2$ and that the trapezoidal rule has the smallest error constant χ_p, cf. (1.3.6), among all A-stable linear multistep methods of order two, we obtain immediately the following result, see also Jeltsch [78a].

<u>(1.7.4) Corollary.</u> *(i) An* I_∞*-stable linear multistep method has order* $p \leq 2$.
(ii) Among all I_∞*-stable linear multistep methods of order* $p = 2$ *the trapezoidal rule has the smallest error constant.*

Example (A.4.7) due to Jeltsch [78a] shows that a *nonlinear* consistent and I_∞-stable method is not necessarily A-stable.

The generalization of Dahlquist's result to nonlinear multistep methods is known as the *Daniel-Moore conjecture*, cf. Daniel and Moore [70]. It was proved by Wanner, Hairer, and Norsett [78a]. For the presentation we recall that the characteristic polynomial,

(1.7.5) $$\pi(\zeta,\eta) = \textstyle\sum_{i=0}^{k}\sigma_i(\eta)\zeta^i = \sum_{j=0}^{\ell}\rho_j(\zeta)\eta^j,$$

is always assumed to be irreducible, cf. Section 1.1.

<u>(1.7.6) Theorem</u>. *Let the k-step ℓ-derivative method satisfy* $\sigma_k(0) \neq 0$ *and* $(\partial\pi/\partial\zeta)(0,1) \neq 0$.

(i) If the method is A-stable then $p \leq 2\ell$ *and* $\mathrm{sgn}(\chi_p) = (-1)^\ell$ *for* $p = 2\ell$.

(ii) The error constant χ_p *of an A-stable method of order* $p = 2\ell$ *satisfies*

$$|\chi_p| \geq |\chi_p^*|, \quad \chi_p^* = (-1)^\ell(\ell!)^2/[(2\ell)!(2\ell+1)!], \qquad p = 2\ell.$$

(iii) Among all A-stable methods of order $p = 2\ell$ *the diagonal Padé approximants, cf.* (1.5.1) *and* (1.5.3), *have the smallest error constants,* χ_p^*.

Theorem (1.7.2) and (1.7.6) yield immediately the following generalization of Corollary (1.7.4):

<u>(1.7.7) Corollary</u>. *(i) An* I_∞*-stable k-step ℓ-derivative method has order* $p \leq 2\ell$.

(ii) Among all I_∞*-stable k-step ℓ-derivative methods of order* $p = 2\ell$ *the diagonal Padé approximants have the smallest error constants.*

After having stated the results concerning implicit methods let us now turn to explicit methods. As concerns linear methods, the only consistent and explicit single-step method is the explicit Euler method, (4.2.1) with $\omega = 0$, which is not I_r-stable for any $r > 0$. For $\ell = 1$ and $k = 2$ the *leap-frog method* of order $p = 2$ with the polynomial

$$\pi(\zeta,\eta) = \zeta^2 - 1 - 2\eta\zeta$$

has the largest stability interval

$$I_r = \{i\eta, -r < \eta < r\} \subseteq S$$

with $r = 1$ as the following result of Jeltsch and Nevanlinna [81] reveals:

<u>(1.7.8) Theorem</u>. *If* (1.1.3) *is an explicit convergent k-step ℓ-derivative method then* $I_\ell \not\subset \overline{S}$ *or* $\overline{I}_\ell = \overline{S}$ *and the characteristic polynomial* $\pi(\zeta,\eta)$ *has a factor*

$$\pi_*(\zeta,\eta) = \zeta^2 - 2i^\ell T_\ell(-i\eta/\ell)\zeta + (-1)^\ell$$

where $T_\ell(\xi) = \cos\ell\arccos\xi$ *is the Tschebyscheff polynomial of degree* ℓ.

Observe that $T_{2\nu}(0) = (-1)^\nu$ hence these methods are not convergent for even ℓ because $0 \notin S$ in these cases.

The next result concerns the case $\ell = 1$ and $k = 3,4$, and is proved by an explicit construction of the desired methods; see Jeltsch and Nevanlinna [81].

<u>(1.7.9) Theorem.</u> *For every* $r \in [0, 1)$ *and* $k = 3,4$ *there exists an explicit linear* I_r*-stable k-step method of order* $p = k$.

<u>(1.7.10) Theorem.</u> *An explicit linear k-step method of order* $p = k$ *cannot be* I_r*-stable if* $k = 1 \bmod 4$.

Proof. See Jeltsch and Nevanlinna [81].

However, for any $r < \sqrt{3}$ there exists an implicit linear I_r-stable 4-step method of order $p = 6$; cf. Dougalis [79] and Lambert [73 , pp. 38, 39].

On the other side, for explicit single-step multiderivative methods van der Houwen [77] has proved:

<u>(1.7.11) Theorem.</u> *If an explicit single-step* ℓ*-derivative method is* I_r*-stable then* $r \leq 2[\ell/2]$. *The equality sign is attained for* ℓ *odd.*

Finally, the question for methods with maximum stability interval on the imaginary axis is answered completely for $\ell = 1,2$ by the following result of Jeltsch and Nevanlinna [82a, 82b]. Here, the methods are not necessarily explicit again.

<u>(1.7.12) Theorem.</u> *(i) If a k-step* ℓ*-derivative method is* I_r*-stable and* $p > 2\ell$ *then*

$$r \leq r_{\ell,\text{opt}} = \begin{cases} \sqrt{3}, & \ell = 1, \\ \sqrt{15}, & \ell = 2. \end{cases}$$

(ii) If $p = 2\ell$ *and* $I_r \subseteq S$ *with* $r > r_{\ell,\text{opt}}$ *then the error constant* χ_p *satisfies*

$$|\chi_p| \geq \begin{cases} (1 - (3/r^2))/12, & \ell = 1, \\ (1 - (15/r^2))/720, & \ell = 2. \end{cases}$$

(iii) The only method with $\ell = 1$ *and* $I_{\sqrt{3}} \subset S$ *is the Milne-Simpson method* (1.2.16). *The only method with* $\ell = 2$ *and* $I_{\sqrt{15}} \subset \bar{S}$ *is the method with the polynomial*

$$\pi(\zeta,\eta) = (\zeta - 1)^2 - \frac{3}{8}\eta(\zeta^2 - 1) + \frac{1}{24}\eta^2(\zeta^2 - 8\zeta + 1).$$

For $\ell = 1$ this result was proved by Dekker [81] in an independent way.

II. Direct Multistep Multiderivative Methods for Differential Systems of Second Order

2.1. Multistep Methods for Conservative Differential Systems

In Section (1.4) the initial value problem (1.4.1) was transformed into a first order problem of twice as large dimension before the numerical treatment which then has provided an approximation of the solution of the original problem and of its derivative simultaneously. In this chapter we consider direct approximation schemes without a-priori transformation. For the general initial value problem of second order with conservative differential equation,

(2.1.1) $y'' = f(t,y),\ t > 0,\ y(0) = y_0,\ y'(0) = y_0^*$,

a multistep method can be written formally in the same form as in Section 1.1. However, we prefer a slightly modified representation in which the even and odd total derivatives $f^{(j)}$ of f are summed up separately:

$$(2.1.2)\quad \begin{aligned} &\textstyle\sum_{j=0}^{\ell}\rho_j(T)(\Delta t^2\Theta^2)^j f_n^{(-2)}(v_n) + \Delta t\sum_{j=0}^{\ell}\rho_j^*(T)(\Delta t^2\Theta^2)^j f_n^{(-1)}(v_n) \equiv \\ &\textstyle\sum_{i=0}^{k}\sigma_i(\Delta t^2\Theta^2)T^i f_n^{(-2)}(v_n) + \Delta t\sum_{i=0}^{k}\sigma_i^*(\Delta t^2\Theta^2)^j T^i f_n^{(-1)}(v_n) = 0,\quad n = 0,1,\ldots \quad . \end{aligned}$$

Here we have to insert

$$(2.1.3)\quad \begin{aligned} &\rho_j(\zeta) = \textstyle\sum_{i=0}^{k}\alpha_{ji}\zeta^i,\ \rho_j^*(\zeta) = \sum_{i=0}^{k}\alpha_{ji}^*\zeta^i, && j = 0,\ldots,\ell, \\ &\sigma_i(\eta) = \textstyle\sum_{j=0}^{\ell}\alpha_{ji}\zeta^j,\ \sigma_i^*(\eta) = \sum_{j=0}^{\ell}\alpha_{ji}^*\zeta^j, && i = 0,\ldots,k, \end{aligned}$$

and $\Theta = \partial/\partial t$ is again the differential operator. Furthermore, $f^{(-2)}(v_n) \equiv v_n$, and $f^{(-1)}(v_n) \equiv w_n$ plays the role of an approximation to y_n'. The total derivatives of f are again to be expressed by partial derivatives of f using the recurrence formula (1.1.2).

In order to overcome the deficiency that the scheme (2.1.2) is only one recurrence formula for the two unknown sequences $\{v_n\}_{n=k}^{\infty}$ and $\{w_n\}_{n=k}^{\infty}$ we have three possibilities:

(i) Put $\rho_j^*(\zeta) \equiv 0$, $j = 0,\ldots,\ell$, i.e., $\sigma_i^*(\eta) \equiv 0$, $i = 0,\ldots,k$.

(ii) Introduce a further scheme of the same type such that - besides other conditions described below - (v_{n+k}, w_{n+k}) can be computed from (v_ν, w_ν), $\nu = n,\ldots,n+k-1$.

(iii) Choose a finite difference approximation $\Delta t^{-1}\tau_i(T)y_n$ to y_{n+i}' and replace w_{n+i} by $\Delta t^{-1}\tau_i(T)v_n$, $i = 0,\ldots,k$.

Rather few is known on the third way hence it shall not be discussed here although it allows a simple generalization to nonconservative differential systems. The second way leads to numerical schemes of Nyström type which are considered in Section 2.4. In this and the next two sections we study the first case, i.e., numerical schemes of the form

(2.1.4)
$$\pi(T,\Delta t^2\Theta^2)f_n^{(-2)}(v_n) \equiv \sum_{j=0}^{\ell}\rho_j(T)(\Delta t^2\Theta^2)^j f_n^{(-2)}(v_n)$$
$$\equiv \sum_{i=0}^{k}\sigma_i(\Delta t^2\Theta^2)T^i f_n^{(-2)}(v_n) = 0, \qquad n = 0,1,\ldots \quad .$$

For $\ell = 1$ we obtain linear multistep methods for conservative differential systems of second order which have been proposed for the solution of dynamic finite element equations e.g. by Bathe and Wilson [76], Dougalis [79], and Gekeler [76 , 80]. However, if $\ell > 1$ then the numerical approximation w_n to $y'(n\Delta t)$ does not appear in (2.1.4) only if the initial value problem is of the form

(2.1.5) $\quad y'' = Ay + c(t),\ t > 0,\ y(0) = y_0,\ y'(0) = y_0^*,$

where A is a constant matrix. Therefore the applicability of multistep *multiderivative* methods is restricted to this special case; cf. e.g. Baker et al. [79].

In analogy to (1.1.5) we assume henceforth that

(2.1.6) $\quad \alpha_{0k} \neq 0,\ \rho_\ell(\zeta) \not\equiv 0,$ and $\sigma_0(\eta) \not\equiv 0,$

and we suppose again without loss of generality that the *characteristic polynomial* $\pi(\zeta,\eta^2)$ defined by (2.1.4) is irreducible with respect to ζ and η^2.

The discretization error of the method (2.1.4) is now

(2.1.7) $\quad d(\Delta t,y)(t) = \sum_{j=0}^{\ell}\rho_j(T)\Delta t^{2j}y^{(2j)}(t)$

and the method is consistent if there exists a constant Γ not depending on Δt such that

$$\|d(\Delta t,u)(t)\| \le \Gamma\Delta t^{p+2} \qquad \forall\ u \in C^{p_*}(\mathbb{R};\mathbb{R}^m)$$

for a $p \in \mathbb{N}$ and $p_* = \max\{p+2,2\ell\}$, the maximum p being the order of the method.

<u>(2.1.8) Lemma</u>. *If the method* (2.1.4) *is consistent of order* p *then*

$$\|d(\Delta t,u)(t)\| \le \Gamma\Delta t^{p+1}\int_t^{t+k\Delta t}\|u^{(p+2)}(\tau)\|d\tau + \sum_{j=[(p+3)/2]}^{\ell}\Delta t^{2j}\|\rho_j(T)u^{(2j)}(t)\| \ \forall\, u\in C^{p_*}(\mathbb{R};\mathbb{R}^m)$$

where Γ *does not depend on* t, Δt, u, *and* m.

Proof. It suffices to prove the assertion for $p \geq 2\ell - 1$. In the same way as in Lemma (1.1.8) we substitute into (2.1.7) the Taylor expansions

$$u^{(2j)}(t+i\Delta t) = \sum_{\nu=0}^{p+1-2j} \frac{(i\Delta t)^\nu}{\nu!} u^{(2j+\nu)}(t) + \frac{1}{(p+1-2j)!} \int_0^{i\Delta t} (i\Delta t-\tau)^{p+1-2j} u^{(p+2)}(t+\tau)d\tau$$

and obtain

$$\begin{aligned} d(\Delta t,u)(t) &= \sum_{j=0}^{\ell}\sum_{\nu=0}^{p+1-2j}\left(\sum_{i=0}^{k}\alpha_{ji}\frac{i^\nu}{\nu!}\right)\Delta t^{2j+\nu}u^{(2j+\nu)}(t) \\ &+ \sum_{j=0}^{\ell}\sum_{i=0}^{k}\alpha_{ji}\frac{\Delta t^{2j}}{(p+1-2j)!}\int_0^{i\Delta t}(i\Delta t-\tau)^{p+1-2j}u^{(p+2)}(t+\tau)d\tau. \end{aligned} \tag{2.1.9}$$

The assumption that the method is consistent of order p implies

$$\begin{aligned} &\sum_{j=0}^{\ell}\sum_{\nu=0}^{p+1-2j}\left(\sum_{i=0}^{k}\alpha_{ji}\frac{i^\nu}{\nu!}\right)\Delta t^{2j+\nu}u^{(2j+\nu)}(t) \\ &\equiv \sum_{\mu=0}^{p+1}\sum_{j=0}^{\min\{[\mu/2],\ell\}}\sum_{i=0}^{k}\alpha_{ji}\frac{i^{\mu-2j}}{(\mu-2j)!}\Delta t^\mu u^{(\mu)}(t) = 0 \end{aligned} \tag{2.1.10}$$

where $[\mu]$ denotes the largest integer not greater than μ. This yields

$$d(\Delta t,u)(t) = \int_0^{k\Delta t}\left[\sum_{j=0}^{\ell}\sum_{i=0}^{k}\alpha_{ji}\frac{\Delta t^{2j}}{(p+1-2j)!}(i\Delta t-\tau)_+^{p+1-2j}\right]u^{(p+2)}(t+\tau)d\tau$$

or

$$\|d(\Delta t,u)(t)\| \leq \left(\sum_{j=0}^{\ell}\sum_{i=0}^{k}|\alpha_{ji}|\frac{i^{p+1-2j}}{(p+1-2j)!}\right)\Delta t^{p+1}\int_t^{t+k\Delta t}\|u^{(p+2)}(\tau)\|d\tau$$

which proves the assertion.

From (2.1.10) we immediately obtain:

<u>(2.1.11) Lemma</u>. *The method* (2.1.4) *is consistent of order* p *iff*

$$\sum_{j=0}^{\min\{[\mu/2],\ell\}}\sum_{i=0}^{k}\alpha_{ji}\frac{i^{\mu-2j}}{(\mu-2j)!} = 0, \qquad \mu = 0,\dots,p+1.$$

In particular, the method (2.1.4) is consistent iff the following conditions for the consistence of *linear* multistep methods are fulfilled,

$$\rho_0(1) = \rho_0'(1) = \rho_0''(1) + 2\rho_1(1) = 0. \tag{2.1.12}$$

The analogue to Lemma (1.1.15) now reads as follows:

<u>(2.1.13) Lemma</u>. *The method* (2.1.4) *is consistent of order* p *iff the characteristic*

polynomial $\pi(\zeta,\eta^2)$ *satisfies*

$$\pi(e^{\Delta t},\Delta t^2) = \chi_p \Delta t^{p+2} + \mathcal{O}(\Delta t^{p+3}),\ \Delta t \to 0,\ \chi_p \neq 0.$$

Again a study of the trivial equation $y'' = \kappa$ shows that we must suppose that $\rho_0''(1) = -2\rho_1(1) \neq 0$ and we may stipulate again that (1.1.14) holds, i.e., $\rho_1(1) = -1$. The conditions (2.1.12) and (1.1.14) together imply that $\zeta = 1$ is a root of $\pi(\zeta,\eta^2)$ for $\eta^2 = 0$ which has exactly multiplicity *two*. Therefore $\eta^2 = 0$ is no longer contained in the stability region S defined by (1.2.7). For this reason we have to weaken the concept of absolute stability appropriately in direct multistep methods for differential systems of second order.

<u>(2.1.14) Definition</u>. *The stability region* S *of a method* (2.1.4) *consists of the* $\eta^2 \in \bar{\mathbb{C}}$ *with the following properties:*

(i) $\sigma_k(\eta^2) \neq 0$, $\eta^2 \in S \cap \mathbb{C}$,

(ii) all roots $\zeta_i(\eta)$ *of* $\pi(\zeta,\eta^2)$ *satisfy* $|\zeta_i(\eta)| \leq 1$,

(iii) all roots $\zeta_i(\eta)$ *of* $\pi(\zeta,\eta^2)$ *with* $|\zeta_i(\eta)| = 1$ *have multiplicity not greater than two.*

As in Chapter I a method with $0 \in S$ is called *zero-stable*. Henrici [62 , Theorem 6.6] has proved that consistent and zero-stable *linear* multistep methods are convergent in the classical sense. A generalization of this result to nonlinear methods is not difficult for the restricted class of applications (2.1.5) therefore consistent and zero-stable methods are again called *convergent* below. Observe that then $0 \in \partial S$ by Lemma (1.3.3).

For the roots $\zeta_i(\eta)$ of the polynomial $\pi(\zeta,\eta^2)$ we obtain by implicit differentiation and Theorem (A.1.4):

Case (i). If $\zeta_i(0)$ is a simple root of $\pi(\zeta,0)$ then $(\rho_0'(\zeta_i(0)) \neq 0$ and)

$$\begin{aligned} &\zeta_i(\eta) = \zeta_i(0) + \chi_i\eta^2 + \mathcal{O}(\eta^4), \qquad \eta \to 0, \\ (2.1.15)\quad & \\ &\chi_i = [-\rho_1/\rho_0'](\zeta_i(0)). \end{aligned}$$

Case (ii). If $\zeta_i(0)$ is a double root of $\pi(\zeta,0)$ then $(\rho_0'(\zeta_i(0)) = 0,\ \rho_0''(\zeta_i(0)) \neq 0)$

$$\begin{aligned} &\zeta_{i,i+1}(\eta) = \zeta_i(0) + \hat{\chi}_i(\pm\eta) + \frac{\tilde{\chi}_i}{2}\eta^2 + \mathcal{O}(\eta^3), \qquad \eta \to 0, \\ (2.1.16)\quad & \\ &\hat{\chi}_i^2 = [-2\rho_1/\rho_0''](\zeta_i(0)), \end{aligned}$$

and if $\hat{\chi}_i \neq 0$ then

(2.1.17) $\tilde{\chi}_i = [(2\rho_0'''\rho_1 - 6\rho_0''\rho_1')/3\rho_0''^2](\zeta_i(0))$.

Accordingly, a convergent method has exactly two roots, $\zeta_1(\eta)$ and $\zeta_2(\eta) = \zeta_1(-\eta)$, called again the *principal roots* which have the property

(2.1.18) $\zeta_{1,2}(\eta) = 1 \pm \eta + \mathcal{O}(\eta^2)$, $\qquad \eta \to 0$.

(2.1.19) Lemma. *Let* $0 \in S$ *then the method defined by* $\pi(\zeta,\eta^2)$ *is consistent of order* p *iff*

(2.1.20) $\zeta_1(\eta) = e^\eta - \chi_p\eta^{p+1} + \mathcal{O}(\eta^{p+2})$, $\chi_p \neq 0$, $\qquad \eta \to 0$.

Proof. In a sufficiently small neighborhood of $\eta = 0$ we define

(2.1.21) $\tilde{\pi}(\zeta,\eta^2) = \pi(\zeta,\eta^2)/[(\zeta - \zeta_1(\eta))(\zeta - \zeta_1(-\eta))]$

and then obtain

(2.1.22) $\tilde{\pi}(1,0) = \rho_0''(1) = -2\rho_1(1) \neq 0$

because by (2.1.12)

$$\pi(\zeta,0) = \rho_0(1) + \rho_0'(1)(\zeta - 1) + \frac{\rho_0''(1)}{2}(\zeta - 1)^2 + \mathcal{O}((\zeta - 1)^3)$$
$$= \frac{\rho_0''(1)}{2}(\zeta - 1)^2 + \mathcal{O}((\zeta - 1)^3), \qquad \eta \to 0.$$

If the method is consistent of order p then Lemma (2.1.13) yields by a substitution of e^η for ζ in $\pi(\zeta,\eta^2)$

(2.1.23) $\pi(e^\eta,\eta^2) = (e^\eta - \zeta_1(\eta))(e^\eta - \zeta_1(-\eta))\tilde{\pi}(e^\eta,\eta^2) = \chi_p\eta^{p+2} + \mathcal{O}(\eta^{p+3})$, $\qquad \eta \to 0$.

But, by (2.1.18),

$$e^\eta - \zeta_1(-\eta) = (1 + \eta) + \mathcal{O}(\eta^2) - (1 - \eta) + \mathcal{O}(\eta^2), \qquad \eta \to 0,$$

therefore (2.1.22) and (2.1.23) prove (2.1.20). On the other side, if (2.1.20) holds then a substitution into the first equation of (2.1.23) yields

$$\pi(e^\eta,\eta^2) = \tilde{\pi}(e^\eta,\eta^2)(e^\eta-\zeta_1(-\eta))(\chi_p\eta^{p+1}+\mathcal{O}(\eta^{p+2})) = \tilde{\pi}(e^\eta,\eta^2)(2\chi_p\eta^{p+2} + \mathcal{O}(\eta^{p+3})), \; \eta \to 0,$$

hence the method has order p by Lemma (2.1.13).

Note that $\chi_p = \chi/2$ where χ is the error constant introduced by Henrici [62 , p.296]; see also Jeltsch and Nevanlinna [81 , (2.8) and subsequent remark]. Moreover, Lemma (2.1.19) and Lemma (1.3.5) have exactly the same form although (or better because) S is defined in two different ways. The approximation properties of $\zeta_1(\eta)$ are thus the same with respect to η in both times.

By (2.1.18) we obtain the same equation as in the proof of Lemma (1.3.12),

$$|\zeta_1(\eta)| = 1 + \mathrm{Re}(\eta) + \mathcal{O}(|\eta|^2), \qquad \eta \to 0.$$

Substituting again $\eta = \rho(1 + e^{i\phi})$, $0 \leq \phi < 2\pi$, we find that - with respect to η^2 - the domain

$$\mathcal{D} = \{\eta^2 \in \mathbb{C},\ \eta^2 = \rho^2(1 + e^{i\phi})^2,\ 0 \leq \rho < \rho_0,\ 0 \leq \phi < 2\pi\}$$

satisfies $\mathcal{D} \subset \overline{\mathbb{C}} \setminus S$ for some $\rho_0 > 0$ sufficiently small. But $\mathbb{C}\setminus\mathcal{D}$ contains no angular domain $\{\eta^2 \in \mathbb{C},\ |\beta - \arg(\eta^2)| \leq \alpha\}$ for a positive α and for the angle $\beta \neq \pi$ not even a half-line in a neighborhood of zero:

<u>(2.1.24) Lemma</u>. *The stability region S of a convergent method defined by* $\pi(\zeta,\eta^2)$ *does not contain in a neighborhood of zero a domain* $\{\eta^2 \in \mathbb{C},\ |\beta - \arg(\eta^2)| \leq \alpha\}$ *with* $\alpha \geq 0$ *for* $\beta \neq \pi$ *and* $\alpha > 0$ *for* $\beta = \pi$.

This result corresponds directly to the third assertion of Lemma (1.3.12) via the mapping $\eta \to \eta^2$. The first two assertions of this lemma have here a somewhat more complicated form. Let

$$\chi_i^* = \chi_i/\zeta_i(0),\ \hat{\chi}_j^* = \hat{\chi}_j/\zeta_j(0),\ \tilde{\chi}_j^* = \tilde{\chi}_j/\zeta_j(0)$$

be the growth parameters of the simple unimodular roots $\zeta_i(0)$ and the double unimodular roots $\zeta_j(0)$ of $\pi(\zeta,0)$ respectively where the constants χ_i, $\hat{\chi}_j$, and $\tilde{\chi}_j$ are defined in (2.1.15), (2.1.16), and (2.1.17).

<u>(2.1.25) Lemma</u>. *Let the method defined by* $\pi(\zeta,\eta^2)$ *be convergent and let all growth parameters* χ_i^* *and* $\hat{\chi}_j^*$ *be nonzero. Then*
(i) S *contains the negative real* η^2*-line in a neighborhood of zero iff all simple and all double unimodular roots of* $\pi(\zeta,0)$ *satisfy*

$$\mathrm{Re}(\chi_i^*) > 0,\ \mathrm{Im}(\hat{\chi}_j^*) = 0,\ \mathrm{Re}(\tilde{\chi}_j^*) - (\hat{\chi}_j^*)^2 > 0,$$

(ii) $\mathbb{C} \setminus$ S *contains the negative real line in a neighborhood of zero iff some simple*

or some double unimodular roots of $\pi(\zeta,0)$ *satisfy*

$$\mathrm{Re}(\chi_i^*) < 0 \text{ or } \mathrm{Im}(\hat{\chi}_j^*) \neq 0 \text{ or } \mathrm{Re}(\tilde{\chi}_j^*) - (\hat{\chi}_j^*)^2 < 0.$$

Proof. With respect to simple roots the statement is the same as in Lemma (1.3.12) with η replaced by η^2; cf. (2.1.15). For the double unimodular roots we observe that (1.3.10) holds, i.e.,

$$|\zeta_{j,j+1}(\eta)| = 1 + \mathrm{Re}(\hat{\chi}_j^*)\mathrm{Re}(\pm\eta) - \mathrm{Im}(\hat{\chi}_j^*)\mathrm{Im}(\pm\eta) + \mathcal{O}(|\eta|^2), \qquad \eta \to 0,$$

and, furthermore, if $\mathrm{Im}(\hat{\chi}_j^*) = 0$ and $\eta^2 < 0$,

$$\begin{aligned}|\zeta_{j,j+1}(\eta)| &= |1 + \hat{\chi}_j^*(\pm\eta) + \frac{\tilde{\chi}_i^*}{2}\eta^2| + \mathcal{O}(|\eta|^3)\\ &= 1 + 2^{-1}(\mathrm{Re}(\tilde{\chi}_j^*) - (\hat{\chi}_j^*)^2)\eta^2 + \mathcal{O}(|\eta|^3), \qquad \eta \to 0.\end{aligned}$$

From these two relations the remaining assertions follow immediately.

More explicitely, writing id for the identity mapping then we have for $\mathrm{Im}(\hat{\chi}_j^*) = 0$

$$\mathrm{Re}(\tilde{\chi}_j^*) - (\hat{\chi}_j^*)^2 = \mathrm{Re}\{[((2\rho_0'''\rho_1 - 6\rho_0''\rho_1')\mathrm{id} + 6\rho_1\rho_0'')/(3\rho_0''^2(\mathrm{id})^2)](\zeta_j(0))\}.$$

The growth parameters χ_i^* and $\hat{\chi}_j^*$ are nonzero if all unimodular roots of $\rho_0(\zeta)$ are not roots of $\rho_1(\zeta)$; cf. (2.1.15) and (2.1.16). This condition is fullfilled at least in two cases:

(i) If the method is linear, i.e., if $\pi(\zeta,\eta^2) = \rho_0(\zeta) + \eta^2\rho_1(\zeta)$ because $\pi(\zeta,\eta^2)$ is irreducibel.

(ii) If $\zeta_{1,2}(0) = 1$ is the unique double unimodular root of $\pi(\zeta,0)$ because a convergent method satisfies $\rho_1(1) \neq 0$ by (2.1.12) and $0 \in S$.

A solution of the test equation $y'' = -\lambda^2 y$, $0 \neq \lambda^2 \in \mathbb{R}$, neither decreases nor increases with increasing t but oscillates. In the long-range solution of a problem (2.1.15) the numerical scheme should have a similar behavior therefore Lambert and Watson [76] have introduced the following notation:

<u>(2.1.26) Definition.</u> *A convergent method defined by* $\pi(\zeta,\eta^2)$ *has the periodicity interval* $[-s, 0]$, $0 < s$, *if all roots* $\zeta_i(\eta)$ *of* $\pi(\zeta,\eta^2)$ *satisfy* $|\zeta_i(\eta)| = 1$ *for* $\eta^2 \in [-s, 0]$.

A method with the polynomial $\pi(\zeta,\eta^2) = \sum_{j=0}^{\ell}\rho_j(\zeta)\eta^{2j} = \sum_{i=0}^{k}\sigma_i(\eta^2)\zeta^i$ is called *symmetric* if

$$(2.1.27)\quad \rho_j(\zeta) = \zeta^k\rho_j(\zeta^{-1}), \qquad j = 0,\ldots,\ell,$$

then $\sigma_i(\eta^2) = \sigma_{k-i}(\eta^2)$, $i = 0,\dots,k$, and $\pi(\zeta,\eta^2) = \zeta^k\pi(\zeta^{-1},\eta^2)$. Lambert and Watson [76] have shown that a *linear* multistep method (2.1.4) with a periodicity interval is necessarily symmetric. Moreover, if the principal root $\zeta_{1,2}(0) = 1$ is the only double root of $\pi(\zeta,0) = \rho_0(\zeta)$ then this necessary condition is also sufficient. The following lemma generalizes this result and simplificates at the same time Lemma (2.1.25) essentially for linear symmetric methods.

(2.1.28) Lemma. (Jeltsch [78b].) *Let the linear multistep method defined by* $\pi(\zeta,\eta^2)$ *be convergent and symmetric. Then it has a periodicity interval iff all growth parameters* $\hat{\chi}_j^*$ *of the double unimodular roots* $\zeta_j(0)$ *satisfy* $(\hat{\chi}_j^*)^2 > 0$.

Proof. We quote the proof because it gives some insight in symmetric methods and proceed in several steps assuming throughout convergence and symmetry.

(i) Let ζ^* be a root of $\rho_0(\zeta)$ then ζ^{*-1} is also a root of $\rho_0(\zeta)$ because $\rho_0(\zeta) = \zeta^k\rho_0(\zeta^{-1})$. Accordingly, because $0 \in S$ we obtain that *all* roots of $\rho_0(\zeta)$ are unimodular, i.e., $\eta^2 = 0$ belongs to the periodicity interval.

(ii) Let $\zeta_j(0)$ be a *simple* (unimodular) root of $\rho_0(\zeta)$ and consider $\zeta_j(\eta)$ near $\eta = 0$, $\eta^2 < 0$. Then $\overline{\zeta_j}(\eta)$ is a root of $\pi(\zeta,\eta^2)$ because $\pi(\zeta,\eta^2)$ has real coefficients and consequently $\overline{\zeta_j}(\eta)^{-1}$ is also a root of $\pi(\zeta,\eta^2)$ because of the symmetry. Let

$$\zeta_m(\eta) = 1/\overline{\zeta_j}(\eta)$$

then we obtain from $|\zeta_j(0)| = 1$, i.e., $\zeta_j(0) = e^{i\phi}$ and $\overline{\zeta_j}(0)^{-1} = e^{i\phi}$, that $\zeta_m(0) = \zeta_j(0)$. If $m \neq j$ then $\zeta_j(0)$ is a double unimodular root which is a contradiction hence $m = j$ and, consequently, $\zeta_j(\eta) = 1/\overline{\zeta_j}(\eta)$ or $|\zeta_j(\eta)| = 1$. Therefore the roots emanating from simple unimodular roots of $\pi(\zeta,0)$ satisfy $|\zeta_j(\eta)| = 1$ for some $\eta_0^2 \le \eta^2 \le 0$, $\eta_0^2 < 0$.

(iii) Let $\zeta_j(0)$ be a *double* (unimodular) root of $\pi(\zeta,0)$ then $\zeta_j(\eta)$ is analytic in a neighborhood of zero by Lemma (A.1.3) and we obtain

$$\zeta_{j,j+1}(\eta) = \zeta_j(0)(1 + \hat{\chi}_j^*(\pm\eta) + \mathcal{O}(\eta^2)), \qquad \eta \to 0,$$

with the growth parameter $\hat{\chi}_j^* \neq 0$. By the Implicit Function Theorem (cf. e.g. Dieudonné [60, § 10.2] we find that $\eta \to \zeta_j(\eta)$ is invertible and an implicit differentiation of $\pi(\zeta,\eta^2(\zeta))$ with respect to ζ reveals that in a neighborhood of $\zeta_j(0) \equiv \zeta_j$

$$(2.1.29)\quad \eta^2(\zeta) = \omega(\zeta) = \kappa(\zeta - \zeta_j)^2 + \mathcal{O}((\zeta - \zeta_j)^3), \qquad \zeta \to \zeta_j,$$

where $\kappa = 1/\hat{\chi}_j{}^2$. $\zeta_j(\eta)$ satisfies $|\zeta_j(\eta)| = 1$ near $\eta^2 = 0$, $\eta^2 < 0$, iff the function $\phi \to \omega(\zeta_j e^{i\phi})$ maps a sufficiently small interval $(-\phi_0, \phi_0)$, $\phi_0 > 0$, onto an interval $[-\eta_1^2, 0]$. But from $\rho_0(\zeta) + \eta^2\rho_1(\zeta) = 0$ we obtain

(2.1.30) $\eta^2 = \omega(\zeta) = -\rho_0(\zeta)/\rho_1(\zeta)$

hence $\omega(\zeta_j e^{i\phi}) = \omega(e^{i\psi}), \zeta_j = e^{i\phi_0}$, $\psi = \phi + \phi_0$, is a real function in ϕ because of the symmetry. Moreover, $\phi = 0$ is a double root of $\omega(\zeta_j e^{i\phi})$ hence it is negative in a real neighborhood of $\phi = 0$ iff the second derivative with respect to ϕ is negative in $\phi = 0$. But by (2.1.29) or (2.1.30) we obtain

(2.1.31) $\partial^2\omega(\zeta_j e^{i\phi})/\partial\phi^2|_{\phi=0} = -2\kappa\zeta_j^2 = -2/(\hat{\chi}_j^*)^2 < 0$

by assumption which proves the assertion.

In the attempt to generalize Lemma (2.1.28) to nonlinear multistep methods (2.1.4) we obtain again a function ω which satisfies (2.1.29) and (2.1.31). But (2.1.30) is naturally fulfilled no longer hence it is more difficult to show that $\omega(\zeta_j e^{i\phi})$ is a real function near $\phi = 0$.

A polynomial $\pi(\zeta,\eta^2)$ can have multiple roots only in a finite number of values η^2; cf. e.g. Ahlfors [53 , § 6.2]. Therefore, if a convergent method has a periodicity interval $[-s_1, 0]$, $s_1 > 0$, then always $[-s_2,0] \subset \partial S$ for some $0 < s_2 \leqq s_1$. It would however be interesting to know wether the case $[-s_1, 0] = \partial S$ can be characterized algebraically.

After these preliminary results on the stability region S we now turn to initial value problems of second order and postpone a study of properties of S in special methods until Section 2.5. We consider the initial value problem

(2.1.32) $y'' = A^2y + c(t) + h(t),\ t > 0,\ y(0) = y_0,\ y'(0) = y_0^*$,

where the notation A^2 instead of A is used only for convenience. As in the first chapter, the perturbation h(t) is neglected in the numerical approximation (2.1.4) and we obtain the computational device

(2.1.33) $\sum_{i=0}^k \sigma_i(\Delta t^2A^2)T^i v_n = -\sum_{i=0}^k T^i \sum_{j=1}^{\ell} \sigma_{ij}(\Delta t^2A^2)\Delta t^{2j} c_n^{(2j-2)}$, $\qquad n = 0,1,\ldots$.

For the definition of the polynomials $\sigma_{ij}(\eta)$ see (1.2.3). The error estimation corresponding to Theorem (1.2.12) has here the following form.

<u>(2.1.34) Theorem</u>. *(i) Let the (m,m)-matrix* A^2 *in* (2.1.32) *be diagonable,* $A^2 = X\Lambda^2X^{-1}$, *and let the solution* y *be (p+2)-times continuously differentiable.*
(ii) Let the method (2.1.33) *be consistent of order* $p \geqq 2\ell - 1$ *with the stability region* S.
(iii) Let $Sp(\Delta t^2A^2) \subset R \subseteq S$ *where* R *is closed in* $\overline{\mathbb{C}}$.
Then for n = k,k+1,...,

$$|X^{-1}(y_n - v_n)| \le \kappa_R |X^{-1}| n\Delta t\Big[\Delta t^{-1}|Y_{k-1} - V_{k-1}| + \Delta t^p \int_0^{n\Delta t} |y^{(p+2)}(\tau)|d\tau + n\Delta t \max_{0\le i\le 2\ell-2} |||h^{(i)}|||_n\Big].$$

Proof. Exactly in the same way as in Theorem (1.2.12) we obtain

(2.1.35) $E_n = F_\pi(\Delta t^2 A^2)E_{n-1} + D_n - H_n, \qquad n = k,k+1,\ldots,$

where $E_n = ((y_{n-k+1} - v_{n-k+1}),\ldots,(y_n - v_n))^T$,

(2.1.36) $D_n = (0,\ldots,0,\sigma_k(\Delta t^2 A^2)^{-1} d(\Delta t,y)_{n-k})^T$,

and now

(2.1.37) $H_n = (0,\ldots,0,\sigma_k(\Delta t^2 A^2)^{-1}\sum_{i=0}^k T^i \sum_{j=1}^\ell \sigma_{ij}(\Delta t^2 A^2)\Delta t^{2j} h_{n-k}^{(2j-2)})^T$.

The Uniform Boundedness Theorem (cf. Appendix) yields by assumption (iii) in the present case

$$|n^{-1}F_\pi(\Delta t^2 \Lambda^2)^n| = \max_{1\le\mu\le m}|n^{-1}F_\pi(\Delta t^2\lambda_\mu^2)^n| \le \kappa_R$$

therefore we obtain

$$|X^{-1}E_n| \le \kappa_R n\Delta t\Big[\Delta t^{-1}|X^{-1}E_{k-1}| + \Gamma|X^{-1}|\Delta t^p \int_0^{n\Delta t} |y^{(p+2)}(\tau)|d\tau + \Delta t^{-1}\textstyle\sum_{\nu=k}^n |X^{-1}H_\nu|\Big].$$

Now the assertion follows in the same way as in Theorem (2.1.12) observing that $\sigma_k(\eta^2)$ is a polynomial of exact degree ℓ in η^2 if $R \subset S$ is unbounded.

This error estimation differs from that in Theorem (1.2.12) by the multiplication factor $n\Delta t$. Moreover, the initial error $|Y_{k-1} - V_{k-1}|$ is multiplied by Δt^{-1} therefore the method for the computation of the start vectors $v_0,\ldots,v_{k-1}$ must be of order p+1 in order that global order p is obtained. This effect is due to the occurence of double unimodular roots in convergent methods. However, if the principal roots, $\zeta_1(\eta)$ and $\zeta_1(-\eta)$, are the only roots coalescing to double unimodular roots in the stability interval [-s, 0] then the error estimation with respect to the initial error can be replaced by a more diversified one which seems somewhat more satisfactory. This is achieved by a splitting off of one of the principal roots from the characteristic polynomial and observing that the resulting modified error equation has the same properties as in multistep methods for *first* order systems.

<u>(2.1.38) Definition</u>. *Let* $\pi^*(\zeta,\eta) = \pi(\zeta,\eta^2)/(\zeta - \zeta_1(\eta))$ *then a method* (2.1.4) *is strongly D-stable in* [-s, 0] *if* [-s, 0] $\subset$ S *and if all roots* $\zeta_i(\eta)$ *of* $\pi^*(\zeta,\eta)$ *with*

$|\zeta_i(\eta)| = 1$ *are simple roots of* $\pi^*(\zeta,\eta)$ *for* $\eta^2 \in [-s, 0]$.

The polynomial $\pi^*(\zeta,\eta)$ can be chosen on the entire real η^2-line with exception of possible poles (which cannot lie in $[-s, 0]$) as a fixed polynomial of degree k-1 in ζ with coefficients that are continuous in η independently of possible real branching points of the algebraic function $\zeta(\eta)$ defined by $\pi(\zeta(\eta),\eta^2) = 0$.

(2.1.39) Theorem. *Let the assumption of Theorem* (2.1.34) *be fulfilled but let* $R = [-s, 0]$ *and let the method* (2.1.4) *be strongly D-stable in* R. *Then for* $n = k,k+1,\ldots,$

$$|X^{-1}(y_n - v_n)| \le |X^{-1}E^0_{k-2}| + \kappa_S|X^{-1}|n\Delta t\Big[|A||E^0_{k-2}| + \Delta t^{-1}|E^0_{k-1} - E^0_{k-2}| + \Delta t^p\int_0^{n\Delta t}|y^{(p+2)}(\tau)|d\tau + n\Delta t \max_{0\le i\le 2\ell-2}|||h^{(i)}|||_n\Big]$$

where $E^0_n = (y_{n-k+2} - v_{n-k+2}, \ldots, y_n - v_n)^T$.

Proof. Using the polynomial $\pi^*(\zeta,\eta)$ instead of $\pi(\zeta,\eta^2)$ the error equation writes as

$$\pi(T,\Delta t^2A^2)e_n = \pi^*(T,\Delta tA)(T - \zeta_1(\Delta tA))e_n = d(\Delta t,y)_n - \textstyle\sum_{i=0}^{k}T^i\sum_{j=1}^{\ell}\sigma_{ij}(\Delta t^2A^2)\Delta t^{2j}h_n^{(2j-2)}$$

therefore we can write the corresponding single-step equation in the following way:

$$(2.1.40)\quad E^0_n - \zeta_1(\Delta tA)E^0_{n-1} = F_{\pi^*}(\Delta tA)(E^0_{n-1} - \zeta_1(\Delta tA)E^0_{n-2}) + D^0_n - H^0_n,\ n = k,k+1,\ldots .$$

D^0_n and H^0_n are here block vectors of the same form as in the proof of Theorem (2.1.34) but with block dimension k-1. By the Uniform Boundedness Theorem we now obtain

$$|F_{\pi^*}(\Delta tA)| \le \sup_{-s\le\eta^2\le 0}\sup_{n\in\mathbb{N}}|F_{\pi^*}(\eta)| \le \kappa_R.$$

An estimation of (2.1.40) by means of this bound yields

$$(2.1.41)\quad \begin{aligned}|X^{-1}E^0_n - \zeta_1(\Delta t\Lambda)X^{-1}E^0_{n-1}| \le \kappa_R\Big[&|X^{-1}E^0_{k-1} - \zeta_1(\Delta t\Lambda)X^{-1}E^0_{k-2}|\\ &+ r|X^{-1}|\Delta t^{p+1}\int_0^{n\Delta t}|y^{(p+2)}(\tau)|d\tau + |X^{-1}|n\Delta t^2\max_{0\le i\le 2\ell-2}|||h^{(i)}|||_n\Big].\end{aligned}$$

But $|\zeta_1(\Delta t\Lambda)| \le 1$ by assumption (iii) of Theorem (2.1.34) hence the triangle inquality and a further iteration yield

$$(2.1.42)\quad |X^{-1}E^0_n| \le |X^{-1}E^0_{k-2}| + n\kappa_R[\ldots\ldots]$$

where [......] is the square-bracketed term on the right side of (2.1.41). Finally,

Lemma (A.1.8) yields

$$|X^{-1}E^0_{k-1} - \zeta_1(\Delta t\Lambda)X^{-1}E^0_{k-2}| \le |X^{-1}E^0_{k-1} - X^{-1}E^0_{k-2}| + |X^{-1}(I - \zeta_1(\Delta tA))E^0_{k-2}|$$
$$\le |X^{-1}|[|E^0_{k-1} - E^0_{k-2}| + \Gamma\Delta t|A||E^0_{k-2}|] .$$

A substitution of this estimation into (2.1.42) proves the desired result.

In Theorem (2.1.39) the initial error is multiplied by $|A|$ instead by Δt^{-1} as in Theorem (2.1.34). This phenomenon was already observed by Dupont [73] for two-step methods. The difference quotient of the initial error corresponds to the initial condition $y'(0) = y^*_0$ of the analytic initial value problem (2.1.32).

A two-step method is always strongly D-stable in $S \cap \mathbb{R}$. But in methods with step number $k > 2$ the polynomial $\pi^*(\zeta,\eta)$ is never computed explicitely. However, if a method is strongly D-stable in $\eta^2 = 0$ and if *all* roots (or at least *all unimodular* roots) of the original polynomial $\pi(\zeta,\eta^2)$ are simple in $[-s, 0) \subset S$ then it is strongly D-stable in $[-s, 0]$. So this somewhat stronger condition may be used in the general case.

2.2. Linear Multistep Methods for Differential Systems with Damping

In this section we consider the real initial value problem

(2.2.1) $y'' = A^2y + B(t)y' + c(t),\ t > 0,\ y(0) = y_0,\ y'(0) = y^*_0,$

with a sufficiently smooth matrix $B(t)$ not necessarily commuting with A^2. That case of orthogonal damping is studied in Section 2.3. Here, we cannot expect error estimations of that strong uniform character as derived in Chapter I even if A^2 *and* $B(t)$ are symmetric and negative definite; cf. e.g. Gear [78]. The utmost we can obtain is a error bound which is uniform with respect to the leading matrix A^2 and 'classical' with respect to B. This error bound remains unaffected by a possible definiteness of $B(t)$ therefore the 'damping' matrix $B(t)$ is allowed to be an arbitrary matrix in this section.

Let us first assume that $B(t)$ is a constant matrix, B. Then the original nonlinear multistep method (2.1.33) reads for $n = 0,1,\ldots,$

(2.2.2) $\sum_{i=0}^k \sigma_i(\Delta t^2A^2)T^iv_n = -\sum_{i=0}^k T^i\sum_{j=1}^{\ell}\sigma_{ij}(\Delta t^2A^2)\Delta t^{2j}(Bv^{(2j-1)} + c^{(2j-2)})_n.$

By means of the differential equation $v^{(2j-1)}$ is to be expressed in this scheme for $j > 1$ by v_n and y'_n. After this operation y'_n is to be approximated by an expression in

$v_n,\dots,v_{n+k}$. This can be achieved only ba a *linear* difference formula because no differential equation is available for y'(t). It thus seems of few practical interest to employ nonlinear multistep methods in the approximation (2.2.2) and we restrict ourselves to linear multistep methods in this and the next section.

We write for simplicity

$$\rho_1(\zeta) = \sum_{i=0}^k \beta_i \zeta^i$$

and introduce for every $\beta_i \neq 0$ a further real polynomial

$$\tau_i(\zeta) = \sum_{\mu=0}^k \gamma_\mu^{(i)} \zeta^\mu .$$

Then a linear multistep method for the problem (2.2.1) is a scheme of the form

$$(2.2.3)\quad \rho_0(T)v_n + \Delta t^2 A^2 \rho_1(T)v_n = -\Delta t \sum_{i=0}^k \beta_i B_{n+i} \tau_i(T)v_n - \Delta t^2 \rho_1(T)c_n,\ n = 0,1,\dots .$$

In the literature mostly the notations $\rho(\zeta) \equiv \rho_0(\zeta)$ and $\sigma(\zeta) \equiv -\rho_1(\zeta)$ are used in this context; see e.g. Lambert [73].

The truncation errors of the method (2.2.3) are

$$d(\Delta t,u)(t) = \rho_0(T)u(t) + \Delta t^2 \rho_1(T)u''(t)$$

and

$$d_i(\Delta t,u)(t) = \tau_i(T)u(t) - \Delta t T^i u'(t), \qquad i = 0,\dots,k.$$

<u>(2.2.4) Definition</u>. *The method* (2.2.3) *is consistent if there exists a positive integer* p *such that for all* $u \in C^{p+2}(\mathbb{R};\mathbb{R}^m)$

$$\|d(\Delta t,u)(t)\| \le \Gamma \Delta t^{p+2} \text{ and } \|d_i(\Delta t,u)(t)\| \le \Gamma \Delta t^{p+1}, \quad i = 0,\dots,k,$$

where Γ *does not depend on* Δt. *The maximum* p *is the order of the method.*

The following lemma is a composition of Lemmas (1.1.8) and (2.1.8) hence we omit the proof.

<u>(2.2.5) Lemma</u>. *If the method* (2.2.3) *is consistent of order* p *then the composed discretization error*

$$d^*(\Delta t,u)(t) = d(\Delta t,u)(t) + \Delta t \sum_{i=0}^k \beta_i B(t+i\Delta t) d_i(\Delta t,u)(t)$$

satisfies for all $u \in C^{p+2}(\mathbb{R};\mathbb{R}^m)$

$$|d^*(\Delta t,u)(t)| \le \Gamma \Delta t^{p+1} \int_t^{t+k\Delta t} (|u^{(p+2)}(\tau)| + |||B|||_{t+k\Delta t} |u^{(p+1)}(\tau)|) d\tau$$

where Γ *does not depend on* t, Δt, u, *and the dimension* m.

In particular, a method (2.2.3) is consistent by (2.1.12) and (1.1.13) iff

$$(2.2.6)\quad \rho_0(1) = \rho_0'(1) = \rho_0''(1) + 2\rho_1(1) = 0,$$

and

$$(2.2.7)\quad \tau_i(1) = 0 \text{ and } \tau_i'(1) = 1\,, \qquad i = 0,\dots,k.$$

<u>(2.2.8) Definition</u>. *The method* (2.2.3) *for the problem* (2.2.1) *with damping has the stability region* S *and it is strongly D-stable in* $[-s, 0] \subset S$ *if the corresponding method* (2.1.4) *for the undamped problem has these properties.*

In the following theorem we derive again an error estimation for the initial value problem (2.2.1) with an additional perturbation $h(t)$,

$$(2.2.9)\quad y'' = A^2y + By' + c(t) + h(t),\ t > 0,\ y(0) = y_0,\ y'(0) = y_0^*.$$

The damping matrix B is assumed to be constant and the linear multistep method (2.2.3) is supposed to be explicit with respect to the finite difference approximations $\tau_i(T)v(t)$ of $y'(t+i\Delta t)$. The latter condition is however no serious restriction because - cum grano salis- (k-1)-step difference formulas for the approximation of $y'(t)$ can be found which have the same order as k-step formulas for the approximation of $y''(t)$.

<u>(2.2.10) Theorem</u>. *(i) Let the (m,m)-matrix* A^2 *in* (2.2.9) *be diagonable,* $A^2 = X\Lambda^2X^{-1}$, *and let the solution* y *be (p+2)-times continuously differentiable.*
(ii) Let the method (2.2.3) *be consistent of order* p *with the stability region* S.
(iii) Let $Sp(\Delta t^2A^2) \subset R \subseteq S$ *where* R *is closed in* $\overline{\mathbb{C}}$.
(iv) Let $\gamma_k^{(i)} = 0$, $i = 0,\dots,k$, *and let* $\rho_0(0) + \eta^2\rho_1(0) \equiv \sigma_0(\eta^2) \neq 0\ \forall\ \eta^2 \in R$.
Then for $n = k,k+1,\dots$,

$$|X^{-1}(y_n - v_n)| \leq \kappa_R|X^{-1}|n\Delta t\,\exp\{\kappa_R^*|X^{-1}BX|n\Delta t\}\ \times$$

$$\times\Big[\Delta t^{-1}|Y_{k-1} - V_{k-1}| + \Delta t^p\int_0^{n\Delta t}(|y^{(p+2)}(\tau)| + |By^{(p+1)}(\tau)|)d\tau + n\Delta t|||h|||_n\Big].$$

Proof. The error $e_n = y_n - v_n$ satisfies

$$(2.2.11)\quad \pi(T,\Delta t^2A^2)e_n = -\Delta t\sum_{i=0}^k\beta_iB\tau_i(T)e_n + d^*(\Delta t,y)_n - \Delta t^2\rho_1(T)h_n,\quad n = 0,1,\dots\ .$$

By assumption (iv) we have

$$\sum_{i=0}^k\beta_iB\tau_i(T)e_n = \sum_{i=0}^k\beta_iB\sum_{\mu=0}^{k-1}\gamma_\mu^{(i)}T^\mu e_n = \sum_{\mu=0}^{k-1}(\sum_{i=0}^k\gamma_\mu^{(i)}\beta_iB)T^\mu e_n \equiv \Phi^*(B)^TE_{n+k-1}$$

where

$$\phi^*(B)^T = (\textstyle\sum_{i=0}^k \gamma_0^{(i)} \beta_i B, \ldots, \sum_{i=0}^k \gamma_{k-1}^{(i)} \beta_i B)$$

and $E_n = (e_{n-k+1}, \ldots, e_n)^T$ is the already above introduced block error vector. (2.2.11) is equivalent to the single-step equation

$$X^{-1}E_n = (F_\pi(\Delta t^2 \Lambda^2) + \Delta t \Phi(X^{-1}BX))X^{-1}E_{n-1} + X^{-1}D_n^* - X^{-1}H_n, \qquad n = k, k+1, \ldots,$$

with the notations

$$D_n^* = (0,\ldots,0,\sigma_k(\Delta t^2 A^2)^{-1} d^*(\Delta t, y)_{n-k})^T, \; H_n = (0,\ldots,0,\sigma_k(\Delta t^2 A^2)^{-1} \Delta t^2 \rho_1(T) h_{n-k})^T.$$

Accordingly,

$$\begin{aligned} (2.2.12) \quad |X^{-1}E_n| &\le |(F_\pi(\Delta t^2\Lambda^2) + \Delta t\Phi(X^{-1}BX))^{n-k+1}||X^{-1}E_{k-1}| \\ &+ \textstyle\sum_{\nu=k}^n |(F_\pi(\Delta t^2\Lambda^2) + \Delta t\Phi(X^{-1}BX))^{n-\nu}||X^{-1}D_\nu^* - X^{-1}H_\nu|, \end{aligned}$$

$$(2.2.13) \quad |X^{-1}H_n| \le \Gamma\Delta t^2 |X^{-1}| \, |||h|||_n,$$

and by Lemma (2.2.5)

$$(2.2.14) \quad |X^{-1}D_n^*| \le \Gamma |X^{-1}| \Delta t^{p+1} \int_{(n-k)\Delta t}^{n\Delta t} (|y^{(p+2)}(\tau)| + |By^{(p+1)}(\tau)|) d\tau.$$

The matrix $\Phi(X^{-1}BX)$ in (2.2.12) is a (k,k)-block matrix whose last row is

$$\sigma_k(\Delta t^2\Lambda^2)^{-1} \phi^*(X^{-1}BX)^T$$

and all other rows are zero therefore we obtain

$$(2.2.15) \quad |\Phi(X^{-1}BX)| \le \kappa |X^{-1}BX|.$$

By (2.2.12), (2.2.13), and (2.2.14) the Theorem is proved if we show that

$$(2.2.16) \quad |(F_\pi(\Delta t^2\Lambda^2) + \Delta t\Phi(X^{-1}Bx))^\nu| \le \kappa_R \nu (1 + \kappa_R^* |X^{-1}BX|\Delta t)^\nu, \qquad \nu = k, k+1, \ldots .$$

But the Frobenius matrix $F_\pi(\Delta t^2\Lambda^2)$ is regular because $\sigma_0(\eta^2)$ is the coefficient of ζ^0 in the characteristic polynomial $\pi(\zeta,\eta^2) \equiv \sum_{i=0}^k \sigma_i(\eta^2)\zeta^i$ of this matrix and $\sigma_0(\eta^2) \ne 0$ $\forall\, \eta^2 \in R$ by assumption (iv). Consequently,

$$(2.2.17) \quad |F_\pi(\Delta t^2\Lambda^2)^{-1}| \le \sup_{\eta^2 \in R} |F_\pi(\eta^2)^{-1}| \le \kappa_R.$$

Now, by (2.1.15) and (2.1.17),

$$|(F_\pi(\Delta t^2\Lambda^2) + \Delta t\Phi(X^{-1}BX))^\nu| \le \nu|\nu^{-1}F_\pi(\Delta t^2\Lambda^2)^\nu|(1 + \Delta t|F_\pi(\Delta t^2\Lambda^2)^{-1}||\Phi(X^{-1}BX))^\nu|)$$

$$\le \nu|\nu^{-1}F_\pi(\Delta t^2\Lambda^2)^\nu|(1 + \kappa_R^*|X^{-1}BX|\Delta t)^\nu, \quad \nu = k,k+1,\ldots,$$

and the Uniform Boundedness Theorem yields

$$|\nu^{-1}F_\pi(\Delta t^2\Lambda^2)^\nu| \le \sup_{\eta^2\in R}\ \sup_{n\in\mathbb{N}} |n^{-1}F_\pi(\eta^2)^n| \le \kappa_R$$

which proves (2.2.16).

It is an open question whether the technical assumption (iv) is necessary here. Cursorily spoken , it is introduced because the Kreiss' Matrix Theorem does not apply to $F_\pi(\eta^2)$ in the present situation. In the next theorem we turn again to strongly D-stable methods. Then assumption (iv) can be omitted and the estimation of the propagation of the initial error obtains a form corresponding to Theorem (2.1.39). However, these improvements are only obtained by considerable effort. In particular, the Kreiss' Matrix Theorem supported by the Uniform Boundedness Theorem plays an important role in the estimation of totally implicit methods and systems with time-varying damping.

<u>(2.2.18) Theorem.</u> *(i) Let the (m,m)-matrix* A^2 *in* (2.2.9) *be diagonable,* $A^2 = X\Lambda^2X^{-1}$, *let the damping matrix* B *vary with time, and let the solution* y *be (p+2)-times contiuously differentiable.*
(ii) Let the method (2.2.3) be consistent of order p *with the stability region* S, *let it be strongly D-stable in* $\eta^2 = 0$, *and let all unimodular roots of* $\pi(\zeta,\eta^2)$ *be simple for* $\eta^2 \in [-s, 0) \subset S$.
(iii) Let $\Delta t^2|A^2| \le s$.
(iv) If the method (2.2.3) *is implicit with respect to* B *then let* $\Delta t|||X^{-1}BX|||_n \le \omega$ *where the constant* ω *is defined in* (2.2.36).
Then for $n = k,k+1,\ldots,$

$$|X^{-1}(y_n - v_n)| \le |X^{-1}E^0_{k-1}| + \kappa_s|X^{-1}|n\Delta t\exp\{\kappa_s^*|||X^{-1}BX|||_n n\Delta t\}\times$$

$$\times\Big[|A||E^0_{k-2}| + \Delta t^{-1}|E^0_{k-1} - E^0_{k-2}| + \Delta t^p\int_0^{n\Delta t}(|y^{(p+2)}(\tau)| + |||B|||_n|y^{(p+1)}(\tau)|)d\tau + n\Delta t|||h|||_n\Big]$$

where $E^0_n = (\,y_{n-k+2} - v_{n-k+2}, \ldots, y_n - v_n\,)^T$.

For the proof of this theorem we need several auxiliary results:

<u>(2.2.19) Lemma.</u> *Let assumption* (2.2.18)(ii) *be fulfilled, let* $F_{\pi^*}(\eta)$ *be the Frobenius matrix of the polynomial* $\pi^*(\zeta,\eta)$ *defined by* (2.1.38), *and let*

$$G(\eta) = \begin{bmatrix} F_{\pi^*}(\eta) & 0 \\ F_{\pi^*}(\eta) - I & \zeta_1(\eta)I \end{bmatrix}.$$

Then

$$\sup_{-s\leq\eta^2\leq 0} \sup_{n\in\mathbb{N}} |G(\eta)^n| \leq \kappa_s.$$

Proof. We follow Gekeler [82b], write shortly $F(\eta) = F_{\pi^*}(\eta)$, and define

$$Z(\eta) = (\zeta_1(\eta)I - F(\eta))^{-1}(I - F(\eta)),\ \eta \in [-s, 0),\ Z(0) = I,$$

where $\zeta_1(\eta)$ is the principal root defined in Section 2.1. Then we obtain

$$G(\eta)^n = \begin{bmatrix} F(\eta)^n & 0 \\ (F(\eta)^n - \zeta_1(\eta)^n I)Z(\eta) & \zeta_1(\eta)^n I \end{bmatrix}, \qquad \eta^2 \in [-s, 0],$$

and the Uniform Boundedness Theorem yields

(2.2.20) $$\sup_{-s\leq\eta^2\leq 0} \sup_{n\in\mathbb{N}} (|F(\eta)^n| + |\zeta_1(\eta)|^n) \leq \kappa_s.$$

Therefore it suffices to show that

(2.2.21) $$\sup_{\eta^2\in[-s,0]} \sup_{n\in\mathbb{N}} |(F(\eta)^n - \zeta_1(\eta)^n I)Z(\eta)| \leq \kappa_s.$$

Let

$$[-s, 0] = R_1 \cup R_2,\ R_1 \cap R_2 = \emptyset,\ 0 \notin R_1,$$

such that the closed set R_1 contains in its interior the finite number of points η^2 in which $\zeta_1(\eta)$ is a multiple roots of the characteristic polynomial $\pi(\zeta, \eta^2)$ and $\sup_{\eta^2\in R_1} |\zeta_1(\eta)| < 1$. Then we find for $\eta^2 \in R_1$

$$|(F(\eta)^n - \zeta_1(\eta)^n)Z(\eta)| = |(\textstyle\sum_{j=0}^{n-1} F(\eta)^j \zeta_1(\eta)^{n-1-j})(F(\eta) - I)|$$

$$\leq \kappa_s \textstyle\sum_{j=0}^{\infty} |\zeta_1(\eta)|^j = \kappa_s(1 - |\zeta_1(\eta)|)^{-1} \leq \kappa_s^*.$$

In order to prove (2.2.21) for $\eta^2 \in R_2$ it suffices to show by (2.2.20) that

(2.2.22) $$\sup_{\eta^2\in R_2} |Z(\eta)| \leq \kappa_s.$$

For the verification of this bound we observe that by Kato [66 , Theorem 2.1.5] the function $\eta \mapsto (\zeta_1(\eta)I - F(\eta))^{-1}$ is continuous for $\eta^2 \in \overline{R}_2 \setminus \{0\}$ since no eigenvalues of $F(\eta)$ coincide with $\zeta_1(\eta)$ in this set by definition of R_2. Therefore we have only to show that $Z(\eta)$ is bounded near $\eta = 0$. For this let $U(\eta)$ be a unitary matrix such that

$$U(\eta)F(\eta)U(\eta)^H = R(\eta) = [r_{ij}(\eta)]_{i,j=1}^{k-1}$$

is an upper triangular matrix with $r_{11}(\eta) = \zeta_2(\eta) = \zeta_1(-\eta)$. Then we have near $\eta = 0$

(2.2.23) $|r_{ij}(\eta)| \leq |R(\eta)| = |F(\eta)| < \Gamma$.

Let $e = (1,0,\dots,0)^T$ be a column vector and let

$$Z^*(\eta) = (\zeta_1(\eta)I - R(\eta))(I - \zeta_1(\eta)ee^T)^{-1}.$$

Then, omitting the argument η,

$$UZU^H = I + (1 - \zeta_1)(\zeta_1 I - R)^{-1} = I + (1 - \zeta_1)(I - \zeta_1 ee^T)^{-1}Z^{*-1}$$

and $|(1 - \zeta_1)(I - \zeta_1 ee^T)^{-1}|$ is bounded near $\eta = 0$. In order to prove that $|Z^{*-1}|$ is bounded we write the upper triangular matrix Z^* as

(2.2.24) $Z^* = \text{diag}(Z^*) - R^* = \text{diag}(Z^*)(I - \text{diag}(Z^*)^{-1}R^*)$

where $\text{diag}(Z^*)$ is the diagonal of Z^* and $R^* = R - \text{diag}(R)$. Accordingly, $(R^*)^k \equiv 0$, $|R^*|$ is bounded near $\eta = 0$ by (2.2.23), and $|\text{diag}(Z^*)|$ as well as $|\text{diag}(Z^*)^{-1}|$ are bounded near $\eta = 0$ by assumption and as by (2.1.18)

$$\zeta_1(\eta) - \zeta_1(-\eta) = 2\eta + \mathcal{O}(\eta^2),\ \zeta_1(\eta) - 1 = \eta + \mathcal{O}(\eta^2) \qquad \eta \to 0.$$

Thus, finally, a von Neumann series expansion of Z^{*-1} via (2.2.24) proves that $|Z^{*-1}|$ is bounded near $\eta = 0$, too.

The next lemma is the Matrix Theorem of Kreiss in a somewhat shortened form adapted to our purposes.

(2.2.25) Lemma. Kreiss [62].) *If the assertion of Lemma* (2.2.19) *is true then there exists to every matrix* $G(\eta)$ *a hermitean matrix* $H(\eta)$ *and a uniform constant* Γ_s *such that*

$$G(\eta)^H H(\eta)G(\eta) \leq H(\eta) \text{ and } 0 < \Gamma_s^{-1} I \leq H(\eta) \leq \Gamma_s I \qquad \forall\ \eta^2 \in [-s, 0].$$

Here $P \geq Q$ means again for two (m,m) matrices P and Q that $\text{Re}(x^H(P - Q)x) \geq 0\ \forall\ x \in \mathbb{C}^m$. As a corollary to Lemma (2.2.19) and (2.2.25) we prove

<u>(2.2.26) Lemma</u>. *Under the assumptions of Theorem* (2.2.18) *there exists to every matrix* $G(\Delta t\Lambda)$ *a norm* $\|\cdot\|_G$ *with* $\|G\|_G = \max_{x \neq 0}\|Gx\|_G/\|x\|_G$ *and*

$$\|G(\Delta t\Lambda)\|_G \leq 1,\ \Gamma_s^{-1/2}|x| \leq \|x\|_G \leq \Gamma_s^{1/2}|x| \qquad \forall\, x \in \mathbb{C}^{2\times(k-1)\times m}.$$

Proof. By assumption and Lemma (2.2.19) we have

$$|G(\Delta t\Lambda)^n| \leq \sup_{-s\leq\eta^2\leq 0} |G(\eta)^n| \leq \kappa_s$$

hence by the Kreiss' Matrix Theorem (2.2.25) there exists to every matrix $G(\Delta t\Lambda)$ a Kreiss' matrix $H_G(\Delta t\Lambda)$ with the property

$$|H_G(\Delta t\Lambda)^{1/2}G(\Delta t\Lambda)H_G(\Delta t\Lambda)^{-1/2}| \leq 1.$$

Therefore, the vector norm $\|x\|_G \equiv |H_G(\Delta t\Lambda)^{1/2}x|$ has the desired properties.

For the proof of Theorem (2.2.18) we now observe that the error $e_n = y_n - v_n$ satisfies with the polynomial $\pi^*(\zeta,\eta)$ introduced in (2.1.38)

$$\text{(2.2.27)}\quad \pi^*(T,\Delta tA)(T-\zeta_1(\Delta tA))e_{n-1} = -\Delta t\textstyle\sum_{i=0}^k \beta_i B_{n+i}\tau_i(T)e_n + d^*(\Delta t,y)_n - \Delta t^2\rho_1(T)h_n,$$

$n = 0,1,\ldots$. By (2.2.7) all polynomials $\tau_i(\zeta)$ of a consistent method have the root $\zeta = 1$. We write

$$\text{(2.2.28)}\quad \tau_i(\zeta)/(\zeta-1) = \textstyle\sum_{\mu=0}^{k-1}\delta_\mu^{(i)}\zeta^\mu,\ \delta_{k-1}^{(i)} = \gamma_k^{(i)},$$

then we obtain using the block error vectors E_n^0 of block dimension k-1

$$\begin{aligned}
&\textstyle\sum_{i=0}^k \beta_i B_{n+i}\tau_i(T)e_n = \sum_{i=0}^k \beta_i B_{n+i}\sum_{\mu=0}^{k-1}\delta_\mu^{(i)}(e_{n+\mu+1} - e_{n+\mu})\\
\text{(2.2.29)}\quad &= \textstyle\sum_{\mu=0}^{k-1}(\sum_{i=0}^k \delta_\mu^{(i)}\beta_i B_{n+i})T^\mu(e_{n+1} - e_n)\\
&= \phi(B)_n(e_{n+k} - e_{n+k-1}) + \phi^{**}(B)_n^T(E_{n+k-1}^0 - E_{n+k-2}^0)
\end{aligned}$$

where $\phi(B)_n = \sum_{i=0}^k \gamma_k^{(i)}\beta_i B_{n+i}$, i.e., $\phi(B)_n \equiv 0$ if the method (2.2.2) is explicit with respect to B. Using (2.2.29), (2.2.27) writes as

$$\begin{aligned}
&\pi^*(T,\Delta t\Lambda)(T-\zeta_1(\Delta t\Lambda))X^{-1}e_n = -\Delta t\phi(X^{-1}BX)_n(X^{-1}e_{n+k} - X^{-1}e_{n+k-1})\\
\text{(2.2.30)}\quad &\\
&- \Delta t\phi^{**}(X^{-1}BX)_n^T(X^{-1}E_{n+k-1}^0 - X^{-1}E_{n+k-2}^0) + X^{-1}d^*(\Delta t,y)_n - \Delta t^2\rho_1(T)X^{-1}h_n.
\end{aligned}$$

With the abbreviations

$$E_n^* = X^{-1}E_n^0,\ \zeta = \zeta_1(\Delta t\Lambda),\ F^* = F_{\pi^*}(\Delta t\Lambda)$$

(2.2.30) is equivalent to the single-step equation

$$E_n^* - \zeta E_{n-1}^* = F^*(E_{n-1}^* - \zeta E_{n-2}^*) - \Delta t\Phi_n(E_n^* - E_{n-1}^*) - \Delta t\Psi_n(E_{n-1}^* - E_{n-2}^*) \tag{2.2.31}$$

$$+ X^{-1}D_n^* - X^{-1}H_n, \qquad n = k,k+1,\ldots ,$$

where D_n^* and H_n are the same block vectors as in (2.2.12) but with block dimension k-1. Φ_n and Ψ_n are (k-1,k-1)-block matrices of which only the last row in nonzero and

$$\text{last row of } \Phi_n = (0,\ldots,0,\sigma_k(\Delta t^2\Lambda^2)^{-1}\phi(X^{-1}BX)_n),$$

$$\text{last row of } \Psi_n = \sigma_k(\Delta t^2\Lambda^2)^{-1}\phi^{**}(X^{-1}BX)_n^T.$$

Substituting

$$E_n^* - E_{n-1}^* = (E_n^* - \zeta E_{n-1}^*) - (E_{n-1}^* - \zeta E_{n-2}^*) + \zeta(E_{n-1}^* - E_{n-2}^*) \tag{2.2.32}$$

into (2.2.31) we obtain

$$(I + \Delta t\Phi_n)(E_n^* - \zeta E_{n-1}^*) = (F^* + \Delta t\Phi_n)(E_{n-1}^* - \zeta E_{n-2}^*) \tag{2.2.33}$$

$$- \Delta t(\Psi_n + \zeta\Phi_n)(E_{n-1}^* - E_{n-2}^*) + X^{-1}D_n^* - X^{-1}H_n,$$

and a substitution of (2.2.31) into (2.2.32) yields

$$(I + \Delta t\Phi_n)(E_n^* - E_{n-1}^*) = (F^* - I)(E_{n-1}^* - \zeta E_{n-2}^*) \tag{2.2.34}$$

$$+ (\zeta I - \Delta t\Psi_n)(E_{n-1}^* - E_{n-2}^*) + X^{-1}D_n^* - X^{-1}H_n.$$

(2.2.33) and (2.2.34) together provide the error equation

$$Z_n = (I + \Delta t\Phi_n)^{-1}G(\Delta t\Lambda)Z_{n-1} - \Delta tL_nZ_{n-1} + D_n^{**} - H_n^{**}, \qquad n = k,k+1,\ldots , \tag{2.2.35}$$

where

$$Z_n = (E_n^* - \zeta E_{n-1}^*, E_n^* - E_{n-1}^*)^T,\ D_n^{**} = (I + \Delta t\Phi_n)^{-1}X^{-1}(D_n^*,D_n^*)^T,\ H_n^{**} = (I + \Delta t\Phi_n)^{-1}X^{-1}(H_n,H_n)^T,$$

$$L_n = (I + \Delta t\Phi_n)^{-1}\begin{bmatrix}-\Phi_n & \Psi_n + \zeta\Phi_n \\ 0 & \Psi_n\end{bmatrix},$$

and $G(\Delta t\Lambda)$ is the matrix introduced in Lemma (2.2.19). Recall that $\Phi_n \equiv 0$ if the method (2.2.3) is explicit with respect to the damping matrix B. If $\Phi_n \equiv 0$ and B is constant then the proof of Theorem (2.2.18) follows the lines of Theorem (2.2.10) estimating Z_n by means of (2.2.35) and Lemma (2.2.19) but *not* involving the Kreiss' Matrix Theorem. However, if the method (2.2.3) is implicit with respect to B then the matrix Φ_n does not necessarily commute with the matrix $G(\Delta t\Lambda)$. Hence we are forced in this case to apply Lemma (2.2.26) and similarly in the case where B varies with time.

We now observe that

$$|||\Psi|||_n \leq \kappa|||X^{-1}BX|||_n, \quad |||\Phi|||_n \leq \tilde{\kappa}|||X^{-1}BX|||_n$$

where we set $\tilde{\kappa} = 0$ if the method is explicit with respect to B. With the constant Γ_s defined in Lemma (2.2.26) we suppose that

$$(2.2.36)\quad \Delta t\Gamma_s\tilde{\kappa}|||X^{-1}BX|||_n \leq 1/2,$$

and we denote for the moment by I^* the identity matrix of the same dimension as $G(\Delta t\Lambda)$. Then we have

$$\|L_n\|_G \leq \kappa_s|||X^{-1}BX|||_n$$

and, by Lemma (2.2.26),

$$\|(I + \Delta t\Phi_n)^{-1}G(\Delta t\Lambda)\|_G \leq \|(I + \Delta t\Phi_n)^{-1}I^*\|_G\|G(\Delta t\Lambda)\|_G \leq \|(I + \Delta t\Phi_n)^{-1}I^*\|_G$$

$$\leq (1 - \Delta t\Gamma_s\tilde{\kappa}|||X^{-1}BX|||_n)^{-1} \leq 1 + \Delta t\tilde{\kappa}_s|||X^{-1}BX|||_n.$$

Using these bounds we obtain from (2.2.35)

$$\|Z_n\|_G \leq (1 + \kappa_s^*|||X^{-1}BX|||_n\Delta t)\|Z_{n-1}\|_G + \|D_n^{**} - H_n^{**}\|_G, \qquad n = k,k+1,\ldots,$$

and we thus find by induction after returning to the Euclid norm that

$$|Z_n| \leq \kappa\ \exp\{\kappa_s^*|||X^{-1}BX|||_n n\Delta t\}[|Z_{k-1}| + \textstyle\sum_{\nu=k}^n|D_\nu^{**} - H_\nu^{**}|].$$

$|D_\nu^{**}|$ and $|H_\nu^{**}|$ have the same bounds as $|X^{-1}D_\nu^*|$ and $|X^{-1}H_\nu|$ in (2.2.13) and (2.2.14) with different Γ hence the rest of the proof follows as in Theorem (2.2.10).

As already noted above, The last condition in assumption (2.2.18)(ii) is slightly stronger than strong D-stability in $[-s, 0)$. It claims that in this interval the method is strongly D-stable *and* that the principal root $\zeta_1(\eta)$ does not coalesce to a double unimodular root. This latter condition cannot be released in the proof of Lemma (2.2.19).

2.3. Linear Multistep Methods for Differential Systems with Orthogonal Damping

The estimations of the preceding section contain the norm of the damping matrix B as exponential multiplication factor. The question whether there are problems and methods which provide damped approximations if the exact solution is damped leads again to the class of initial value problems with orthogonal damping which was already considered in Theorem (1.4.7). Perhaps it should be remarked once more that the occupation with these problems is not only a theoretical pastime. At the present state of matrix structural analysis the majority of dynamic finite element equations (1) with damping seems to be of that form. Besides the references given in the Introduction we quote Geradin [74] and Jensen [74] to which we were pointed by Godlewski and Puech-Rauolt [79]. The latter contribution deserves special emphasis here because it deals with linear multistep methods and problems with orthogonal damping in detail.

We reconsider in this section the initial value problem (2.2.9) and linear multistep methods (2.2.3) under the assumption of Theorem (1.4.7) but *without* the condition that the damping is smaller than the critical damping:

<u>(2.3.1) Assumption</u>. *(i) Let the solution* y *of* (2.2.9) *be (p+2)-times continuously differentiable.*
(ii) Let the constant (m,m)-matrices A^2 *and* B *be real symmetric and negative semidefinite with the same system* X *of eigenvectors,* $A^2 = X\Lambda^2X^T$, $B = X\Omega X^T$, $X^TX = I$.

The following lemma describes orthogonal damping completely in an algebraic way.

<u>(2.3.2) Lemma</u>. *Let* A *and* B *be two diagonable (m,m)-matrices. Then the following three conditions are equivalent:*
(i) $AB = BA$,
(ii) there exists a diagonable matrix C *and polynomials* $p(x)$ *and* $q(x)$ *such that* $A = p(C)$ *and* $B = q(C)$,
(iii) there exists a nonsingular matrix X *such that* $X^{-1}AX$ *and* $X^{-1}BX$ *are both diagonal.*

Proof. See Householder [64, p. 30].

As the damping matrix B is constant throughout we introduce the polynomial

$$\tau(\zeta) = \sum_{i=0}^{k}\beta_i\tau_i(\zeta) = \sum_{i=0}^{k}\beta_i\sum_{\mu=0}^{k}\gamma_\mu^{(i)}\zeta^\mu = \sum_{\mu=0}^{k}\left(\sum_{i=0}^{k}\beta_i\gamma_\mu^{(i)}\right)\zeta^\mu = \sum_{\mu=0}^{k}\chi_\mu\zeta^\mu$$

then the linear multistep method (2.2.3) writes as

$$(2.3.3)\quad \rho_0(T)v_n + \Delta t^2A^2\rho_1(T)v_n + \Delta tB\tau(T)v_n = -\Delta t^2\rho_1(T)c_n, \qquad n = 0,1,\ldots,$$

and the *characteristic polynomial* is now

$$\pi(\zeta,\eta^2,\mu) = \pi(\zeta,\eta^2) + \mu\tau(\zeta) = \rho_0(\zeta) + \eta^2\rho_1(\zeta) + \mu\tau(\zeta).$$

Substituting the exact solution y of the initial value problem (2.2.1) into the scheme (2.3.3) we obtain again the discretization error $d^*(\Delta t,y)(t)$ of the method,

$$\begin{aligned} d^*(\Delta t,y)_n &= \rho_0(T)y_n + \Delta t^2A^2\rho_1(T)y_n + \Delta tB\tau(T)y_n + \Delta t^2\rho_1(T)c_n \\ &= \rho_0(T)y_n + \Delta t^2\rho_1(T)y_n'' + \Delta tB(\tau(T)y_n - \Delta t\rho_1(T)y_n'), \end{aligned}$$

or

(2.3.4) $d^*(\Delta t,y)(t) = d(\Delta t,y)(t) + \Delta tB\tilde{d}(\Delta t,y)(t)$

where $d(\Delta t,y)$ is the discretization error (2.1.7) of the corresponding undamped problem (2.1.5) and

(2.3.5) $\tilde{d}(\Delta t,y)(t) = \tau(T)y(t) - \Delta t\rho_1(T)y'(t).$

If the method (2.3.3) is consistent of order p after Definition (2.2.4) then for all $u \in C^{p+2}(\mathbb{R};\mathbb{R}^m)$

(2.3.6) $\|d(\Delta t,u)(t)\| \le \Gamma\Delta t^{p+2}$ and $\|\tilde{d}(\Delta t,u)(t)\| \le \Gamma\Delta t^{p+1}$

where Γ does not depend on Δt but the contrary is not true in general because the polynomial $\tau(\zeta)$ does not define the polynomials $\tau_i(\zeta)$, $i = 0,\dots,k$, in a unique way. Nevertheless Lemma (2.2.5) remains true if (2.2.4) is replaced by (2.3.6) because, by (1.1.6) for $\ell = 1$, $-\tilde{d}(\Delta t,u)$ is the discretization error of a linear multistep method (1.1.3) with $\rho_0(\zeta) = -\tau(\zeta)$ for a *first order* differential system and Lemma (2.2.5) is a composition of Lemma (1.1.8) and Lemma (2.1.8). So we may define:

<u>(2.3.7) Definition.</u> *The method* (2.3.3) *is consistent if there exists a positive integer* p *such that* (2.3.6) *holds for all* $u \in C^{p+2}(\mathbb{R};\mathbb{R}^m)$. *The maximum* p *is the order of the method.*

In particular, the method (2.3.3) is consistent iff

(2.3.8) $\rho_0(1) = \rho_0'(1) = \rho_0''(1) + 2\rho_1(1) = \tau(1) = \tau'(1) - \rho_1(1) = 0.$

Recall now that by Lemma (1.1.12) a polynomial $\rho_1(\zeta)$ of degree p defines in a unique way a polynomial $-\tau(\zeta)$ such that (2.3.5) fulfils (2.3.6). Starting from a method (2.1.4) of order p for the undamped problem we so derive easily a method (2.3.3) of order p if the polynomial $\rho_1(\zeta)$ has degree p.

As a further tool for the construction of methods (2.3.3) we quote the following result of Godlewski and Puech-Raoult [79].

<u>(2.3.9) Lemma</u>. *The method* (2.3.3) *has order* p *iff*
(i) the corresponding method for the undamped problem has order p,
(ii) the method (1.1.3) *with* $\pi(\zeta,\eta) = \rho_0(\zeta) + \eta\tau(\zeta)$ *has order* p+1.

Proof. Let $d(\Delta t,u)$ be defined by (2.1.7) for $\ell = 1$ and let $\hat{d}(\Delta t,u) = \rho_0(T)u + \Delta t\tau(T)u'$ be the discretization error of the linear method (1.1.3) for first order differential systems with the polynomial $\rho_0(\zeta) + \eta\tau(\zeta)$. Then

$$d(\Delta t,u) = -\Delta t\tilde{d}(\Delta t,u') + \hat{d}(\Delta t,u)$$

and hence (2.3.4) proves the result.

After having studied the consistence let us now consider the stability of the method (2.3.3). The error equation (2.2.11) has here the form

$$(2.3.10)\quad [\rho_0(T) + \Delta t^2A^2\rho_1(T) + \Delta tB\tau(T)]e_n = d^*(\Delta t,y)_n - \Delta t^2\rho_1(T)h_n, \quad n = 0,1,\ldots\ .$$

Under Assumption (2.3.1) this is equivalent to the single-step equation

$$(2.3.11)\quad X^TE_n = F_\pi(\Delta t^2\Lambda^2,\Delta t\Omega)X^TE_{n-1} + X^TD_n^* - X^TH_n, \qquad n = k,k+1,\ldots,$$

where now $F_\pi(\eta^2,\mu)$ is now the Frobenius matrix with the characteristic polynomial $\pi(\zeta,\eta^2,\mu)$ and the following further notations are used:

$$\sigma_k^*(\Delta t^2\Lambda^2,\Delta t\Omega) = \alpha_{0k}I + \beta_k\Delta t^2\Lambda^2 + \chi_k\Delta t\Omega,$$

$$X^TD_n^* = (0,\ldots,0,\sigma_k^*(\Delta t^2\Lambda^2,\Delta t\Omega)^{-1}X^Td^*(\Delta t,y)_{n-k})^T,$$

$$X^TH_n = (0,\ldots,0,\sigma_k^*(\Delta t^2\Lambda^2,\Delta t\Omega)^{-1}\Delta t^2\rho_1(T)X^Th_{n-k})^T.$$

As above, let $\overline{\mathbb{C}}$ be the complex plane extended in the usual sense, and let $\overline{\mathbb{C}^2} = \overline{\mathbb{C}}^2 \cup \{\infty,\infty\}$ be extended likewise in the usual sense; moreover, let $\pi(\zeta,\infty,\mu) = \rho_1(\zeta)$, $\mu \in \mathbb{C}$, $\pi(\zeta,\eta^2,\infty) = \tau(\zeta)$, $\eta^2 \in \mathbb{C}$, and $\pi(\zeta,\infty,\infty) = \rho_1(\zeta) + \tau(\zeta)$. For simplicity we *define* here the *two-dimensional* region of stability S^2 as follows, cf. also Godlewski and Puech-Raoult [79]:

<u>(2.3.12) Definition</u>. *The two-dimensional stability region* S^2 *of the method* (2.3.3) *consists of the pairs* $(\eta^2,\mu) \in \overline{\mathbb{C}^2}$ *with the following properties:*
(i) $\sigma_k^*(\eta^2,\mu) \neq 0$, $(\eta^2,\mu) \in S^2 \cap \mathbb{C}^2$,

(ii) all roots $\zeta_i(\eta^2,\mu)$ *of* $\pi(\zeta,\eta^2,\mu)$ *satisfy* $|\zeta_i(\eta^2,\mu)| \le 1$,
(iii) all roots $\zeta_i(\eta^2,\mu)$ *of* $\pi(\zeta,\eta^2,\mu)$ *with* $|\zeta_i(\eta^2,\mu)| = 1$ *have multiplicity not greater than two.*

With this notation, the error estimation corresponding to Theorem (2.1.34) and Theorem (2.2.10) is then

<u>(2.3.13) Theorem.</u> *(i) Let Assumption* (2.3.1) *be fulfilled.*
(ii) Let the method (2.3.3) *be consistent of order* p *with the stability region* S^2.
(iii) Let $Sp(\Delta t^2 A^2) \times Sp(\Delta t B) \subset R \subseteq S^2$ *where* R *is closed in* $\mathbb{C}^2$.
Then for $n = k, k+1, \ldots,$

$$|y_n - v_n| \le \kappa_R n\Delta t\Big[\Delta t^{-1}|Y_{k-1} - V_{k-1}| + \Delta t^p \int_0^{n\Delta t} (|y^{(p+2)}(\tau)| + |By^{(p+1)}(\tau)|)d\tau + n\Delta t|||h|||_n\Big].$$

If supplementary the method is implicit with respect to B *and the discretization error* $\tilde{d}(\Delta t,u)$ *defined by* (2.3.5) *has order* p+1 *then the assertion holds with* $|By^{(p+1)}(\tau)|$ *cancelled.*

Proof. Under the assumption we find by the Uniform Boundedness Theorem again that

$$\sup_{n\in\mathbb{N}} |n^{-1}F_\pi(\Delta t^2\Lambda^2,\Delta t\Omega)^n| \le \kappa_R$$

hence the first assertion follows as in Theorem (2.1.34) using Lemma (2.2.5) instead of Lemma (2.1.8). The second assertion follows in the same way because under the present assumption

$$|\sigma_k^*(\Delta t^2\Lambda^2,\Delta t\Omega)^{-1}\Delta t\Omega| \le \tilde{\kappa}$$

and accordingly

$$\begin{aligned}|\sigma_k^*(\Delta t^2\Lambda^2,\Delta t\Omega)^{-1}X^T d^*(\Delta t,y)(t)| &\le \kappa|d(\Delta t,y)(t)| + |\sigma_k^*(\Delta t^2\Lambda^2,\Delta t\Omega)^{-1}\Delta t\Omega||\tilde{d}(\Delta t,y)(t)| \\ &\le \kappa\Gamma\Delta t^{p+1}\int_t^{t+k\Delta t}|y^{(p+2)}(\tau)|d\tau + \tilde{\kappa}\Gamma\Delta t^{p+1}\int_t^{t+k\Delta t}|y^{(p+2)}(\tau)|d\tau.\end{aligned}$$

The principal roots $\zeta_1(\eta,\mu)$ and $\zeta_2(\eta,\mu) = \zeta_1(-\eta,\mu)$ now depend on η and μ but otherwise the situation is the same as in Section 2.1. The concept of strong D-stability can be generalized to the present case in a likewise simple way:

<u>(2.3.14) Definition.</u> *Let* $\pi^*(\zeta,\eta,\mu) = \pi(\zeta,\eta^2,\mu)/(\zeta - \zeta_1(\eta,\mu))$ *then a method* (2.3.3) *is strongly D-stable in* $[-s, 0] \times [-r, 0]$ *if* $[-s, 0] \times [-r, 0] \subseteq S^2$ *and if for all* $(\eta^2,\mu) \in [-s, 0] \times [-r, 0]$ *the roots* $\zeta_i(\eta)$ *of* $\pi^*(\zeta,\eta,\mu)$ *with* $|\zeta_i(\eta,\mu)| = 1$ *are simple*

roots of $\pi^*(\zeta,\eta)$.

Finally, the following result is the analogue to Theorem (2.1.39) and Theorem (2.2.18).

(2.3.15) Theorem. *Let the assumptions of Theorem* (2.3.13) *be fulfilled but let* R = [-s, 0] × [-r, 0] *and let the method* (2.3.3) *be strongly D-stable in* R. *Then for* n = k,k+1,...,

$$|y_n - v_n| \le |E^0_{k-2}| + \kappa_R n\Delta t\Big[(|A| + |B|)|E^0_{k-2}| + \Delta t^{-1}|E^0_{k-1} - E^0_{k-2}| + \Delta t^p \int_0^{n\Delta t} (|y^{(p+2)}(\tau)| + |By^{(p+1)}(\tau)|)d\tau + n\Delta t|||h|||_n\Big]$$

where $E^0_n = (y_{n-k+2} - v_{n-k+2}, \ldots, y_n - v_n)^T$. *If supplementary the method is implicit with respect to* B *and the discretization error* $\tilde{d}(\Delta t,u)$ *defined by* (2.3.5) *has order* p+1 *then the assertion holds with* $|By^{(p+1)}(\tau)|$ *cancelled.*

Proof. Using Lemma (A.1.13) instead of Lemma (A.1.8) the first assertion is proved as in Theorem (2.1.39) and the second assertion is proved as in the preceding theorem.

Both error bounds of this section contain no longer an exponential increasing multiplication factor but, on the other side, numerical damping is derived neither. The verification of exponential damping in the error bounds however follows rather closely the lines of Section 1.3. It is therefore omitted in order to avoid to many repetitions.

2.4. Nyström Type Methods for Conservative Differential Systems

As announced at the outset of the chapter we turn in this section to the second way of approximating conservative differential systems of second order directly by multistep multiderivative methods. Supplementary to the polynomials $\rho_j(\zeta)$, $\rho^*_j(\zeta)$, $j = 0,\ldots,\ell$, and the polynomials $\sigma_i(\eta)$, $\sigma^*_i(\eta)$, $i = 0,\ldots,k$, defined by (2.1.3) we introduce the polynomials

$$\tau_j(\zeta) = \textstyle\sum_{i=0}^k \beta_{ji}\zeta^i, \quad \tau^*_j(\zeta) = \sum_{i=0}^k \beta^*_{ji}\zeta^i,$$

and

$$\chi_i(\eta) = \textstyle\sum_{j=0}^{\ell} \beta_{ji}\eta^j, \quad \chi^*_i(\eta) = \sum_{j=0}^{\ell} \beta^*_{ji}\eta^j.$$

Then a general multistep multiderivative method of Nyström type for conservative systems

$$y'' = f(t,y)$$

is a scheme formally consisting of two formulas (2.1.2), i.e.,

$$\sum_{j=0}^{\ell}\rho_j(T)(\Delta t^2\Theta^2)^j f_n^{(-2)}(v_n) + \Delta t\sum_{j=0}^{\ell}\rho_j^*(T)(\Delta t^2\Theta^2)^j f_n^{(-1)}(v_n)$$

$$\equiv \sum_{i=0}^{k}\sigma_i(\Delta t^2\Theta^2)T^i f_n^{(-2)}(v_n) + \Delta t\sum_{i=0}^{k}\sigma_i^*(\Delta t^2\Theta^2)T^i f_n^{(-1)}(v_n) = 0,$$

(2.4.1)

$$\sum_{j=0}^{\ell}\tau_j(T)(\Delta t^2\Theta^2)^j f_n^{(-2)}(v_n) + \Delta t\sum_{j=0}^{\ell}\tau_j^*(T)(\Delta t^2\Theta^2)^j f_n^{(-1)}(v_n)$$

$$\equiv \sum_{i=0}^{k}\chi_i(\Delta t^2\Theta^2)T^i f_n^{(-2)}(v_n) + \Delta t\sum_{i=0}^{k}\chi_i^*(\Delta t^2\Theta^2)T^i f_n^{(-1)}(v_n) = 0, \; n = 0,1,\ldots \quad .$$

Here we have to write $f_n^{(-2)}(v_n) = v_n$ and $f_n^{(-1)}(v_n)$ is to be replaced by w_n which plays the role of an approximation to y_n'. In order to guarantee that these formulas provide a unique solution (v_{n+k}, w_{n+k}) for sufficiently small time steps Δt and to remove a certain arbitrariness in the choice of the coefficients we stipulate that

(2.4.2) $\alpha_{0k} = 1$ and $\beta_{0k}^* = 1$

and that the matrix

$$\begin{bmatrix}\alpha_{0k} & \alpha_{0k}^* \\ \beta_{0k} & \beta_{0k}^*\end{bmatrix} = \begin{bmatrix}1 & \alpha_{0k}^* \\ \beta_{0k} & 1\end{bmatrix}$$

is regular. Moreover, in analogy to (2.1.6) it is supposed that

$$\begin{bmatrix}\rho_\ell(\zeta) & \rho_\ell^*(\zeta) \\ \tau_\ell(\zeta) & \tau_\ell^*(\zeta)\end{bmatrix} \not\equiv 0 \text{ and } \begin{bmatrix}\sigma_0(\eta) & \sigma_0^*(\eta) \\ \chi_0(\eta) & \chi_0^*(\eta)\end{bmatrix} \not\equiv 0.$$

Finally, let

$$\tilde{\rho}_j(\zeta) \equiv \sum_{i=0}^{k}\tilde{\alpha}_{ji}\zeta^i = \begin{cases}\rho_{j/2}(\zeta) & \text{if } j \text{ even} \\ \rho^*_{(j-1)/2}(\zeta) & \text{if } j \text{ odd}\end{cases},$$

$$\tilde{\tau}_j(\zeta) \equiv \sum_{i=0}^{k}\tilde{\beta}_{ji}\zeta^i = \begin{cases}\tau_{j/2}(\zeta) & \text{if } j \text{ even} \\ \tau^*_{(j-1)/2}(\zeta) & \text{if } j \text{ odd}\end{cases}.$$

Then the discretization error of the method (2.4.1) is

$$D(\Delta t,u)(t) = (d_\rho(\Delta t,u)(t), d_\tau(\Delta t,u)(t))^T, \qquad u \in C^{2\ell+1}(\mathbb{R};\mathbb{R}^m),$$

where

(2.4.3) $d_\rho(\Delta t,u)(t) = \sum_{j=0}^{2\ell+1} \Delta t^j \tilde{\rho}_j(T)u^{(j)}(t)$, $d_\tau(\Delta t,u)(t) = \sum_{j=0}^{2\ell+1} \Delta t^j \tilde{\tau}_j(T)u^{(j)}(t)$.

(2.4.4) Definition. *The method (2.4.1) is consistent if there exists a positive integer* p *such that for all* $u \in C^{p+2}(\mathbb{R};\mathbb{R}^m)$

$$\|D(\Delta t,u)(t)\| \leq \Gamma \Delta t^{p+2}$$

where Γ *does not depend on* Δt. *The maximum* p *is the order of the method.*

By the formulas (2.4.3) we find in the same way as in Lemma (1.1.12)

(2.4.5) Lemma. *The method (2.4.1) is consistent of order* p *iff*

$$\sum_{j=0}^{\min\{\mu,2\ell+1\}} \sum_{i=0}^{k} \tilde{\alpha}_{ji} \frac{i^{\mu-j}}{(\mu-j)!} = \sum_{j=0}^{\min\{\mu,2\ell+1\}} \sum_{i=0}^{k} \tilde{\beta}_{ji} \frac{i^{\mu-j}}{(\mu-j)!} = 0,\ \mu = 0,1,\ldots,p+1.$$

In particular, the method (2.4.1) is consistent iff

(2.4.6) $\tilde{\rho}_0(1) = \tilde{\rho}_0'(1) + \tilde{\rho}_1(1) = \tilde{\rho}_0''(1) + \tilde{\rho}_0'(1) + 2\tilde{\rho}_1'(1) + 2\tilde{\rho}_2(1) = 0$

and

(2.4.7) $\tilde{\tau}_0(1) = \tilde{\tau}_0'(1) + \tilde{\tau}_1(1) = \tilde{\tau}_0''(1) + \tilde{\tau}_0'(1) + 2\tilde{\tau}_1'(1) + 2\tilde{\tau}_2(1) = 0$.

For simplicity we consider in the sequel only nonlinear single-step methods, i.e., methods (2.4.1) with k = 1. Furthermore, we assume that the method is at most semi-implicit in the sense that

$$\sigma_1^*(\eta) = x_1(\eta) \equiv 0,$$

and with respect to the computational efficiency we suppose that

$$\sigma_1(\eta) = x_1^*(\eta) = \sigma(\eta).$$

With these restrictions, the approximation of the initial value problem (2.1.5),

(2.4.8) $y'' = A^2y + c(t),\ t > 0,\ y(0) = y_0,\ y'(0) = y_0^*$,

yields the following scheme which is at once written in explicit form:

$$(2.4.9)\quad \begin{aligned} v_{n+1} &= -\,\sigma(\Delta t^2A^2)^{-1}[\sigma_0(\Delta t^2A^2)v_n + \sigma_0^*(\Delta t^2A^2)\Delta tw_n] - \sigma(\Delta t^2A^2)^{-1}c_n^* \\ \Delta tw_{n+1} &= -\,\sigma(\Delta t^2A^2)^{-1}[\chi_0(\Delta t^2A^2)v_n + \chi_0^*(\Delta t^2A^2)\Delta tw_n] - \sigma(\Delta t^2A^2)^{-1}c_n^{**},\ n = 0,1,\ldots \end{aligned}$$

where

$$c_n^* = \sum_{j=1}^{\ell}\{\sigma_{0j}(\Delta t^2A^2)\Delta t^{2j}c_n^{(2j-2)} + \sigma_{0j}^*(\Delta t^2A^2)\Delta t^{2j+1}c_n^{(2j-1)} + \sigma_{1j}(\Delta t^2A^2)\Delta t^{2j}c_{n+1}^{(2j-2)}\},$$

$$c_n^{**} = \sum_{j=1}^{\ell}\{\chi_{0j}(\Delta t^2A^2)\Delta t^{2j}c_n^{(2j-2)} + \chi_{0j}^*(\Delta t^2A^2)\Delta t^{2j+1}c_n^{(2j-1)} + \chi_{1j}(\Delta t^2A^2)\Delta t^{2j+1}c_{n+1}^{(2j-1)}\},$$

$$\chi_{ij}(\eta) = \sum_{m=j}^{\ell}\beta_{mi}\eta^{m-j},\quad \chi_{ij}^*(\eta) = \sum_{m=j}^{\ell}\beta_{mi}^*\eta^{m-j},\qquad j = 0,\ldots,\ell,$$

and $\sigma_{ij}(\eta)$, $\sigma_{ij}^*(\eta)$ are defined in (1.2.3).

By (2.4.2), (2.4.6), and (2.4.7) the method (2.4.9) is consistent iff

$$(2.4.10)\quad \alpha_{00} = \alpha_{00}^* = -1,\ \alpha_{10} + \alpha_{11} = -1/2,\ \beta_{00} = 0,\ \beta_{00}^* = \beta_{10} = -1.$$

The method (2.4.9) and the corresponding Runge-Kutta-Nyström method differ only in the treatment of the time-dependent right side c(t). Accordingly, both methods can be studied together as concerns problems of stiffness and absolute stability since these concepts are defined a-priori only for systems (2.4.8) with constant matrix A^2. Moreover, the discretization error of multiderivative methods can be estimated more easily whereas in Runge-Kutta method an ill-conditioned matrix A^2 can affect the discretization error in a negative way; see Chapter V.

Let now

$$V_n = (v_n, \Delta tw_n)^T$$

and

$$G(\eta) = -\,\sigma(\eta)^{-1}\begin{bmatrix}\sigma_0(\eta) & \sigma_0^*(\eta)\\ \chi_0(\eta) & \chi_0^*(\eta)\end{bmatrix} = \begin{bmatrix} g_1(\eta) & g_2(\eta)\\ g_3(\eta) & g_4(\eta)\end{bmatrix}.$$

Then we can write instead of (2.4.9)

$$(2.4.11)\quad V_{n+1} = G(\Delta t^2A^2)V_n - C_n,\qquad n = 0,1,\ldots,$$

where $C_n = \sigma(\Delta t^2A^2)^{-1}(c_n^*, c_n^{**})^T$. Obviously, we have

$$G(0) = \begin{bmatrix}1 & 1\\ 0 & 1\end{bmatrix}$$

and the eigenvalues of $G(\eta)$ are

(2.4.12) $\omega_{1,2}(\eta) = [g_1(\eta) + g_4(\eta) \pm ((g_1(\eta) - g_4(\eta))^2 + 4g_2(\eta)g_3(\eta))^{1/2}]/2.$

(2.4.13) Definition. *The stability region S of the method* (2.4.9) *consists of the* $\eta^2 \in \overline{\mathbb{C}}$ *with* $\mathrm{spr}(G(\eta^2)) \le 1$. *The method* (2.4.9) *is strongly D-stable in* $[-s, 0] \subseteq S$ *if all eigenvalues of modulus one of* $G(\eta^2)$ *are simple eigenvalues for* $\eta^2 \in [-s, 0)$.

Because of (2.4.12) the elements $g_i(\eta^2)$ of $G(\eta^2)$ must be bounded for $\eta^2 \in S$ therefore $\sigma(\eta^2)$ in (2.4.9) must be a polynomial of exact degree ℓ if S is unbounded in $\overline{\mathbb{C}}$.

We reconsider the initial value problem (2.1.32) and quote the following result:

(2.4.14) Theorem. *(i) Let the (m,m)-matrix* A^2 *in* (2.1.32) *be diagonable,* $A^2 = X\Lambda^2X^{-1}$, *and let the solution* y *be (p+2)-times continuously differentiable.*
(ii) Let the method (2.4.9) *be consistent of order* $p \ge 2\ell - 1$ *with the stability region* S.
(iii) Let $\mathrm{Sp}(\Delta t^2A^2) \subset R \subseteq S$ *where* R *is closed in* $\overline{\mathbb{C}}$.
Then for $n = 1,2,\ldots,$

$$|X^{-1}E_n| \le \kappa_R|X^{-1}|n\Delta t\left[\Delta t^{-1}|E_0| + \Delta t^p \int_0^{n\Delta t} |y^{(p+2)}(\tau)|d\tau + n\Delta t \max_{0\le i\le 2\ell-1} |||h^{(i)}|||_n\right]$$

where $E_n\ (y_n - v_n, \Delta t(y'_n - w_n))^T$.

This theorem is proved in the same way as Theorem (2.1.34) by application of the Uniform Boundedness Theorem. However, we note that in the present case the crucial estimation

$$\sup_{\eta^2\in R}\ \sup_{n\in\mathbb{N}} |n^{-1}G(\eta^2)^n| \le \kappa_R$$

can be carried out in a straightforward way by a Jordan canonical decomposition; see Gekeler [77].

In order to derive the optimum order of convergence with respect to the initial error we need the following auxiliary result:

(2.4.15) Lemma. *Let the method* (2.4.9) *be strongly D-stable in* $[-s, 0] \equiv R \subseteq S$ *and let*

$$G(\eta^2)^n = \begin{bmatrix} g_1(\eta^2;n) & g_2(\eta^2;n) \\ g_3(\eta^2;n) & g_4(\eta^2;n) \end{bmatrix}.$$

Then

$$|g_{1,4}(\eta^2;n)| \le \kappa_R(1 + n|\eta|),\ |g_2(\eta^2;n)| \le \kappa_R n,\ |g_3(\eta^2;n)| \le \kappa_R n|\eta^2|.$$

Proof. The assertion is true for $\eta^2 = 0$. Let $\eta^2 \neq 0$ and observe that for a consistent method

(2.4.16) $|g_{1,4}(\eta^2)| \le 1 + \kappa_R|\eta^2|,\ |g_2(\eta^2)| \le \kappa_R,\ |g_3(\eta^2)| \le \kappa_R|\eta^2|.$

We consider two cases:

If $\omega_1(\eta^2) = \omega_2(\eta^2)$ in (2.4.12) then $|\omega_1(\eta^2)| < 1$ and this value is bounded away from one because the method is strongly D-stable in [-s, 0]. Hence the elements of $G(\eta^2)^n$ are bounded and it suffices to consider $g_3(\eta^2;n)$. By a Jordan canonical decomposition of $G(\eta^2)^n$ we find directly that

$$g_3(\eta^2;n) = g_3(\eta^2)n\omega_1(\eta^2)^{n-1}$$

hence the assertion follows by (2.4.16).

If $\omega_1(\eta^2) \neq \omega_2(\eta^2)$ then let

$$\theta_n(\eta^2) = (\omega_1(\eta^2)^n - \omega_2(\eta^2)^n)/(\omega_1(\eta^2) - \omega_2(\eta^2)).$$

A Jordan canonical decomposition of $G(\eta^2)^n$ yields

$$g_1(\eta^2;n) = \omega_1(\eta^2)^n - (\omega_1(\eta^2) - g_1(\eta^2))\theta_n(\eta^2),\ g_2(\eta^2;n) = g_2(\eta^2)\theta_n(\eta^2),$$

$$g_4(\eta^2;n) = \omega_2(\eta^2)^n + (\omega_1(\eta^2) - g_1(\eta^2))\theta_n(\eta^2),\ g_3(\eta^2;n) = g_3(\eta^2)\theta_n(\eta^2).$$

But

$$|\theta_n(\eta^2)| \le n$$

because $|\omega_{1,2}(\eta^2)| \le 1$ by assumption hence we obtain from (2.4.16) that the assertion is true for $g_2(\eta^2;n)$ and $g_3(\eta^2;n)$. Furthermore,

$$|\omega_1(\eta^2) - g_1(\eta^2)| \le |\omega_1(\eta^2)| + |g_1(\eta^2)| \le \kappa_R^*$$

and by (2.4.12)

$$|\omega_1(\eta^2) - g_1(\eta^2)| \le \kappa_R^{**}(|\eta^2| + (|\eta^2|^2 + |\eta^2|)^{1/2} \le \kappa_R^{***}(1 + |\eta|)|\eta|,\ \eta^2 \to 0,$$

because $g_1(\eta^2) - g_4(\eta^2) = \mathcal{O}(\eta^2),\ \eta^2 \to 0$. Accordingly,

$$|\omega_1(\eta^2) - g_1(\eta^2)| \le |\eta|\min\{\kappa_R^*/|\eta|,\ \kappa_R^{***}(1 + |\eta|)\} \le \kappa_R|\eta|$$

which proves the assertion for $g_{1,4}(\eta^2;n)$, too.

(2.4.17) Theorem. *Let the assumption of Theorem* (2.4.14) *be fulfilled but let the method* (2.4.9) *be strongly D-stable in* $R \equiv [-s, 0]$. *Then for* $n = 1,2,\dots$,

$$|X^{-1}(y_n - v_n)| \le \kappa_s |X^{-1}|[(1 + n\Delta t|A|)|y_0 - v_0| + n\Delta t|y_0' - w_0| + n\Delta t\Sigma_n]$$

$$|X^{-1}(y_n' - w_n)| \le \kappa_s |X^{-1}|[n\Delta t|A^2||y_0 - v_0| + (1 + n\Delta t|A|)|y_0' - w_0| + n\Sigma_n]$$

where

$$\Sigma_n = \Delta t^p \int_0^{n\Delta t} |y^{(p+2)}(\tau)|d\tau + n\Delta t \max_{0\le i\le 2\ell-1} |||h^{(i)}|||_n .$$

Proof. The error $E_n = (y_n - v_n, \Delta t(y_n' - w_n))^T$ satisfies

(2.4.18) $X^{-1}E_n = G(\Delta t^2\Lambda^2)^n X^{-1}E_0 + \sum_{\nu=0}^{n-1} G(\Delta t^2\Lambda^2)^{n-\nu-1}(X^{-1}D_\nu^* - X^{-1}H_\nu), \quad n = 0,1,\dots,$

where H_n is the same vector as C_n in (2.4.11) with c_n replaced by h_n and

$$D_n^* = \sigma(\Delta t^2A^2)^{-1}(d_\rho(\Delta t,y)_n, d_\tau(\Delta t,y)_n)^T.$$

Applying Lemma 2.1.8 to the components $d_\rho(\Delta t,y)$ and $d_\tau(\Delta t,y)$ defined by (2.4.3) we obtain

$$|X^{-1}D_n^*| \le \Gamma|X^{-1}|\Delta t^{p+1} \int_{n\Delta t}^{(n+1)\Delta t} |y^{(p+2)}(\tau)|d\tau,$$

and

$$|X^{-1}H_n| \le \kappa_s |X^{-1}|\Delta t^2 \max_{0\le i\le 2\ell-1} |||h^{(i)}|||_n$$

follows in the same way as in Theorem (1.2.12) using again the fact that $\sigma(\eta)$ must be a polynomial of exact degree ℓ if R is unbounded in $\overline{\mathbb{C}}$. The Uniform Boundedness Theorem or a direct verification yields again

$$|\eta^{-1}G(\Delta t^2\Lambda^2)^n| \le \sup_{-s\le\eta^2\le 0} |\eta^{-1}G(\eta^2)^n| \le \kappa_s$$

and so we deduce from (2.4.18)

$$|X^{-1}(y_n - v_n)| \le |g_1(\Delta t^2\Lambda^2;n)||X^{-1}(y_0 - v_0)| + |g_2(\Delta t^2\Lambda^2;n)|\Delta t|X^{-1}(y_0' - w_0)| + \kappa_s|X^{-1}|n\Delta t\Sigma_n$$

$$|\Delta tX^{-1}(y_n' - w_n)| \le |g_3(\Delta t^2\Lambda^2;n)||X^{-1}(y_0 - v_0)| + |g_4(\Delta t^2\Lambda^2;n)|\Delta t|X^{-1}(y_0' - w_0)| + \kappa_s|X^{-1}|n\Delta t\Sigma_n.$$

By assumption we have $Sp(\Delta t^2\Lambda^2) \subset [-s, 0]$ and $\Delta t^2\Lambda^2$ is a diagonal matrix therefore we can substitute $\Delta t^2\Lambda^2$ for η^2 in Lemma (2.4.15) which proves the Theorem.

By and large, the error bound of $X^{-1}(y_n - v_n)$ in this Theorem is the same as that of Theorem (2.1.25) for the original multistep multiderivative methods (2.1.33). Additionally, Nyström type procedures provide an approximation of y_n' in a similar way as indirect methods do. Roughly spoken we can say that $\Delta t X^{-1}(y_n' - w_n)$ has the same error bound as $X^{-1}(y_n - v_n)$ but with the initial error multiplied by $\Delta t|A|$.

In this section, we have adopted the definition of the consistence order p of Section 2.1 for both formula (2.4.1) representing the Nyström type method. Then, in Theorem (2.4.14) and (2.4.17), the order of convergence is p with respect to the solution y of (2.4.8) and p-1 with respect to its derivative y'. Hence we cannot speak of convergence with respect to the approximation of y' if $p = 1$ therefore the consistence order is defined sometimes for Nyström methods as $p^* = p-1$ so that $p^* \geq 1$ implies convergence of at least order p^* for *both* sequences $\{v_n\}$ and $\{w_n\}$.

For examples of Nyström type methods and Runge-Kutta-Nyström methods we refer to Appendix A.5.

2.5. Stability on the Negative Real Line

In Section 1.4 the linear problem

(2.5.1) $y'' = A^2y + c(t),\ t > 0,\ y(0) = y_0,\ y'(0) = y_0^*,$

was transformed into a first order problem and then approximated by a k-step ℓ-derivative method (1.1.3). If we restrict ourselves to the test equation $y'' = \lambda^2 y$ and define the method (1.1.3) by its characteristic polynomial,

$$\pi(\zeta,\eta) = \textstyle\sum_{j=0}^{\ell}\sum_{i=0}^{k}\alpha_{ji}\zeta^i\eta^j \equiv \sum_{j=0}^{\ell}\rho_j(\zeta)\eta^j,$$

then this indirect procedure yields the computational device

(2.5.2)
$$\textstyle\sum_{j=0}\rho_{2j}(T)\Delta t^{2j}\lambda^{2j}v_n + \sum_{j=0}\rho_{2j+1}(T)\Delta t^{2j+1}\lambda^{2j}w_n = 0,$$
$$\textstyle\sum_{j=0}\rho_{2j}(T)\Delta t^{2j}\lambda^{2j}w_n + \sum_{j=0}\rho_{2j+1}(T)\Delta t^{2j+1}\lambda^{2j+2}v_n = 0, \qquad n = 0,1,\ldots,$$

where $\alpha_{ji} = 0$ for $j > \ell$. Here we can eliminate the terms containing the approximation w_n to y_n' and obtain the following scheme

(2.5.3) $[(\sum_{j=0}\rho_{2j}(T)\Delta t^{2j}\lambda^{2j})^2 - \Delta t^2\lambda^2(\sum_{j=0}\rho_{2j+1}(T)\Delta t^{2j}\lambda^{2j})^2]v_n = 0,\quad n = 0,1,\ldots,$

or

$$\pi(T,\Delta t\lambda)\pi(T,-\Delta t\lambda)v_n = 0, \qquad n = 0,1,\ldots .$$

Accordingly, the transformation

$$(2.5.4) \quad \Sigma:\ \pi(\zeta,\eta) \to \pi(\zeta,\eta)\pi(\zeta,-\eta) = \tilde{\pi}(\zeta,\eta^2)$$

defines a 2k-step method (2.1.4) for the problem (2.5.1) involving the first $\ell - 1$ even derivatives of the right side of (2.5.1). See Jeltsch and Nevanlinna [82b]. For $\ell = 1$ the transformation (2.5.4) yields a linear multistep method for the nonlinear problem (2.1.1). Below a method (2.1.4) given by the polynomial

$$\pi(\zeta,\eta^2) = \sum_{j=0}^{\ell}\sum_{i=0}^{k}\alpha_{ji}\zeta^i\eta^{2j}$$

is also briefly called a (k,ℓ)-method as the k-step ℓ-derivative methods (1.1.3) for first order systems.

If the starting values are chosen suitably then the schemes (2.5.2) and (2.5.3) produce the same sequence $\{v_n\}_{n=0}^{\infty}$ if rounding errors are not regarded. Therefore the method (1.1.3) and its Σ-*transformation* defined by (2.5.4) have the same order p and the same error constant. For the sequel we recall that the stability regions S are defined in different ways for first order and second order problems, cf. Definitions (1.2.7) and (2.1.14). If the method (1.1.3) has the stability interval $I_r = \{i\eta,\ -r < \eta < r\}$ on the imaginary axis then all unimodular roots of $\pi(\zeta,\eta)$ are simple for $\eta \in I_r$ by Definition (1.2.7). But if $\zeta_j(\eta)$ is a unimodular root and $\eta \in I_r$ then $\zeta_j(-\eta)$ is a unimodular root, too, because $\pi(\zeta,\eta)$ is a real polynomial. Hence the method (2.1.4) given by the Σ-transformation has at most double unimodular roots for $\eta \in I_r$ and therefore has the real stability interval $(-r^2, 0] \subset S$. If I_r is maximum for the method (1.1.3) then $(-r^2, 0]$ is maximum for the Σ-transformation. Using this fact some results from Section 1.7 on the stability on the imaginary axis of methods (1.1.3) can be transformed into results on the stability on the negative real line of methods (2.1.4). In these latter methods only the stability interval $(-s, 0]$, $s > 0$, is of numerical interest because the test equation $y'' = \lambda^2 y$ has bounded solutions only for negative real λ^2.

The following result corresponds to Corollary (1.7.7) and has been proved by Hairer [79] and Jeltsch and Nevanlinna [82b].

(2.5.5) Theorem. *(i) A (k,ℓ)-method* (2.1.4) *with* $(-\infty, 0] \subset S$ *has order* $p \le 2\ell$. *(ii) Let* $\pi_\ell(\zeta,\eta)$ *be the polynomial* (1.5.1) *of the diagonal Padé approximant. Among all (k,ℓ)-methods of order* $p = 2\ell$ *with* $(-\infty, 0] \subset S$ *the method with the polynomial* $\Sigma\pi_\ell(\zeta,\eta)$ *has the smallest error constant.*

As concerns explicit methods, the result corresponding to Theorem (1.7.8) reads:

(2.5.6) Theorem. (Jeltsch and Nevanlinna [82b].) *If* (2.1.4) *is an explicit convergent* (k,ℓ)*-method then either* $[-4\ell^2, 0] \not\subset \bar{S}$ *or* $[-4\ell^2, 0] = \bar{S}$ *and* $\pi(\zeta,\eta^2)$ *has the factor*

$$\pi_*(\zeta,\eta^2) = \zeta^2 - 2i^{2\ell}T_{2\ell}(-i\eta/2\ell)\zeta + (-1)^{2\ell}$$

where $T_\ell(\xi) = \cos\ell\arccos\xi$ *is the Tschebyscheff polynomial of degree* ℓ.

Recall that Tschebyscheff polynomials $T_\ell(\xi)$ are even for ℓ even and odd for ℓ odd. For $\ell = 1$ we have

$$\pi_*(\zeta,\eta^2) = \zeta^2 - 2\zeta + 2 - \eta^2\zeta$$

which defines Störmer's method of order two with $[-4, 0] = S$.

Methods (2.1.4) with $(-\infty, 0] \subset S$ are necessarily implicit. The result corresponding to Corollary (1.7.4) has been proved by Dahlquist [78a]:

(2.5.7) Theorem. *(i) A linear multistep method* (2.1.4) *with* $(-\infty, 0] \subset S$ *has order* $p \le 2$.
(ii) Let $\pi^\dagger(\zeta,\eta) = (\zeta - 1) - (\eta/2)(\zeta + 1)$ *be the polynomial of the trapezoidal rule. Among all linear methods* (2.1.4) *of order* $p = 2$ *with* $(-\infty, 0] \subset S$ *the method given by* $\Sigma\pi^\dagger(\zeta,\eta)$ *has the smallest error constant.*

Finally, the result corresponding to Theorem (1.7.12) has been proved again by Jeltsch and Nevanlinna [82b] for linear methods:

(2.5.8) Theorem. *(i) If* (2.1.4) *is a linear method of order* $p > 2$ *with* $(-s, 0] \subset S$ *then* $s \le 6$.
(ii) If $p = 2$ *and* $s > 6$ *then the error constant* χ_p *satisfies*

$$|\chi_p| \geq (1 - (6/s))/12.$$

(iii) The only linear method (2.1.4) *with* $s = 6$ *is the method of Cowell (or Numerov) given by*

$$(2.5.9) \quad \pi(\zeta,\eta^2) = \Sigma\pi^\dagger(\zeta,\eta) + \frac{\eta^2}{6}(\zeta - 1)^2 = (\zeta^2 - 2\zeta + 1) - \frac{\eta^2}{12}(\zeta^2 + 10\zeta + 1).$$

Cowell's method has order four. For every $s < 6$ there exists a linear 4-step method (2.1.4) with order six and $S = [-s, 0]$, cf. Lambert and Watson [76] , Jeltsch [78b], and Dougalis [79].

Baker, Dougalis, and Serbin [79, 80] have developed high order single-step methods (2.1.4) with $(-\infty, 0] \subset S$ for the homogeneous linear problem

$$y'' = A^2 y,\ t > 0,\ y(0) = y_0,\ y'(0) = y_0^*$$

which are based on rational approximations to the cosinus function.

2.6. Examples of Linear Multistep Methods

Examples of Nyström methods are given in Appendix A.5 because of their strong relationship to Runge-Kutta methods. The general 2-step method (2.1.4) with $\ell = 2$ is considered in Appendix A.6. In concluding this chapter we give in this section some examples of linear mutistep methods for differential systems of second order without and with damping:

The general linear 2-step method (2.1.4) of order $p \geq 2$ has the polynomial

$$(2.6.1)\quad \pi(\zeta,\eta^2) = \zeta^2 - 2\zeta + 1 - \eta^2(\omega\zeta^2 + (1-2\omega)\zeta + \omega), \qquad \omega \in \mathbb{R}.$$

For $\omega = 1/12$ this method has order 4, cf. (2.5.9), and order 2 else. The stability region for η^2 is $S = [-s, 0]$ where $s = 4/(1-4\omega)$ for $0 \leq \omega < 1/4$ and $s = \infty$ for $\omega \geq 1/4$. $\pi(\zeta,\eta^2)$ has double unimodular roots in $\eta^2 = 0$ and only for $0 \leq \omega \leq 1/4$ in $\eta^2 = -s$. Accordingly the method is strongly D-stable in $[-\infty, 0]$ for $\omega > 1/4$.

The 3-step method of order 2 with the polynomial

$$(2.6.2)\quad \pi(\zeta,\eta^2) = 2\zeta^3 - 5\zeta^2 + 4\zeta - 1 - \eta^2\zeta^3$$

is strongly D-stable in $[-\infty, 0]$, too, but here we have $(-\infty, 0) \subset \mathring{S}$ hence $[-\infty, 0]$ is not a periodicity interval.

The 4-step methods given by

$$\pi(\zeta,\eta^2) = (\zeta-1)^2(\zeta^2 - 2\zeta\cos\phi + 1)$$

$$- \eta^2[(9+\cos\phi)(\zeta^4+1) + 8(13-3\cos\phi)(\zeta^3+\zeta) + 2(7-97\cos\phi)\zeta^2]/120$$

have optimum order $p = 6$ for $\phi \in [0, \pi]$ and $0 \in S$ holds for $\phi \in (0, \pi]$. The stability region is

$$S = [-s(\phi), 0],\ s(\phi) = 60(1+\cos\phi)/(11+9\cos\phi), \qquad \phi \in (0, \pi)$$

hence there exists for every $0 < s < 6$ a $\phi \in (0, \pi)$ with $s = s(\phi)$; cf. e.g. Jeltsch [78b].

The procedure given by (2.6.2) is a backward differentiation method as only the leading coefficent α_{13} of the polynomial $\rho_1(\zeta)$ is nonzero. Choosing for $\tau_3(\zeta)$ in (2.2.3) the backward differentiation approximation of Table (A.4.3) for $k = p = 3$,

$$(2.6.3)\quad \tau_3(\zeta) = (11\zeta^3 - 18\zeta^2 + 9\zeta - 2)/6,$$

we obtain $d_i(\Delta t,u)(t) \equiv 0$, $i = 0,1,2$, and

$$d_3(\Delta t,u)(t) = \tau_3(T)u(t) - \Delta t T^3 u'(t) = \mathcal{O}(\Delta t^4), \qquad \Delta t \to 0,$$

for the discretization errors with respect to y'. A substitution of (2.6.2) and (2.6.3) into (2.2.3) yields Houbolt's method of order 2 for the problem (2.2.1),

$$(12 - 6\Delta t^2 A^2 - 11\Delta t B_{n+3})v_{n+3}$$

$$= (30 - 18\Delta t B_{n+3})v_{n+2} - (24 - 9\Delta t B_{n+3})v_{n+1} + (6 - 2\Delta t B_{n+3})v_n + 6\Delta t^2 c_{n+3},$$

cf. e.g. Bathe and Wilson [76]. With respect to $\tilde{d}(\Delta t,u)$ the order is 3 and the method is strongly D-stable in $[-\infty, 0]\times[-\infty, 0] \subset S^2$.

Recall now that a linear method for the problem

$$(2.6.4)\quad y'' = A^2 y + By' + c(t),\ t > 0,\ y(0) = y_0,\ y'(0) = y_0^*,$$

with constant matrices A^2 and B is given by the polynomial

$$\pi(\zeta,\eta^2,\mu) = \rho_0(\zeta) + \eta^2\rho_1(\zeta) + \mu\tau(\zeta).$$

If (1.1.3) is a linear method of order p for systems of first order with the polynomial $\pi(\zeta,\eta) = \rho(\zeta) - \eta\sigma(\zeta)$ then

$$(2.6.5)\quad \pi(\zeta,\eta^2,\mu) = \rho(\zeta)^2 - \eta^2\sigma(\zeta)^2 - \mu\sigma(\zeta)\rho(\zeta)$$

defines a method of order p for (2.6.4). This is shown in the same way as in the previous section. Moreover, if the method (1.1.3) is A-stable then the method given by (2.6.5) is strongly D-stable in $[-s, 0]\times[-r, 0]$ for every $0 < s < \infty$, $0 < r < \infty$.

The general 2-step method of order 2 has the polynomial

$$\pi(\zeta,\eta^2,\mu) = (\zeta - 1)^2 - \eta^2(\omega\zeta^2 + (1 - 2\omega)\zeta + \omega) - \mu[(\zeta^2 - 1)/2].$$

For $\omega > 1/4$ the method is strongly D-stable in $[-\infty, 0]\times[-\infty, 0]$.

The general 3-step method of order 3 has the polynomial

$$\pi(\zeta,\eta^2,\mu) = (\omega + \tfrac{1}{2})(\zeta - 1)^3 + (\zeta - 1)^2$$
$$- \eta^2[(\tfrac{1}{24} + \tfrac{\omega}{6} + \theta)(\zeta - 1)^3 + (\tfrac{7}{12} + \omega)(\zeta - 1)^2 + (\tfrac{3}{2} + \omega)(\zeta - 1) + 1]$$
$$- \mu[(\tfrac{\omega}{2} + \tfrac{1}{6})(\zeta - 1)^3 + (\omega + 1)(\zeta - 1)^2 + (\zeta - 1)], \quad \omega,\theta \in \mathbb{R}.$$

For $\theta = -\omega/12$ we obtain

$$\pi(\zeta,\eta^2,\mu) = [(\tfrac{1}{2} + \omega)\zeta + (\tfrac{1}{2} - \omega)][(\zeta^2 - 2\zeta + 1) - \eta^2((\zeta^2 + 10\zeta + 1)/12)]$$
$$- \mu(\zeta - 1)[(\tfrac{1}{6} + \tfrac{\omega}{2})\zeta^2 + \tfrac{2}{3}\zeta + (\tfrac{1}{6} - \tfrac{\omega}{2})].$$

Hence the associated method for conservative systems with the polynomial $\pi(\zeta,\eta^2,0)$ is a trivial modification of Cowell's method in this case and thus has order 4 with the stability region $S = [-6, 0]$, cf. Theorem (2.5.8). For $\theta = -\omega/12$ and $\omega > 0$ the real stability region $S^2 \cap \mathbb{R}^2$ of the general method is the closed triangle with the vertices $(0,0)$, $(-6, 0)$, and $(0, -12\omega)$; cf. Godlewski and Puech-Raoult [79].

III. Linear Multistep Methods and Problems with Leading Matrix $A(t) = a(t)A$

3.1. Differential Systems of First Order and Methods with Diagonable Frobenius Matrix

During the last years great progress has been made in the understanding of the behavior of numerical integration schemes for stiff differential equations in the case where the leading matrix A varies with time and even some interesting results were obtained for general nonlinear systems. In this and the next chapter we consider some of these a-priori error bounds for differential systems of first and second order as far as they are of that special uniform character which is desired in the context with dynamic finite element equations. The generalization of the results of this chapter to nonlinear multistep methods is somewhat involved as concerns the notation but can be derived otherwise in a straightforward way.

Over a long period the majority of contributions to numerical stability dealed only with the test equation $y' = \lambda y$. Nevertheless the study of this trivial equation has revealed to be very successful and a large variety of interesting results and useful new methods has been derived by this way. Similarly, it seems advantageous in the study of time-dependent problems to consider at first the case where the leading matrix A depends in a *scalar* way on time in order to derive the optimum results. So we study in this chapter the initial value problem

$$(3.1.1)\quad y' = a(t)Ay + c(t) + h(t),\ t > 0,\ y(0) = y_0,$$

where a is a scalar-valued function, and the corresponding second-order problem. This special time-dependent form of the leading matrix is necessary in this and the third section by technical reasons but not in Section 3.2.

Using the conventional notations

$$\rho(\zeta) = \textstyle\sum_{i=0}^{k}\alpha_i\zeta^i = \rho_0(\zeta),\ \alpha_k > 0,\ \sigma(\zeta) = \sum_{i=0}^{k}\beta_i\zeta^i = -\rho_1(\zeta),\ \beta_k \geq 0,$$

linear multistep methods have for (3.1.1) the form

$$(3.1.2)\quad \rho(T)v_n - \Delta tA\sigma(T)(av)_n = \Delta t\sigma(T)c_n, \qquad n = 0,1,\ldots,$$

where $\sigma(T)(av)_n = \sum_{i=0}^{k}\beta_i a_{n+i}v_{n+i}$ and the defect $h(t)$ is again omitted in the computational device.

In this section we follow Hackmack [81] and deduce a generalization of Theorem

(1.2.12) to the problem (3.1.1) under the restriction that the associated Frobenius matrix is diagonable in the considered stability (sub-)region $R \subset S$. As in the proof a mean value theorem is applied to the eigenvalues $\zeta_i(\Delta t a(t)\lambda)$ of $F_\pi(\Delta t a(t)\lambda)$ with respect to t, this assumption ensures the necessary smoothness of $\zeta_i(\eta)$.

<u>(3.1.3) Theorem.</u> *(i) Let the (m,m)-matrix* A *in* (3.1.1) *be diagonable,* $A = X\Lambda X^{-1}$, *let* $a(t) \neq 0$, $t > 0$, *and let the solution* y *of* (3.1.1) *be (p+1)-times continuously differentiable.*

(ii) Let the method (3.1.2) *be consistent of order* p *with stability region* S.

(iii) Let $Sp(\Delta t a(t) A) \subset R \subseteq S$, $t > 0$, *where* R *is closed in* $\overline{\mathbb{C}}$ *and convex.*

(iv) *Let the Frobenius matrix* $F_\pi(\eta)$ *of the method* (3.1.2) *be diagonable in* R.

Then for $n = k, k+1, \ldots,$

$$|X^{-1}(y_n - v_n)|$$

$$\leq \kappa_R |X^{-1}| \exp\{\kappa_R^* \Theta_n n\Delta t\} \left[|Y_{k-1} - V_{k-1}| + \Delta t^p \int_0^{n\Delta t} \exp\{-\kappa_R^* \Theta_n \tau\} |y^{(p+1)}(\tau)| d\tau + n\Delta t |||h|||_n \right]$$

where $\Theta_n = \max_{k \leq \nu \leq n} \max_{(\nu-k)\Delta t \leq t \leq \nu\Delta t} |a'(t)/a(\nu\Delta t)|$.

Proof. In the present case the error equation (1.2.10) changes to

$$(3.1.4) \quad E_n = (F_n + R_n)E_{n-1} + D(\Delta t, y)_n - H_n, \qquad n = k, k+1, \ldots,$$

where $F_n = F_\pi(\Delta t a_n A)$ and R_n is a (k,k)-block matrix of which only the last block row is nonzero:

last block row of R_n

$$= \Delta t A(\alpha_k I - \Delta t \beta_k a_n A)^{-1} (\beta_0(a_{n-k} - a_n)I, \ldots, \beta_{k-1}(a_{n-1} - a_n)I).$$

Under the above assumptions we find easily that

$$(3.1.5) \quad |X^{-1} R_n X| \leq \kappa_1 \Theta_n \Delta t$$

and the Frobenius matrix $F_\pi(\eta)$ is diagonable to $Z(\eta)$ in the closed set R, $F_\pi(\eta) = W(\eta)Z(\eta)W(\eta)^{-1}$. Using the abbreviations $W_n = W(\Delta t a_n \Lambda)$ and $Z_n = Z(\Delta t a_n \Lambda)$ we thus can write instead of (3.1.4)

$$(3.1.6) \quad \begin{aligned} W_n^{-1}X^{-1}E_n &= (Z_n + W_n^{-1}X^{-1}R_n X W_n) W_n^{-1} X^{-1} E_{n-1} + W_n^{-1}X^{-1}(D(\Delta t, y)_n - H_n) \\ &= (Z_n + W_n^{-1}X^{-1}R_n X W_n)(W_n^{-1} W_{n-1}) W_{n-1}^{-1} X^{-1} E_{n-1} + W_n^{-1}X^{-1}(D(\Delta t, y)_n - H_n). \end{aligned}$$

By assumption we have $|Z_n| \le 1$ and observing $|W(\Lambda)| = \max_{1\le\mu\le m}|W(\lambda_\mu)|$ we obtain from Lemma (A.2.4) that

(3.1.7) $\quad |W_n| \le \kappa_2, \; |W_n^{-1}| \le \tilde{\kappa}_R.$

Hence if

(3.1.8) $\quad |W_n^{-1}W_{n-1}| \le 1 + \tilde{\kappa}_R^*\Theta_n\Delta t$

then we find from (3.1.6) that

$$|W_n^{-1}X^{-1}E_n| \le (1 + (\kappa_R^*/2)\Theta_n\Delta t)^2|W_{n-1}^{-1}X^{-1}E_{n-1}| + \kappa_R|X^{-1}(D(\Delta t,y)_n - H_n)|,\; n = k,k+1,\ldots,$$

writing

$$(1 + \kappa_1\kappa_2\tilde{\kappa}_R\Theta_n\Delta t)(1 + \tilde{\kappa}_R^*\Theta_n\Delta t) = (1 + (\kappa_R^*/2)\Theta_n\Delta t)^2.$$

Using once more (3.1.7) the assertion follows from this inequality by induction.

In order to prove (3.1.8) we observe that by Lemma (A.2.4)

(3.1.9)
$$\begin{aligned} |W_n^{-1}W_{n-1}| &= \max_{1\le\mu\le m}|W(\Delta t a_n\lambda_\mu)^{-1}W(\Delta t a_{n-1}\lambda_\mu)| \\ &\le 1 + \kappa_R \max_{1\le\mu\le m}\max_{1\le i\le k}|\zeta_i(\Delta t a_n\lambda_\mu) - \zeta_i(\Delta t a_{n-1}\lambda_\mu)|. \end{aligned}$$

In this inequality $\zeta_1(\eta),\ldots,\zeta_k(\eta)$ are the analytic roots of the characteristic polynomial $\pi(\zeta,\eta)$ of the method (3.1.2) in the closed and convex domain R. If $\beta_k = 0$ then R is bounded by Lemma (A.1.3) therefore we obtain by the Mean Value Theorem and assumption (iii)

$$|\zeta_i(\Delta t a_n\lambda_\mu) - \zeta_i(\Delta t a_{n-1}\lambda_\mu)| \le \max_{\eta\in R}|\zeta_i'(\eta)|\Theta_n \mathrm{spr}(\Delta t a_n A)\Delta t \le \tilde{\kappa}_R^*\Theta_n\Delta t$$

which proves together with (3.1.9) the desired inequality (3.1.8). If $\beta_k \ne 0$ then observe that $\zeta_i(\eta)$, $i = 1,\ldots,k$, are the roots of the normed polynomial

$$\textstyle\sum_{i=0}^k \gamma_i(\eta)\zeta^i,\; \gamma_i(\eta) = (\alpha_i - \beta_i\eta)/(\alpha_k - \beta_k\eta).$$

Accordingly we have omitting the argument $\eta = \Delta t a(t)\lambda$

(3.1.10) $$\frac{\partial\zeta_i}{\partial t}(\gamma_0,\ldots,\gamma_k) = \sum_{j=0}^k \frac{\partial\zeta_i}{\partial\gamma_j}(\gamma_0,\ldots,\gamma_k)\frac{\partial\gamma_j}{\partial t}.$$

But

$$\max_{1\le i\le k}\sup_{\eta\in R}|1/\frac{\partial\pi}{\partial\zeta}(\zeta_i(\eta),\eta)| \le \kappa_R$$

because $\pi(\zeta,\eta)$ has no double roots $\zeta_i(\eta)$ in R, and moreover

$$\frac{\partial\pi}{\partial\gamma_j}(\zeta_i) = \frac{\partial\pi}{\partial\zeta}(\zeta_i)\frac{\partial\zeta_i}{\partial\gamma_j} + \zeta_i^j \equiv 0, \qquad i = 1,\dots,k.$$

Consequently,

$$(3.1.11)\quad |\frac{\partial\zeta_i}{\partial\gamma_j}| \le |\zeta_i^j/\frac{\partial\pi}{\partial\zeta}(\zeta_i)| \le \kappa_R.$$

Substituting this bound into (3.1.10) and observing that

$$(3.1.12)\quad |\frac{\partial\gamma_j}{\partial t}(\Delta t a(t)\lambda_\mu)| \le \kappa\Theta_n, \qquad \mu = 1,\dots,m,\ k\Delta t \le t \le n\Delta t,$$

we can estimate the right side of (3.1.9) by $1 + \tilde{\kappa}_R^*\Theta_n\Delta t$ which proves the inequality (3.1.8) for implicit methods.

Theorem (3.1.3) does not need the assumption that the domain R is contained in the open interior $\mathring{S}$ of the stability region S thus, in particular, it is not necessary here that S has a non-empty interior. So, for instance, the Simpson rule of Section 1.2 satisfies the assumption with

$$R = \{\eta = i\tilde{\eta},\ -s \le \tilde{\eta} \le s\},\ 0 < s < \sqrt{3}.$$

On the other side, assumption (iv) is difficult to delete in the above way of proof, and a generalization of the result to differential systems $y' = A(t)y + c(t)$ with general time-dependent matrix A is obstructed by the unavailability of a suitable dimensionfree mean value theorem for functions of matrix-valued functions $f(A(t))$; see e.g. Gekeler [81].

3.2. Differential Systems of First Order and Methods with Non-Empty $\mathring{S}$

It was the merit of LeRoux [79a] to have shown a way of error estimation for the problem studied in Section 3.1 which circumvents a Jordan canonical decomposition of the Frobenius matrix $F_\pi(\eta)$ and avoids so the difficulties arising in this way of proceeding if more general results are desired. We dedicate this section to the work of LeRoux in which linear differential systems with a rather general operator A(t) are considered. However, for the sake of simplicity, we shall confine ourselves to the special non-autonomous differential system (3.1.1) and adopt moreover the assumption of Corollary (1.3.15). The original results of LeRoux concern "strongly $A(\alpha)$-stable" linear multistep methods of which the best-known representatives are the backward differentiation methods given in Appendix A.4. The subsequent generalization to linear multistep methods with non-empty $\mathring{S} \cap \mathbb{R}$ in a neighborhood of zero suggests itself under the stipulated restriction.

Before we formulate the result of this section we have to modify the exponential multiplication factor of Theorem (3.1.3) slightly: Let

$$\tilde{\Theta}_n = \max_{k\le\nu\le n} \max_{(\nu-1)\Delta t\le t\le\nu\Delta t} |a'(t)/(a_{\nu-1}a_\nu)^{1/2}|$$

and

$$\Theta_n^* = \begin{cases} \max\{\Theta_n, \tilde{\Theta}_n\} \\ \tilde{\Theta}_n \text{ if } \beta_0 = \beta_1 = \ldots = \beta_{k-1} = 0 \text{ (cf. (3.1.2))}. \end{cases}$$

<u>(3.2.1) Theorem</u>. (LeRoux [79a].) *(i) Let the (m,m)-matrix* $a(t)A$ *in* (3.1.1) *be hermitean and negative definite,* $a(t)A \le -\alpha I$, $\alpha > 0$, $t > 0$, *and let the solution* y *of* (3.1.1) *be (p+1)-times continuously differentiable.*
(ii) Let the method (3.1.2) *be consistent of order* p *with stability region* S*; let* $0 \in S$, $[-s, 0) \subset \mathring{S}$, $0 < s \le \infty$, *and* $\mathrm{Re}(\chi_i^*) > 0$, $i = 1,\ldots,k_*$.
(iii) Let $\Delta t|a(t)A| \le s$, $t > 0$.
Then for $n = k,k+1,\ldots$,

$$|(y_n - v_n)| \le \kappa_s \exp\{\kappa_s^*\Theta_n^* n\Delta t\}\left[|Y_{k-1} - V_{k-1}| + \Delta t^p \int_0^{n\Delta t} \exp\{-\kappa_s^*\Theta_n^*\tau\}|y^{(p+1)}(\tau)|d\tau + n\Delta t|||h|||_n\right].$$

Proof. Let

$$\Psi_n = F_n + R_n = \Psi(\Delta t, \Delta t a_n A)$$

be the leading matrix in the error equation (3.1.4) then we obtain instead of (1.2.13)

$$|E_n| \le |\Psi_n\cdots\Psi_k||E_{k-1}| + \textstyle\sum_{\nu=k}^{n-1}|\Psi_n\cdots\Psi_{\nu+1}||D(\Delta t,y)_\nu - H_\nu| + |D(\Delta t,y)_n - H_n|.$$

Consequently, if

$$|\Psi_n\cdots\Psi_\nu| \le \kappa_s \exp\{\kappa_s^*\Theta_n^*(n-\nu)\Delta t\}, \qquad \nu = k,\ldots,n,$$

then the assertion follows in the same way as in Chapter I. But, since A is now unitary diagonable we have

$$|\Psi_n\cdots\Psi_\nu| = |\Psi(\Delta t,\Delta t a_n \Lambda)\cdots\Psi(\Delta t,\Delta t a_\nu \Lambda)| = \max_{1\le\mu\le m}|\Psi(\Delta t,\Delta t a_n\lambda_\mu)\cdots\Psi(\Delta t,\Delta t a_\nu\lambda_\mu)| .$$

Writing shortly $\Phi_n = \Psi(\Delta t, \Delta t a_n \lambda)$ we thus have to prove

$$(3.2.2)\quad \max_{0\le t\le n\Delta t}\sup_{-s\le\Delta t a(t)\lambda\le 0}|\Phi_n\cdots\Phi_\nu| \le \kappa_s\exp\{\kappa_s^*\Theta_n^*(n-\nu)\Delta t\}, \qquad \nu = k,\dots,n.$$

The proof of this crucial inequality is derived in several steps. The first one is of rather general character:

<u>(3.2.3) Lemma</u>. *Let* $\mathbb{N}_0 = \mathbb{N}\cup\{0\}$ *and let* P_i *and* Q_i, $i = 0,1,\dots$, *be two sequences of matrices such that*

(i) $\sup_{i\in\mathbb{N}_0, j\in\mathbb{N}_0}|Q_i^j| \le \kappa_0$, $\qquad n = 0,1,\dots,$

(ii) $\max_{0\le i\le n}|P_i - Q_i| \le \kappa_1\Gamma_n$, $\qquad n = 0,1,\dots,$

(iii) $\max_{1\le i\le n, 1\le j\le n}|Q_i^j - Q_{i-1}^j| \le \kappa_2\Gamma_n$, $\qquad n = 1,2,\dots$.

Then

$$|P_n\cdots P_\nu| \le \kappa_3\exp\{\kappa_3\Gamma_n(n-\nu)\}, \qquad \nu = 0,\dots,n,\ n = 0,1,\dots,$$

where $\kappa_3 = \max\{|P_0|, \kappa_0(2+|P_0|), \kappa_0\kappa_1+\kappa_2\}$.

Proof. It suffices to prove the assertion for $\nu = 0$. By LeRoux [79a, Prop. 3] we have for $n = 0,1,\dots,$ the following decomposition writing $P_i\cdots P_k = I$ for $i < k$

$$P_n\cdots P_0 = Q_0^{n+1} + \sum_{i=1}^{n+1}(Q_i^{n+1-i}P_{i-1}\cdots P_0 - Q_{i-1}^{n+2-i}P_{i-2}\cdots P_0)$$

$$= Q_0^{n+1} + \sum_{i=1}^{n}(Q_i^{n+1-i} - Q_{i-1}^{n+1-i})P_{i-1}\cdots P_0 + \sum_{i=1}^{n+1}Q_{i-1}^{n+1-i}(P_{i-1} - Q_{i-1})P_{i-2}\cdots P_0.$$

Writing $z_i = |P_{i-1}\cdots P_0|$ we obtain $z_1 = |P_0| \le \kappa_3$ and

$$z_{n+1} \le \kappa_0 + |Q_0^n||P_0 - Q_o| + \kappa_2\Gamma_n\sum_{i=1}^n z_i + \kappa_0\kappa_1\Gamma_n\sum_{i=2}^{n+1}z_{i-1}$$

or

$$z_{n+1} \le \kappa_3(1 + \Gamma_n\sum_{i=1}^n z_i), \qquad n = 1,2,\dots \ .$$

By induction, finally, we derive from this inequality that

$$z_i \le \kappa_3(1+\kappa_3\Gamma_n)^{i-1}, \qquad i = 1,\dots,n+1,$$

or

$$|P_i\cdots P_0| \le \kappa_3\exp\{\kappa_3\Gamma_n i\}, \qquad i = 0,\dots,n.$$

Under the assumption of Theorem (3.2.1) the Uniform Boundedness Theorem yields

$$\sup_{-s\le\eta\le 0}\sup_{j\in\mathbb{N}}|F_\pi(\eta)^j| \le \kappa_s$$

and (3.1.5) says that

$$|R_i| = |R(\Delta t,\Delta ta_i\lambda)| = |\Phi_i - F_\pi(\Delta ta_i\lambda)| \le \kappa\Theta_n\Delta t, \qquad i = 1,\ldots,n.$$

Writing $P_i = \Phi_i$, $Q_i = F_\pi(\Delta ta_i\lambda)$, and $\Gamma_n = \Theta_n^*\Delta t$, these are the first two assumptions of Lemma (3.2.3). Accordingly, this lemma yields the desired inequality (3.2.2) if the third assumption is fulfilled which now reads as

(3.2.4) $$\max_{1\le i\le n}\max_{1\le j\le n}\sup_{-s\le\Delta ta(t)\lambda\le 0}|F_\pi(\Delta ta_i\lambda)^j - F_\pi(\Delta ta_{i-1}\lambda)^j| \le \kappa_s\tilde{\Theta}_n\Delta t.$$

In order to prove this inequality we need some auxiliary results.

(3.2.5) Lemma.

$$\sum_{j=1}^{n-1}(j(n-j))^{-1/2} \le \pi.$$

Proof. By a standard argument we find that $1/2 = \arg\min_{0<x<1}(x(1-x))^{-1/2}$ therefore we obtain for $j = 1,\ldots,n-1$,

$$n^{-1}[(j/n)(1-(j/n))]^{-1/2} \le \int_{(j-1)/n}^{j/n}(x(1-x))^{-1/2}dx, \qquad j/n \le 1/2,$$

$$n^{-1}[(j/n)(1-(j/n))]^{-1/2} \le \int_{j/n}^{(j+1)/n}(x(1-x))^{-1/2}dx, \qquad j/n \ge 1/2,$$

which leads to the desired result,

$$\sum_{j=1}^{n-1}(j(n-j))^{-1/2} \le \int_0^1(x(1-x))^{-1/2}dx \le \pi.$$

(3.2.6) Lemma. *If* $0 \in S$, $[-s, 0) \subset \mathring{S}$, $0 < s < \infty$, *and* $\mathrm{Re}(\chi_i^*) > 0$, $i = 1,\ldots,k_*$, *then there exist positive constants* κ_s *and* κ_s^* *such that*

$$|F_\pi(\eta)^n| \le \kappa_s\exp\{\kappa_s^*n\eta\}, \qquad -s \le \eta \le 0,\ n = 1,2,\ldots\ .$$

Proof. By Corollary (1.3.13) there exist positive constants δ, κ_1, and κ_1^* such that

(3.2.7) $$|F_\pi(\eta)^n| \le \kappa_1\exp\{\kappa_1^*n\eta\}, \qquad -\delta \le \eta \le 0.$$

Because $[-s, -\delta] \subset \mathring{S}$ there exists a $\mu > 0$ such that $[-s, -\delta] \subset S_\mu$. In the same way as in Theorem (1.2.18) we then obtain

$$|F_\pi(\eta)^n| \leq \tilde{\kappa}_s \exp\{-n\mu\}, \quad - s \leq \eta \leq - \delta.$$

As an immediate consequence we find from this inequality that

$$(3.2.8) \quad |F_\pi(\eta)^n| \leq \tilde{\kappa}_s \exp\{(\mu/s)n\eta\}, \quad - s \leq \eta \leq - \delta.$$

From (3.2.7) and (3.2.8) the assertion now follows with $\kappa_s = \max\{\kappa_1, \tilde{\kappa}_s\}$ and $\kappa_s^* = \min\{\kappa_1^*, \mu/s\}$.

<u>(3.2.9) Lemma</u>. *If* $0 \in S$, $[-\infty, 0) \subset \mathring{S}$, *and* $\mathrm{Re}(\chi_i^*) > 0$, $i = 1,\dots,k_*$, *then there exist positive constants* s_0, κ, *and* κ^* *such that*

(i) $\quad |F_\pi(\eta)^n| \leq \kappa\exp\{\kappa^* n\eta\}, \qquad - s_0 \leq \eta \leq 0,$

(ii) $\quad |F_\pi(\eta)^n| \leq \kappa\exp\{-\kappa^* n\}, \qquad - \infty \leq \eta \leq - s_0,$

(iii) $\quad |F_\pi(\eta)^n - F_\pi(\infty)^n| \leq \kappa\exp\{-\kappa^* n\}/|\eta|, \qquad - \infty \leq \eta \leq - s_0.$

Proof. By assumption we have $\mathrm{spr}(F_\pi(\infty)) < 1$ hence there is a neighborhood $\{\eta \in \overline{\mathbb{C}}, |\eta| \geq s_0\}$ of ∞ such that $\mathrm{spr}(F_\pi(\eta)) < 1 - \mu_0 < 1$ by continuity for $|\eta| \geq s_0$. Now the second assertion follows as in Theorem (1.2.18). The first assertion is that of Lemma (3.2.6) with $s = s_0$. In order to prove the last assertion observe that $\beta_k \neq 0$ by Lemma (A.1.3) and that by a simple estimation

$$(3.2.10) \quad |F_\pi(\eta) - F_\pi(\infty)| \leq \kappa|\eta|^{-1}.$$

For $n > 1$ we have

$$F_\pi(\eta)^n - F_\pi(\infty)^n = \textstyle\sum_{i=0}^{n-1} F_\pi(\eta)^i (F_\pi(\eta) - F_\pi(\infty)) F_\pi(\eta)^{n-1-i}$$

therefore we obtain by (3.2.10) and the second assertion

$$|F_\pi(\eta)^n - F_\pi(\infty)^n| \leq \tilde{\kappa}|\eta|^{-1} \textstyle\sum_{i=0}^{n-1} e^{-\tilde{\kappa}^* i} e^{-\tilde{\kappa}^*(n-1-i)} \leq \kappa|\eta|^{-1} e^{-\kappa^* n}, \quad - \infty \leq \eta \leq - s_0.$$

<u>(3.2.11) Lemma</u>. *Let the assumption of Theorem* (3.2.1) *be fulfilled and let* $x = - a\lambda \geq 0$. *Then, for every fixed* $\varepsilon > 0$, *and* $0 \leq x \leq s$,

(i) $|x^{\varepsilon}F_{\pi}(\Delta tx)^{j}| \le \kappa_s(j\Delta t)^{-\varepsilon}$, $\quad s < \infty$,

(ii) $|x^{\varepsilon}(F_{\pi}(\Delta tx)^{j} - F_{\pi}(\infty)^{j})| \le \kappa(j\Delta t)^{-\varepsilon}$, $\quad s = \infty$, $j = 1,2,\ldots$.

Proof. For $s < \infty$ we have by Lemma (3.2.6) with $\kappa_s^* > 0$

$$|x^{\varepsilon}F_{\pi}(\Delta tx)^{j}| \le \kappa_s x^{\varepsilon}\exp\{-\kappa_s^* j\Delta tx\} = \kappa_s(j\Delta tx)^{\varepsilon}\exp\{-\kappa_s^* j\Delta tx\}(j\Delta t)^{-\varepsilon} \le \kappa_s(j\Delta t)^{-\varepsilon}.$$

For $s = \infty$ we have $spr(F_{\pi}(\infty)) < 1$. Consequently there are by Lemma (3.2.9) positive constants s_0, κ, and κ^* such that for $0 \le x \le s_0/\Delta t$

$$|F_{\pi}(\Delta tx)^{j}| \le \kappa\exp\{-\kappa^* j\Delta tx\},\ |F_{\pi}(\infty)^{j}| \le \kappa\exp\{-\kappa^* j\},$$

and so we obtain by the first assertion

$$|x^{\varepsilon}(F_{\pi}(\Delta tx)^{j} - F_{\pi}(\infty)^{j})| \le \kappa_{s_0}(j\Delta t)^{-\varepsilon} + \kappa x^{\varepsilon}\exp\{-\kappa^* j\}$$

$$\le \kappa_1((j\Delta t)^{-\varepsilon} + (s_0 j)^{\varepsilon}\exp\{-\kappa^* j\}(\Delta tj)^{-\varepsilon}) \le \kappa_2(\Delta tj)^{-\varepsilon}.$$

For $x > s_0/\Delta t$ we have by Lemma (3.2.9)

$$|x^{\varepsilon}(F_{\pi}(\Delta tx)^{j} - F_{\pi}(\infty)^{j})| \le \kappa\exp\{-\kappa^* j\}x^{\varepsilon}/\Delta tx \le \kappa\exp\{-\kappa^* j\}/[(\Delta tx)^{1-\varepsilon}\Delta t^{\varepsilon}]$$

$$\le \kappa\exp\{-\kappa^* j\}/\Delta t^{\varepsilon}s_0^{1-\varepsilon} \le \kappa_1 j^{\varepsilon}\exp\{-\kappa^* j\}/(j\Delta t)^{\varepsilon} \le \kappa_2(j\Delta t)^{-\varepsilon}.$$

<u>(3.2.12) Lemma</u>. *Let the assumption of Theorem* (3.2.1) *be fulfilled. Then for* $i = 1,\ldots$ $\ldots,n$,

(i) $|F_{\pi}(\Delta ta_i\lambda) - F_{\pi}(\Delta ta_{i-1}\lambda)| \le \kappa\tilde{\Theta}_n\Delta t$,

(ii) $|(F_{\pi}(\Delta ta_i\lambda) - F_{\pi}(\Delta ta_{i-1}\lambda))(a_i a_{i-1}\lambda^2)^{-1/2}| \le \kappa\tilde{\Theta}_n\Delta t^2$,

(iii) $|(F_{\pi}(\Delta ta_i\lambda) - F_{\pi}(\Delta ta_{i-1}\lambda))(a_{\ell}\lambda)^{-1/2}| \le \kappa\tilde{\Theta}_n\Delta t^{3/2}$, $\quad \ell = i-1,i$.

Proof. For brevity let $x_i = a_i\lambda$ and $F_i = F_{\pi}(\Delta tx_i)$ then we have

$$|F_i - F_{i-1}| \le k^{1/2}\max_{0\le\nu\le k-1}|(\alpha_k - \beta_k\Delta tx_i)^{-1}(\beta_{\nu}\alpha_k - \beta_k\alpha_{\nu})\Delta t(x_i - x_{i-1})(\alpha_k - \beta_k\Delta tx_{i-1})^{-1}|.$$

Extending by $(\Delta tx_i)^{1/2}$ and $(\Delta tx_{i-1})^{1/2}$ we thus obtain

$$|F_i - F_{i-1}| \le \kappa\Theta_n \Delta t^2 \max_{0\le\nu\le k-1} |(\alpha_k - \beta_k \Delta tx_i)^{-1}(\Delta tx_i)^{1/2}||(\alpha_k - \beta_k \Delta tx_{i-1})^{-1}(\Delta tx_{i-1})^{1/2}|$$

which yields the first assertion observing that $|\Delta tx_i|$ is bounded if $\beta_k = 0$. The second and the third assertion follow in a similar way extending with $(\Delta tx_\ell)^{1/2}$ in the last case.

The first assertion of Lemma (3.2.12) is the inequality (3.2.4) for j = 1. We now turn to the verification of that inequality for j = 2,...,n writing again $x_i = a_i\lambda$ $F_i = F_\pi(\Delta tx_i)$, and $F_\infty = F_\pi(\infty)$:

(i) For $0 < s < \infty$ we choose the partition

$$|F_i^j - F_{i-1}^j| \le \sum_{\nu=0}^{j-1} |F_i^\nu||F_i - F_{i-1}||F_{i-1}^{j-1-\nu}|$$

$$\le (|F_i^j| + |F_{i-1}^j|)|F_i - F_{i-1}| + \sum_{\nu=1}^{j-2} |x_i^{1/2}F_i^\nu||(F_i - F_{i-1})(x_i x_{i-1})^{-1/2}|x_{i-1}^{1/2}F_{i-1}^{j-1-\nu}|.$$

Then the Uniform Boundedness Theorem and Lemma (3.2.12) yield

$$|F_i^j - F_{i-1}^j| \le \kappa_s\tilde{\Theta}_n\Delta t(1 + \Delta t\sum_{\nu=1}^{j-2} |x_i^{1/2}F_i^\nu||x_{i-1}^{1/2}F_{i-1}^{j-1-\nu}|),$$

the first assertion of Lemma (3.2.11) yields for $\varepsilon = 1/2$

$$|F_i^j - F_{i-1}^j| \le \kappa_s\tilde{\Theta}_n\Delta t(1 + \sum_{\nu=1}^{j-2}(\nu(j-1-\nu))^{-1/2}),$$

and thus Lemma (3.2.5) proves the desired result for the case of finite s.

(ii) For $s = \infty$ we choose the partition

$$\begin{aligned} |F_i^j - F_{i-1}^j| \le &\sum_{\nu=1}^{j-2} |F_i^\nu - F_\infty^\nu||F_i - F_{i-1}||F_{i-1}^{j-1-\nu} - F_\infty^{j-1-\nu}| \\ &+ \sum_{\nu=0}^{j-2} |F_\infty^\nu||F_i - F_{i-1}||F_{i-1}^{j-1-\nu} - F_\infty^{j-1-\nu}| \\ &+ \sum_{\nu=1}^{j-1} |F_i^\nu - F_\infty^\nu||F_i - F_{i-1}||F_\infty^{j-1-\nu}| \\ &+ \sum_{\nu=0}^{j-1} |F_\infty^\nu||F_i - F_{i-1}||F_\infty^{j-1-\nu}|. \end{aligned} \tag{3.2.13}$$

Now, Lemma (3.2.11)(ii) for $\varepsilon = 1/2$, Lemma (3.2.12)(ii), and Lemma (3.2.5) yield

$$\begin{aligned} &\sum_{\nu=1}^{j-2} |F_i^\nu - F_\infty^\nu||F_i - F_{i-1}||F_{i-1}^{j-1-\nu} - F_\infty^{j-1-\nu}| \\ &\le \sum_{\nu=1}^{j-2} |x_i^{1/2}(F_i^\nu - F_\infty^\nu)||(F_i - F_{i-1})(x_i x_{i-1})^{-1/2}||x_{i-1}^{1/2}(F_{i-1}^{j-1-\nu} - F_\infty^{j-1-\nu})| \\ &\le \tilde{\kappa}_s\tilde{\Theta}_n\Delta t\sum_{\nu=1}^{j-2}(\nu(j-1-\nu))^{-1/2} \le \kappa_s\tilde{\Theta}_n\Delta t. \end{aligned}$$

Lemma (3.2.9)(ii), Lemma (3.2.11)(iii) for $\varepsilon = 1/2$, Lemma (3.2.12)(iii), and Lemma (3.2.5) yield

$$\textstyle\sum_{\nu=0}^{j-2}|F_\infty^\nu||F_i - F_{i-1}||F_{i-1}^{j-1-\nu} - F_\infty^{j-1-\nu}| \le \sum_{\nu=0}^{j-2}|F_\infty^\nu||(F_i - F_{i-1})x_{i-1}^{-1/2}||x_{i-1}^{1/2}(F_{i-1}^{j-1-\nu} - F_\infty^{j-1-\nu})|$$

$$\le \tilde{\kappa}_s\tilde{\Theta}_n\Delta t\textstyle\sum_{\nu=0}^{j-2}e^{-\kappa^*\nu}(j-1-\nu)^{-1/2} \le \hat{\kappa}_s\tilde{\Theta}_n\Delta t(1 + \sum_{\nu=1}^{j-2}(\nu(j-1-\nu))^{-1/2}) \le \kappa_s\tilde{\Theta}_n\Delta t$$

and analogeously

$$\textstyle\sum_{\nu=1}^{j-1}|F_i^\nu - F_\infty^\nu||F_i - F_{i-1}||F_\infty^{j-1-\nu}| \le \tilde{\kappa}_s\tilde{\Theta}_n\Delta t\sum_{\nu=1}^{j-1}e^{-\kappa^*(j-1-\nu)}\nu^{-1/2}$$

$$\le \hat{\kappa}_s\tilde{\Theta}_n\Delta t(1 + \textstyle\sum_{\nu=1}^{j-2}(\nu(j-1-\nu))^{-1/2}) \le \kappa_s\tilde{\Theta}_n\Delta t.$$

Lemma (3.2.9)(ii) and Lemma (3.2.12)(i) yield

$$\textstyle\sum_{\nu=0}^{j-1}|F_\infty^\nu||F_i - F_{i-1}||F_\infty^{j-1-\nu}| \le \tilde{\kappa}\tilde{\Theta}_n\Delta t\sum_{\nu=0}^{j-1}e^{-\kappa^*(\nu+(j-1-\nu))} \le \hat{\kappa}\tilde{\Theta}_n\Delta t j e^{-\kappa^*(j-1)} \le \kappa\tilde{\Theta}_n\Delta t.$$

A substitution of these bounds into (3.2.13), finally, proves that

$$|F_i^j - F_{i-1}^j| \le \kappa_s\tilde{\Theta}_n\Delta t$$

holds independently of i, j, and $x_i \in [-\infty, 0]$. This is the desired inequality (3.2.4) for $s = \infty$.

3.3. An Error Bound for Differential Systems of Second Order

The exact solution of the linear initial value problem

(3.3.1) $\quad y'' = a(t)^2A^2y + c(t) + h(t),\ t > 0,\ y(0) = y_0,\ y'(0) = y_0^*,$

oscillates without decreasing exponentially if the differential equation is homogeneous and the leading matrix $a(t)^2A^2$ is hermitean and negative definite. This behavior should naturally be carried over to the numerical approximation scheme and, indeed, convergent linear multistep methods with a periodicity interval $[-s, 0]$, cf. Definition (2.1.26), inherit this property from the analytic problem under the spectral condition $|a(t)^2A^2| \le s$. But the error estimation of the preceding section utilizes in an essential way that the spectrum of the leading matrix is contained in the *interior* of S, cf. Lemma (3.2.6) and Lemma (3.2.9), and the periodicity interval obviously is a subset of the *boundary* of S. Therefore the way of proceeding proposed by LeRoux [79a] fails here. It is an open question whether the above ingenious technique can be generalized to linear multistep methods for second order problems if the stability region has a non-empty interior of a suitable form. The main obstacle for a straightforward adaption seems to be the fact that now $\pi(\zeta,\eta^2)$ has two principal roots coalescing to one for $\eta^2 \to 0$ therefore the Uniform Boundedness Theorem yields correctly

$$|F_\pi(\eta^2)^n| \le \kappa_S n, \qquad -s \le \eta^2 \le 0,$$

which cannot play longer the role of the first assumption in the basic Lemma (3.2.3) without further ado.

Because of the two coalescing principal roots the error estimation of Section 3.1 does not hold longer, too. The bound (3.1.7) for the inverse of the Vandermonde matrix W holds only if all roots of $\pi(\zeta,\eta^2)$ are bounded away from each other. In the present situation this implies that Δt must be bounded away from zero in the case where the leading matrix is negative definite and the principal roots are the sole coalescing pair coinciding only in $\eta^2 = 0$. However, this drawback is only of technical character. Subsequently a more thorough study of the Vandermonde matrix leads to error bounds which correspond to those of the first section in a satisfactory way.

Using the notations introduced in Section 3.1, a linear multistep method for the problem (3.3.1) with general time-dependent matrix $A(t)^2$ has the form

$$(3.3.2)\quad \rho(T)v_n - \Delta t^2\sigma(T)(A^2v)_n = \Delta t^2\sigma(T)c_n, \qquad n = 0,1,\dots \;.$$

Let the exponential multiplication factor be once more modified slightly,

$$\Theta_n^{**} = \max\begin{cases}\Theta_n = \max_{k\le\nu\le n}\max_{(\nu-k)\Delta t\le t\le\nu\Delta t}|a'(t)/a_\nu| \\ \hat{\Theta}_n = \max_{k\le\nu\le n}\max_{(\nu-k)\Delta t\le t\le\nu\Delta t}|(a(t)^2)'/a_\nu^2|\end{cases} \qquad a_\nu = a(\nu\Delta t),$$

and recall that the essential roots $\zeta_i(\eta)$ of $\pi(\zeta,\eta^2)$, i.e. the roots with $|\zeta_i(0)| = 1$, are numbered by $i = 1,\dots,k_*$, without loss of generality. Then the following result is due to Hackmack [81, Theorem 4.8].

<u>(3.3.3) Theorem.</u> *(i) Let the (m,m)-matrix* $a(t)^2A^2$ *in* (3.3.1) *be hermitean and negative definite,* $a(t)^2A^2 \le -\alpha I$, $\alpha > 0$, $t > 0$, *and let the solution* y *of* (3.3.1) *be (p+1)-times continuously differentiable.*
(ii) Let the method (3.3.2) *be consistent of order* p *with stability region* S, *and let* $[-s, 0] \subset S$, $0 < s \le \infty$.
(iii) Let the Frobenius matrix $F_\pi(\eta^2)$ *be diagonable for* $\eta^2 \in [-s, 0)$.
Then for $n = k,k+1,\dots,$

$$|y_n - v_n| \le \kappa_S(1 + \alpha^{-1})\exp\{\kappa_S^*\Theta_n^{**}n\Delta t\}\Big[|E_{k-1}| + \Delta t^{-1}\max_{1\le i\le k_*}|E_{k-1}^0 - \zeta_i(0)E_{k-2}^0| +$$

$$+ \Delta t^p \int_0^{n\Delta t} \exp\{-\kappa_s^* \Theta_n^{**} \tau\} |y^{(p+2)}(\tau)| d\tau + n\Delta t |||h|||_n \Big].$$

where $E_n^0 = (y_{n-k+2} - v_{n-k+2}, \ldots, y_n - v_n)^T$ *and* $E_n = (y_{n-k+1} - v_{n-k+1}, E_n^0)^T$.

Proof. As in Section 3.1 let $F_\pi(\eta) = W(\eta)Z(\eta)W(\eta)^{-1}$ be the Jordan canonical decomposition of $F_\pi(\eta)$. Let $\Lambda^2 = (\lambda_1^2, \ldots, \lambda_m^2)$ be the diagonal matrix of the eigenvalues of A^2, and write briefly $W_n = W(\Delta t a_n \Lambda)$ and $Z_n = Z(\Delta t a_n \Lambda)$. Then the error equation (3.1.6) is the same as in Section 3.1,

$$(3.3.4) \quad W_n^{-1}X^{-1}E_n = (Z_n + W_n^{-1}X^{-1}R_n X W_n)W_n^{-1}W_{n-1}X^{-1}E_{n-1} + W_n^{-1}X^{-1}(D(\Delta t, y)_n - H_n),$$

where R_n is again a (k,k)-block matrix of which only the last block row in nonzero:

$$(3.3.5) \quad \text{last block row of } R_n = (\Delta t^2 A^2(\alpha_k I - \beta_k \Delta t^2 a_n^2 A^2)^{-1}(\beta_0(a_{n-k}^2 - a_n^2)I, \ldots, \beta_{k-1}(a_{n-1}^2 - a_n^2)I).$$

$D(\Delta t, y)_n$ and H_n are the same vectors as in (2.1.37) and (2.1.38) hence we obtain in the same way as in the proof of Theorem (2.1.34) for $\ell = 1$ that

$$|X^{-1}(D(\Delta t, y)_n - H_n)| \leq \kappa_s (\Delta t^{p+1} \int_{(n-k)\Delta t}^{n\Delta t} |y^{(p+2)}(\tau)| d\tau + \Delta t^2 |||h|||_n).$$

For the estimation of the multiplication factor we observe that by assumption (iii)

$$|Z_n| \leq 1, \; |W_n| \leq \kappa,$$

and consequently

$$(3.3.6) \quad |(Z_n + W_n^{-1}X^{-1}R_n X W_n)(W_n^{-1}W_{n-1})| \leq (1 + \kappa |W_n^{-1}X^{-1}R_n X|)|W_n^{-1}W_{n-1}|.$$

In order to estimate $|W_n^{-1}W_{n-1}|$ we observe that

$$(3.3.7) \quad |W_n^{-1}W_{n-1}| = |W(\Delta t a_n \Lambda)^{-1} W(\Delta t a_{n-1}\Lambda)| = \max_{1 \leq \mu \leq m} |W(\Delta t a_n \lambda_\mu)^{-1} W(\Delta t a_{n-1}\lambda_\mu)|.$$

Now we consider two cases:

(i) If $s < \infty$ then Lemma (A.2.6) yields for $-s \leq \Delta t^2 a(t)^2 \lambda^2 \leq 0$

$$(3.3.8) \quad |W(\Delta t a_n \lambda)^{-1} W(\Delta t a_{n-1}\lambda)| \leq 1 + \kappa_s |\Delta t a_n \lambda|^{-1} |\Delta t a_n \lambda - \Delta t a_{n-1}\lambda| \leq 1 + \kappa_s \Theta_n \Delta t.$$

(ii) If $s = \infty$ then Lemma (A.2.6) yields for $-s \leq \Delta t^2 a(t)^2 \lambda^2 \leq 0$ and $|\Delta t a_n \lambda| \geq 1$

$$|W(\Delta t a_n\lambda)^{-1}W(\Delta t a_{n-1}\lambda)| \leq 1 + \kappa_s \max_{1\leq i\leq k}|\zeta_i(\Delta t a_n\lambda) - \zeta_i(\Delta t a_{n-1}\lambda)|.$$

But now we have $\beta_k \neq 0$ by Lemma (A.1.3) hence the same estimation as in the corresponding part in the proof of (3.1.8) shows that

$$|\zeta_i(\Delta t a_n\lambda) - \zeta_i(\Delta t a_{n-1}\lambda)| \leq \kappa_s \hat{\Theta}_n \Delta t.$$

For $s = \infty$ and $|\Delta t a_n\lambda| < 1$ Lemma (A.2.6) yields for $-s \leq \Delta t^2 a(t)^2\lambda^2 \leq 0$

$$(3.3.9)\quad |W(\Delta t a_n\lambda)^{-1}W(\Delta t a_{n-1}\lambda)| \leq 1 + \kappa_s|\Delta t a_n\lambda|^{-1}\max_{1\leq i\leq k}|\zeta_i(\Delta t a_n\lambda) - \zeta_i(\Delta t a_{n-1}\lambda)|.$$

In this case, too, the estimation follows the lines of the proof of (3.1.8) but (3.1.12) is to be modified in the following way:

$$\left|\frac{\partial\gamma_j}{\partial t}(\Delta t^2 a(t)^2\lambda^2)\right| = \left|\frac{\partial}{\partial t}\frac{\alpha_j - \beta_j\Delta t^2 a(t)^2\lambda^2}{\alpha_k - \beta_k\Delta t^2 a(t)^2\lambda^2}\right| \leq \kappa\left|\frac{\partial}{\partial t}(\Delta t^2 a(t)^2\lambda^2)\right|,\ j = 0,\dots,k-1.$$

Therefore we obtain from (3.1.10)

$$|\zeta_i(\Delta t a_n\lambda) - \zeta_i(\Delta t a_{n-1}\lambda)| \leq \kappa_s|\Delta t^2 a_n^2\lambda^2|\hat{\Theta}_n\Delta t$$

and accordingly by (3.3.9), because $|\Delta t a_n\lambda| < 1$ in the present case,

$$|W(\Delta t a_n\lambda)^{-1}W(\Delta t a_{n-1}\lambda)| \leq 1 + \kappa_s|\Delta t a_n\lambda|^{-1}|\Delta t^2 a_n^2\lambda^2|\hat{\Theta}_n\Delta t \leq 1 + \kappa_s\hat{\Theta}_n\Delta t.$$

Now we have shown that

$$|W(\Delta t a_n\lambda)^{-1}W(\Delta t a_{n-1}\lambda)| \leq 1 + \kappa_s\hat{\Theta}_n\Delta t, \qquad -s \leq \Delta t^2 a(t)^2\lambda^2 \leq 0,\ 0 < s \leq \infty,$$

and a substitution of this bound into (3.3.7) yields

$$(3.3.10)\quad |W_n^{-1}W_{n-1}| \leq 1 + \kappa_s\hat{\Theta}_n\Delta t.$$

In order to estimate $|W_n^{-1}X^{-1}R_nX|$ we observe that

$$(3.3.11)\quad |W_n^{-1}X^{-1}R_nX| = |W(\Delta t a_n\Lambda)^{-1}R(\Delta t a_n\Lambda)| = \max_{1\leq\mu\leq m}|W(\Delta t a_n\lambda_\mu)^{-1}R(\Delta t a_n\lambda_\mu)|.$$

We omit the argument $\Delta t a_n\lambda_\mu$ and write

$$W^{-1} = [w^*_{ij}]_{i,j=1}^k,\ R = [r_{ij}]_{i,j=1}^k,\ Z = [z_{ij}]_{i,j=1}^k \equiv W^{-1}R,$$

then

(3.3.12) $z_{ij} = w^*_{ik} r_{kj}$

because the first k-1 rows of R are zero. Now we have by (3.3.5)

$$|r_{kj}| = |\Delta t^2\lambda^2(\alpha_k - \beta_k\Delta t^2 a_n^2\lambda^2)^{-1}\beta_{j-1}(a^2_{n-k+j-1} - a_n^2)|$$

(3.3.13)

$$\leq \kappa|\Delta t^2 a_n^2\lambda^2||(\alpha_k - \beta_k\Delta t^2 a_n^2\lambda^2)^{-1}|\Theta_n\Delta t,$$

and by (A.2.5) we have ($s_0 = 1$)

(3.3.14) $$|w^*_{ik}| = |\prod_{\nu\neq i}(\zeta_\nu(\Delta t a_n\lambda) - \zeta_i(\Delta t a_n\lambda))^{-1}| \leq \kappa_s|\zeta_i(\Delta t a_n\lambda) - \zeta_{i+1}(\Delta t a_n\lambda)|^{-1}$$

because all roots $\zeta_\nu(\eta)$ with $\nu \neq i+1$ are bounded away from $\zeta_i(\eta)$ for $\eta^2 \in [-s, 0]$ by assumption (iv) and because $0 \in S$.

(i) If $s < \infty$ then we obtain from (3.3.14) and Lemma (A.1.10)(ii)

(3.3.15) $$|w^*_{ik}| \leq \kappa_s \max\{1, |\Delta t a_n\lambda|^{-1}\}$$

and $|\alpha_k - \beta_k\eta^2|^{-1}$ is uniformly bounded for $\eta^2 \in [-s, 0]$ independently of finite or infinite s. Thus (3.3.12), (3.3.13), and (3.3.15) yield together

$$|z_{ij}| \leq \tilde{\kappa}_s\max\{1, |\Delta t a_n\lambda|^{-1}\}|\Delta t a_n\lambda|^2\hat{\Theta}_n\Delta t \leq \tilde{\kappa}_s\max\{\sqrt{s}, s\}\hat{\Theta}_n\Delta t \leq \kappa_s\hat{\Theta}_n\Delta t,\ 0 < s < \infty.$$

(ii) If $s = \infty$ and $|\Delta t a_n\lambda| \geq 1$ then we obtain by Lemma (A.1.10)(ii) and (3.3.14) that $|w^*_{ik}| \leq \kappa$ hence, because now $\beta_k \neq 0$, (3.3.12) and (3.3.13) yield

$$|z_{ij}| \leq \tilde{\kappa}|\Delta t^2 a_n^2\lambda^2(\alpha_k - \beta_k\Delta t^2 a_n^2\lambda^2)^{-1}|\hat{\Theta}_n\Delta t \leq \kappa\hat{\Theta}_n\Delta t.$$

If $s = \infty$ and $|\Delta t a_n\lambda| < 1$ then we obtain by Lemma (A.1.10)(ii) again (3.3.15) and therefore, by (3.3.12),

$$|z_{ij}| \leq \kappa|\Delta t a_n\lambda|^{-1}|\Delta t a_n\lambda|^2\hat{\Theta}_n\Delta t \leq \kappa\hat{\Theta}_n\Delta t.$$

These bounds together imply that under the assumption of the Theorem

$$|W(\Delta t a_n\lambda)^{-1}R(\Delta t a_n\lambda)| \leq \kappa_s\hat{\Theta}_n\Delta t$$

and a substitution into (3.3.11) yields

$$|W_n^{-1}X^{-1}R_nX| \leq \kappa_s\hat{\Theta}_n\Delta t.$$

The error equation (3.3.4), (3.3.6), the bound (3.3.10) and this bound now yield

$$|W_n^{-1}E_n| \le (1+\kappa_S^*\Theta_n^{**}\Delta t)^{2(n-k+1)}|W_{k-1}^{-1}E_{k-1}| + \sum_{\nu=k}^{n}(1+\kappa_S^*\Theta_n^{**}\Delta t)^{2(n-\nu)}|W^{-1}(D(\Delta t,y)_\nu - H_\nu)|.$$

But $|W_n^{-1}E_n| \ge |W_n|^{-1}|E_n| \ge \kappa^{-1}|E_n|$ and

$$|W_\nu^{-1}(D(\Delta t,y)_\nu - H_\nu)| \le \Delta t^{-1}\alpha^{-1}|(D(\Delta t,y)_\nu - H_\nu)|$$

follows in the same way as the estimation of (3.3.11) because the first k-1 block elements of the vectors D and H are zero. The remaining part of the proof follows the repeatedly described way estimating $|W_{k-1}^{-1}E_{k-1}|$ by Lemma (A.2.10).

In concluding this section it should be remarked that obviously the following bounds for the initial error can be inserted in Theorem (3.3.3):

$$|E_{k-1}| + \Delta t^{-1}\max_{1\le i\le k_*}|E_{k-1}^0 - \zeta_i(0)E_{k-2}^0| \le \kappa\Delta t^{-1}|E_{k-1}|$$

and if the method (3.3.2) is strongly D-stable in $0 \in S$ then

$$\max_{1\le i\le k_*}|E_{k-1}^0 - \zeta_i(0)E_{k-2}^0| = |E_{k-1}^0 - E_{k-2}^0|.$$

Thus, with exception of the technical assumption (iv), Theorem (3.3.3) agrees well with Theorem (2.1.34) and Theorem (2.1.39) respectively.

IV. Linear Multistep Methods and Nonlinear Differential Systems of First Order

4.1. An Error Bound for Stable Differential Systems

As early as 1959 Dahlquist [59] has derived a-priori error bounds for linear multistep methods and nonlinear differential systems but he introduced the concept of absolute stability for numerical methods not before 1963 in a very celebrated communication. The contributions of Dahlquist and the book of Henrici [62] constitute the foundations of modern numerical analysis of initial value problems. In this section we shall combine one of Dahlquist's former ideas with a recent result of LeRoux mentioned already above and concerning linear multistep methods and *linear* initial value problems,

(4.1.1) $y' = A(t)y + c(t),\ t > 0,\ y(0) = y_0.$

<u>(4.1.2) Theorem.</u> (LeRoux [79a].) *(i) For* $t > 0$ *let the (m,m)-matrix* $A(t)$ *in* (4.1.1) *be hermitean and negative definite,* $A(t) \leq -\alpha I$, $\alpha > 0$, *and let the solution* y *of* (4.1.1) *be (p+1)-times continuously differentiable.*
(ii) Let there exist two positive constants, Θ *and* ε, $0 < \varepsilon < 1$, *such that for* $t > 0$

$$|A(t+\Delta t)^{-\varepsilon}(A(t+\Delta t) - A(t))A(t)^{-1+\varepsilon}| \leq \Theta\Delta t,$$

and if $\sigma(\zeta) \neq \beta_k\zeta^k$ *in* (3.1.2) *then*

$$|A(t)^{-1}(A(t) - A(\tau))| \leq \Theta|t-\tau|, \qquad t - k\Delta t \leq \tau \leq t.$$

(iii) Let the method (3.1.2) *be consistent of order* p *with the stability region* S, *let* $0 \in S$, $[-\infty, 0) \subset \overset{\circ}{S}$, *and* $\mathrm{Re}(\chi_i^*) > 0$, $i = 1,\dots,k_*$.
Then there exist two positive constants, κ *and* κ^*, *depending only on* α, ε, Δt_0, *and the the data of the method* (3.1.2) *such that for* $n = k, k+1, \dots,$ *and* $0 < \Delta t \leq \Delta t_0$

$$|y_n - v_n| \leq \kappa e^{\kappa^*\Theta n\Delta t}[|Y_{k-1} - V_{k-1}| + \Delta t^p \int_0^{n\Delta t} e^{-\kappa^*\Theta\tau}|y^{(p+1)}(\tau)|d\tau].$$

The proof of this theorem follows the lines of Theorem (3.2.1) but the necessary auxiliary results are more difficult to derive therefore we refer here to the original contribution of LeRoux [79a].

We now follow Hackmack [81] and turn to nonlinear initial value problems,

(4.1.3) $y' = f(t,y),\ 0 < t < T,\ y(0) = y_0,$

$f : [0, T] \times \mathbb{R}^m \to \mathbb{R}^m$, and linear multistep methods,

(4.1.4) $\rho(\tau)v_n - \Delta t\sigma(\tau)f_n(v_n) = 0, \qquad n = 0,1,\ldots,$

or

$$\alpha_k v_{n+k} - \beta_k \Delta t f_{n+k}(v_{n+k}) = -\sum_{i=0}^{k-1}\alpha_i v_{n+i} + \Delta t\sum_{i=0}^{k-1}\beta_i f((n+i)\Delta t,\ v_{n+i}),$$

$n = 0,1,\ldots$. Naturally, if $\beta_k \neq 0$ then Δt must be chosen sufficiently small here such that v_{n+k} exists for all n.

With respect to the problem (4.1.3) the following notations are used in this section:

(4.1.5) $A(t) = (\partial f/\partial y)(t,y(t)),$

$$\kappa_f(\Omega) = \sup_{0<t<T}\sup_{w\in\Omega}|(\partial^2 f/\partial y^2)(t,w)|,$$

where y is the exact solution of (4.1.3). Ω denotes a sufficiently large convex domain such that y and the numerical approximation v are contained in Ω. An interdependence of this form appears always in the error estimation of genuine nonlinear problems. It is however not a serious restriction because it can be assumed under sufficient smoothness of f that this function is replaced by a function $\tilde{f}$ having bounded second derivatives with respect to y such that f and $\tilde{f}$ coincide inside a tube in $[0, T] \times \mathbb{R}^m$ containing the exact solution y.

For the following auxiliary result we refer to Ortega and Rheinboldt [70, § 3.3].

<u>(4.1.6) Lemma</u>. *Let* $\Xi \subset \mathbb{R}^m$ *be convex and let* $f \in C^2(\Xi;\mathbb{R}^m)$ *then*

$$|f(v) - f(w) - f'(w)(v - w)| \leq \sup_{0\leq\xi\leq 1}|f''(w + \xi(v - w))||v - w|^2$$

where f' *denotes the Jacobi matrix and* f'' *the* (m,m^2)*-matrix of the second derivatives of* f.

The next lemma is the announced result proposed by Dahlquist [59] for the error estimation of nonlinear problems.

<u>(4.1.7) Lemma</u>. *Let* $\Gamma > 0$ *be a fixed constant. Let* Φ *and* ξ *be two nonnegative continuous functions such that*

$$\Phi(0) = \xi(0) = 0,\ \xi(\Delta t) \leq \Phi(\Delta t) + \Gamma\xi(\Delta t)^2,\ 0 < \Delta t < \Delta t_0,\ 4\Gamma\Phi(\Delta t_0) = 1,$$

and let Φ *be strictly monotonically increasing for* $0 \le \Delta t \le \Delta t_0$. *Then*

$$\xi(\Delta t) \le 2\Phi(\Delta t), \qquad 0 \le \Delta t \le \Delta t_0.$$

Proof. We consider $\Delta t \in (0, \Delta t_0)$ and obviously have $0 < 4\Gamma\Phi(\Delta t) < 1$ hence the equation

$$\zeta = \Phi(\Delta t) + \Gamma\zeta^2$$

has two solutions, $0 < \zeta_0(\Delta t) < \zeta_1(\Delta t)$, and $\lim_{\Delta t \to 0}\zeta_1(\Delta t)$ is positive. From the assumption that $\Gamma\xi(\Delta t)^2 - \xi(\Delta t) + \Phi(\Delta t) \ge 0$ we obtain that $\xi(\Delta t) \le \zeta_0(\Delta t)$ or $\xi(\Delta t) \ge \zeta_1(\Delta t)$. But $\xi(\Delta t) < \zeta_1(\Delta t)$ for $\Delta t > 0$ sufficiently small therefore $\xi(\Delta t) \le \zeta_0(\Delta t)$ by the continuity of ξ. Now the assertion follows observing that the smaller root ζ_0 of $\Gamma\zeta^2 - \zeta + \Phi$ satisfies by Vieta's root criterion

$$0 < \zeta_0 = \Phi + \Gamma\zeta_0^2 \le \Phi + \Gamma\zeta_0\zeta_1 = 2\Phi.$$

With (4.1.5) the error equation is now written as

$$\rho(T)e_n - \Delta t\sigma(T)(A_n e_n) = \Delta t\sigma(T)(f_n(y_n) - f_n(v_n) - A_n e_n) + d(\Delta t, y)_n \equiv \tilde{d}_n, \; n = 0,1,\ldots,$$

where again $e_n = y_n - v_n$ and $d(\Delta t, y)_n$ is the discretization error estimated in Lemma (1.1.8). By means of Lemma (4.1.6) we obtain

(4.1.8) $$|\sigma(T)(f_n(y_n) - f_n(v_n) - A_n e_n| \le \kappa\kappa_f(\Omega)(\beta_k|E_{n+k}|^2 + |E_{n+k-1}|^2),$$

and if the initial value problem

(4.1.9) $$\tilde{y}' = (\partial f/\partial y)(t,y(t))\tilde{y}, \; 0 < t < T, \; \tilde{y}(0) = y_0,$$

satisfies the assumption of Theorem (4.1.2) then a slight modification of this Theorem with respect to the discretization error yields

$$|E_n| \le \kappa e^{\kappa^*\Theta n\Delta t}[|E_{k-1}| + \textstyle\sum_{\nu=k}^{n} e^{-\kappa^*\Theta\nu\Delta t}|\tilde{D}_\nu|], \qquad n = k,k+1,\ldots,$$

with $\tilde{D}_\nu = (0,\ldots,0,(\alpha_k I - \beta_k\Delta t A_n)^{-1}\tilde{d}_{\nu-k})^T$. Therefore we obtain

$$e^{-\kappa^*\Theta n\Delta t}|E_n| \le \kappa[|E_{k-1}| + \textstyle\sum_{\nu=k}^{n} e^{-\kappa^*\Theta\nu\Delta t}(\Delta t\kappa_f(\Omega)(|E_\nu|^2 + |E_{\nu-1}|^2) + d(\Delta t,y)_{\nu-k})]$$

or

$$e^{-\kappa^*\Theta n\Delta t}|E_n| \le \Phi(\Delta t) + 2\Delta t\kappa\kappa_f(\Omega)\sum_{\nu=k}^{n} e^{\kappa^*\Theta\nu\Delta t}(e^{-\kappa^*\Theta\nu\Delta t}|E_\nu|)^2, \quad n = k,k+1,\ldots,$$

where

$$\Phi(\Delta t) \le \kappa[|E_{k-1}| + \kappa_f(\Omega)|E_{k-1}|^2 + \Delta t^p \int_0^{n\Delta t} e^{-\kappa^*\Theta\tau}|y^{(p+1)}(\tau)|d\tau].$$

This inequality remains true if on the right side the step number, n, is kept fixed and on the left side n is replaced by μ, $k \le \mu \le n$. Then we obtain

$$\text{(4.1.10)} \quad \max_{k\le\mu\le n}\{e^{-\kappa^*\Theta\mu\Delta t}|E_\mu|\} \le \Phi(\Delta t) + \Gamma(\max_{k\le\mu\le n}\{e^{-\kappa^*\Theta\mu\Delta t}|E_\mu|\})^2$$

where

$$\Gamma = \kappa\kappa_f(\Omega)e^{\kappa^*\Theta k\Delta t}\,\frac{e^{\kappa^*\Theta T} - 1}{\kappa^*\Theta} \ge \Delta t\kappa\kappa_f(\Omega)\sum_{\nu=k}^{n} e^{\kappa^*\Theta\nu\Delta t}.$$

Here, κ, κ^*, and Θ are the constants of Theorem (4.1.2) with respect to the problem (4.1.9). Now we assume that $|E_{k-1}| = \mathcal{O}(\Delta t)$, $\Delta t \to 0$ then $\Phi(\Delta t)$ and

$$\xi(\Delta t) = \max_{k\le\mu\le n}\{e^{-\kappa^*\Theta\mu\Delta t}|E_\mu|\}$$

fulfil the assumption of Lemma (4.1.7) for a certain $\Delta t_0 > 0$ defined by $4\Gamma\Phi(\Delta t_0) = 1$. Consequently, (4.1.10) yields

$$|E_\mu| \le 2\Phi(\Delta t)e^{\kappa^*\Theta\mu\Delta t}, \qquad \mu = k,\ldots,n.$$

The result may be summarized in the following theorem.

<u>(4.1.11) Theorem</u>. (Hackmack [81].) *(i) Let the solution* y *of the problem* (4.1.3) *exist uniquely and be (p+1)-times continuously differentiable; let* f *be two-times continuously differentiable with respect to* y *in a sufficiently large domain* $[0, T] \times \Omega$.
(ii) Let the initial value problem (4.1.9) *and the method* (4.1.4) *satisfy the assumption of Theorem* (4.1.2) , *and let* κ, κ^*, *and* Θ *be the constants of Theorem* (4.1.2) *with respect to* (4.1.9).
(iii) Let the initial values $v_0,\ldots, v_{k-1}$, *satisfy* $|Y_{k-1} - V_{k-1}| = \mathcal{O}(\Delta t)$, $\Delta t \to 0$.
Then there exists a Δt_0 *such that for* $0 < \Delta t \le \Delta t_0$, $n = k,k+1,\ldots, n\Delta t \le T$,

$$|y_n - v_n| \le \kappa e^{\kappa^*\Theta n\Delta t}[|Y_{k-1} - V_{k-1}| + \kappa_f(\Omega)|Y_{k-1} - V_{k-1}|^2 + \Delta t^p \int_0^{n\Delta t} e^{-\kappa^*\Theta\tau}|y^{(p+1)}(\tau)|d\tau].$$

For a further error estimation of nonlinear problems and linear multistep methods we refer to LeRoux [80].

4.2. The Modified Midpoint Rule

In chapter III and the previous section it was shown that for stable problems and suitable linear multistep methods an ill condition of the leading matrix A does not affect the error propagation seriously even if this matrix varies with time. But the bounds are here no longer of that uniform character as it was obtained in the first two chapters for systems with constant matrix because the exponential multiplication factor depends now on $A'(t)$ in some way. Certainly, this perturbation effect is partly due to the fact that a first or second order differential equation is approximated by a higher order difference scheme whose characteristic polynomial has some 'spurious' roots besides the principal ones. However, for some single step multiderivative methods very similar results were derived, too, by Nassif and Descloux [77]. All these error bounds seem to be optimum if no further conditions are imposed on the analytic problem. Therefore, if one seeks for numerical schemes with error bounds being independent of the data of the differential equation one would consider first of all single step single derivative and single step single stage methods.

The general consistent single step single derivative method has for (4.1.3) the form

$$(4.2.1)\quad v_{n+1} = v_n + \Delta t[\omega f_{n+1}(v_{n+1}) + (1-\omega)f_n(v_n)], \qquad n = 0,1,\dots,$$

and the general consistent singe step single stage method for (4.1.3) is

$$v_{n,1} = v_n + \Delta t\omega f((n+\omega)\Delta t, v_{n,1})$$

$$v_{n+1} = v_n + \Delta t f((n+\omega)\Delta t, v_{n,1}), \qquad n = 0,1,\dots,$$

or

$$(4.2.2)\quad v_{n+1} = v_n + \Delta t f((n+\omega)\Delta t, \omega v_{n+1} + (1-\omega)v_n), \qquad n = 0,1,\dots,$$

cf. also the following chapter. For $\omega = 0$ and $\omega = 1$ we obtain in both cases the explicit and the implicit Euler method whereas for $\omega = 1/2$ (4.2.1) yields the trapezoidal rule and (4.2.2) the midpoint rule. The discretization error of the method (4.2.1) follows immediately from Section 1.1 or by direct verification. Both methods, (4.2.1) and (4.2.2), are of order one and for $\omega = 1/2$ of order two and they are both A-stable for $1/2 \leq \omega \leq 1$. The trapezoidal rule is the unique A(0)-stable linear k-step method of order k+1 (Widlund [67]) and among all A-stable linear multistep methods of (maximum reachable) order two it has the smallest error constant (Dahlquist [63]).

In the remaining part of this chapter the initial value problem (4.1.3) with $T = \infty$ is supposed to satisfy the rather general stability condition

(4.2.3) $(v - w)^T(f(t,v) - f(t,w)) \le -\alpha|v - w|^2 \qquad \forall\, v,w \in \mathbb{R}^m,\ \alpha > 0.$

As two arbitrary solutions y and y^* of $y' = f(t,y)$ satisfy

$$\frac{d|y(t) - y^*(t)|^2}{dt} = 2(y(t) - y^*(t))^T(f(t,y(t)) - f(t,y^*(t))),$$

this monotonicity assumption implies for $\alpha > 0$ that $|y(t) - y^*(t)|$ decreases exponentially for increasing t.

If (4.2.3) holds then the error of the implicit Euler method satisfies

(4.2.4) $|y_{n+1} - v_{n+1}| \le (1 + \alpha\Delta t)^{-1}(|y_n - v_n| + |d(\Delta t,y)_n|), \qquad n = 0,1,\ldots,$

which is optimum because in this *first order* method the propagation factor is a *first order* approximation of the propagation factor of the analytic problem,

$$e^{-\alpha\Delta t} = (1 + \alpha\Delta t)^{-1} + \mathcal{O}(\Delta t^2), \qquad \Delta t \to 0.$$

Let us now turn to the case of consistence order two. The characteristic polynomial of the method (4.2.1), $\pi(\zeta,\eta) = \zeta - 1 - \eta(\omega\zeta + (1 - \omega))$, has for $\omega = 1/2$ and $\eta = \infty$ the root $\zeta = -1$ and it is well-known that for A-stable methods the property $\infty \in \partial S$ leads to undue oscillations of the numerical approximations if $\eta = \Delta t\lambda$ is very large. Naturally, also the midpoint rule suffers from this drawback because both methods, trapezoidal rule and midpoint rule, coincide for the test equation $y' = \lambda y$. If ω is chosen greater than 1/2 then these oscillations disappear but also the optimum order two is lost since

$$d(\Delta t,y)(t) = (\tfrac{1}{2} - \omega)\Delta t^2 y''(t) + \mathcal{O}(\Delta t^3), \qquad \Delta t \to 0,$$

is the discretization error of the method (4.2.1). So it suggests itself to choose ω *slightly* greater than 1/2 and Kreth [81] has proposed to choose $\omega = 1/2 + \mathcal{O}(\Delta t) > 1/2$ in order to preserve the optimum order. It seems also in other methods very promising to choose some parameters in a suitable dependence of the steplength Δt. For the midpoint rule modified by this way we then obtain an error propagation which resembles that of the implicit Euler method (4.2.4) but now holds for a method of order two:

<u>(4.2.5) Theorem.</u> (Kreth [81].) *If the problem* (4.1.3) *satisfies* (4.2.3) *and*

$$\omega \ge [-1 + \Delta t\alpha + (1 + \Delta t^2\alpha^2)^{1/2}]/2\Delta t\alpha$$

then the error of the modified midpoint rule (4.2.2) *satisfies*

$$|y_{n+1} - v_{n+1}| \le \frac{1 - (1 - \omega)\Delta t\alpha}{1 + \omega\Delta t\alpha}|y_n - v_n| + \frac{1}{1 + \omega\Delta t\alpha}|d(\Delta t,y)_n|, \; n = 0,1,\ldots \quad .$$

Proof. By an application of the monotonicity condition (4.2.3) we obtain from (4.2.2) writing $e_n = y_n - v_n$ and $d_n = d(\Delta t,y)_n$

$$(e_{n+1} - e_n, \omega e_{n+1} + (1-\omega)e_n) + \Delta t\alpha|\omega e_{n+1} + (1-\omega)e_n|^2 \le (d_n, \omega e_{n+1} + (1-\omega)e_n)$$

which yields after some simple transformations

$$|e_{n+1} + \frac{1-2\omega(1-(1-\omega)\Delta t\alpha)}{2\omega(1+\omega\Delta t\alpha)} e_n|^2$$

$$\le [\frac{(1-\omega)(1-(1-\omega)\Delta t\alpha)}{\omega(1+\omega\Delta t\alpha)} + (\frac{1-2\omega(1-(1-\omega)\Delta t\alpha)}{2\omega(1+\omega\Delta t\alpha)})^2]|e_n|^2 + \frac{(d_n, \omega e_{n+1} + (1-\omega)e_n)}{\omega(1+\omega\Delta t\alpha)} .$$

Now we observe that

$$\frac{(1-\omega)(1-(1-\omega)\Delta t\alpha)}{\omega(1+\omega\Delta t\alpha)} + (\frac{1-2\omega(1-(1-\omega)\Delta t\alpha)}{2\omega(1+\omega\Delta t\alpha)})^2 = \frac{1}{4\omega^2(1+\omega\Delta t\alpha)^2}$$

hence

$$|e_{n+1} + \frac{1-2\omega(1-(1-\omega)\Delta t\alpha)}{2\omega(1+\omega\Delta t\alpha)} e_n|^2$$

$$\le \frac{1}{4\omega^2(1+\omega\Delta t\alpha)^2}|e_n|^2 + (\frac{d_n}{1+\omega\Delta t\alpha}, e_{n+1} + \frac{1-2\omega(1-(1-\omega)\Delta t\alpha)+1}{2\omega(1+\omega\Delta t\alpha)} e_n)$$

or

(4.2.6)
$$|e_{n+1} + \frac{1-2\omega(1-(1-\omega)\Delta t\alpha)}{2\omega(1+\omega\Delta t\alpha)} e_n - \frac{1}{2(1+\omega\Delta t\alpha)} d_n|^2 \le |\frac{1}{2\omega(1+\omega\Delta t\alpha)} e_n + \frac{1}{2(1+\omega\Delta t\alpha)} d_n|^2 .$$

If

(4.2.7) $1 - 2\omega(1-(1-\omega)\Delta t\alpha) \le 0$

or

$$\omega \ge [-1 + \Delta t\alpha + (1 + \Delta t^2\alpha^2)^{1/2}]/2\Delta t\alpha$$

then the last inequality yields

$$|e_{n+1}| \le [\frac{2\omega(1-(1-\omega)\Delta t\alpha)-1}{2\omega(1+\omega\Delta t\alpha)} + \frac{1}{2\omega(1+\omega\Delta t\alpha)}]|e_n| + \frac{1}{1+\omega\Delta t\alpha}|d_n|, \; n = 0,1,\ldots,$$

which is the assertion.

The choice of

$$(4.2.8)\quad \omega = [-1 + \Delta t\alpha + (1 + \Delta t^2\alpha^2)^{1/2}]/2\Delta t\alpha$$

yields

$$\frac{1-(1-\omega)\Delta t\alpha}{1+\omega\Delta t\alpha} = \frac{1}{(1+(\Delta t\alpha)^2)^{1/2} + \Delta t\alpha} = 1 - \Delta t\alpha + \frac{(\Delta t\alpha)^2}{2} + \mathcal{O}((\Delta t\alpha)^3).$$

We thus can state that the *second order* method (4.2.2) with the parameter (4.2.8) is optimum with respect to the stability because the propagation factor is a *second order* approximation of that of the analytic problem, $e^{-\alpha\Delta t}$. On the other side, if we only require that the method is A-stable, i.e. $1/2 \le \omega \le 1$, and not (4.2.7) then (4.2.6) yields

$$(4.2.9)\quad |e_{n+1}| \le \frac{1-\omega}{\omega}|e_n| + |d_n|, \qquad n = 0,1,\ldots .$$

For $\omega = 1/2$ the damping disappears here completely in agreement with the above remark on the case $\infty \in \partial S$.

In the proof of Theorem (4.2.5) the inequality (4.2.7) was obtained in a purely algebraic way. We conclude this section with a more heuristic foundation of this condition and consider the initial value problem

$$(4.2.10)\quad y' = Ay,\ t > 0,\ y(0) = y_0,$$

with diagonable matrix A, $A = X\Lambda X^{-1}$. Let $\Lambda = (\lambda_1,\ldots,\lambda_m)$ be the diagonal matrix of the eigenvalues of A such that $\lambda_m \le \ldots \le \lambda_2 \le \lambda_1 = -\alpha < 0$, and let the eigenvectors of A, i.e. the columns of X, be denoted by $x_1,\ldots,x_m$. Then the exact solution of (4.2.10) is

$$y(t) = \sum_{\mu=1}^m \chi_\mu e^{\lambda_\mu t} x_\mu,\quad y_0 = \sum_{\mu=1}^m \chi_\mu x_\mu .$$

Here, all components $\chi_\mu x_\mu$ of $y(0) = y_0$ are damped at least as strong as the first component $\chi_1 x_1$. If we require the same property also for the numerical approximation (4.2.2) being here identical with (4.2.1) then we obtain the condition

$$\sup_{-\infty\le\eta\le-\Delta t\alpha}\left|\frac{1+(1-\omega)\eta}{1-\omega\eta}\right| \le \frac{1-(1-\omega)\Delta t\alpha}{1+\omega\Delta t\alpha}, \qquad 0 \le \omega \le 1.$$

This demand is very similar to the concept of strong stability introduced by Grigorieff [72 , p. 98] and a short calculation shows that it is equivalent to

$$\frac{1-\omega}{\omega} \le \frac{1-(1-\omega)\Delta t\alpha}{1+\omega\Delta t\alpha}$$

which yields the inequality (4.2.7).

4.3. G-Stability and A-Stability

The midpoint rule is not really a method of the class indicated in the title of this chapter but a single step single stage method. Nevertheless we have studied it in the previous section because of its interesting properties and its strong relationship to the trapezoidal rule. With respect to the stability, Theorem (4.2.5) may serve as a model for the error propagation in other methods but the discretization error was not estimated at all in this result as the reader has certainly remarked. Here we are faced with an entirely different situation in genuine single stage and in multistage methods therefore we postpone a study to the following chapter on Runge-Kutta methods.

In order to obtain error bounds for linear multistep methods comparable to that of Theorem (4.2.5) Dahlquist [75] has introduced the concept of G-stability. But he has considered only 'one-leg' methods, i.e., methods of the form

$$(4.3.1)\quad \rho(T)v_n - \Delta t f(\sigma(T)t_n, \sigma(T)v_n) = 0,\ t_n = n\Delta t,\ \sigma(1) = 1, \qquad n = 0,1,\dots,$$

of which the midpoint rule is an example. These methods are genuine multistep single stage methods because they can be written in the form

$$v_{n,1} = (\sigma(T) - \rho(T))v_n + \Delta t f(\sigma(T)t_n, v_{n,1})$$

$$\rho(T)v_n = \Delta t f(\sigma(T)t_n, v_{n,1}), \qquad n = 0,1,\dots,$$

So they suffer from the same difficulties concerning a data-free estimation of the discretization error as Runge-Kutta methods. However, after some essential pre-work of Dahlquist [78b], Nevanlinna and Odeh [81] have recently shown how to apply the concept of G-stability in multistep *single derivative* methods.

We reconsider linear k-step methods (4.1.4) for the initial value problem (4.1.3) under the general monotonicity condition (4.2.3). Recalling that ρ and σ are real polynomials of degree not greater than k the following definition is the above mentioned G-stability due to Dahlquist [75].

<u>(4.3.2) Definition</u>. *The method* (4.1.4) *is G-stable if there exists a real symmetric and positive definite matrix* $G = [g_{ij}]_{i,j=1}^{k}$ *such that for all* $w_i \in \mathbb{C}$, $w_{i+1} = Tw_i$, $i = 0,\dots,k-1$, and $W_n = (w_{n-k+1},\dots,w_n)^T$,

$$W_k^H G W_k - W_{k-1}^H G W_{k-1} \le 2\mathrm{Re}(\overline{\sigma(T)w_0}\rho(T)w_0).$$

If $w_i \in \mathbb{C}^m$ then let

$$W_k^H G W_k = \sum_{i,j=1}^{k} g_{ij} w_i^H w_j .$$

With this notation the following result shows that a G-stable method satisfies

$$W_k^H G W_k - W_{k-1}^H G W_{k-1} \leq 2\,\mathrm{Re}((\sigma(T)w_0)^H \rho(T)w_0)$$

for arbitrary *vectors* $w_i \in \mathbb{C}^m$.

(4.3.3) Lemma. *The real symmetric matrix* $Q = [q_{ij}]_{i,j=1}^{k+1}$ *is positive semidefinite iff*

$$\sum_{i,j=1}^{k+1} q_{ij} z_i^H z_j \geq 0 \qquad \forall\, z_i \in \mathbb{C}^m .$$

Proof. The condition is obviously sufficient for the matrix Q to be positive semidefinite. On the other side, we have $Q = X\Omega X^T$, $X^T X = I$, by assumption where Ω is the diagonal matrix of the eigenvalues of Q. Hence, if Q is positive semidefinite then

$$\sum_{i,j=1}^{k+1} q_{ij} z_i^H z_j = \sum_{i=1}^{k+1} \omega_i \Big|\sum_{j=1}^{k+1} x_{ij} z_j\Big|^2 \geq 0 .$$

Recall now that a method (4.1.4) is A-stable iff

(4.3.4) $\{\eta \in \mathbb{C},\ \mathrm{Re}(\eta) \leq 0\} \subset S$ and $\{\infty\} \in S$.

This is a slight modification of the frequently used original version of Dahlquist [63] who defined a method to be A-stable iff

(4.3.5) $\{\eta \in \mathbb{C},\ \mathrm{Re}(\eta) < 0\} \subset \mathring{S}$.

Obviously, (4.3.4) implies (4.3.5) but the converse is also true since by remark (i) after (A.1.25) no point of $\bar{S} \setminus S$ can lie on the straight line $i\mathbb{R} \cup \{\infty\}$.

(4.3.6) Theorem. (Dahlquist [75].) *G-stability implies A-stability.*

Proof. With respect to the test equation $y' = \lambda y$, G-stability yields for the method (4.1.4)

$$V_n^H G V_n - V_{n-1}^H G V_{n-1} \leq 2\mathrm{Re}((\sigma(T)v_{n-k})^H \Delta t \lambda \sigma(T) v_{n-k}) = 2\Delta t \mathrm{Re}(\lambda) |\sigma(T) v_{n-k}|^2 .$$

Hence the sequence $\{v_n\}_{n=0}^{\infty}$ defined by $\rho(T)v_n - \Delta t \lambda \sigma(T) v_n = 0$ is bounded for $\mathrm{Re}(\lambda) \leq 0$ and arbitrary start values. This implies A-stability.

It is rather astonishing that the conversion of Theorem (4.3.6) is also true. We state this result in a somewhat more general context:

(4.3.7) Definition. (Dahlquist [78b].) *A rational function* ϕ *is an A-function if* $|\zeta| > 1$ *implies* $\mathrm{Re}(\phi(\zeta)) > 0$.

This definition is equivalent to the statement that $|\arg\phi(\zeta)| < \pi/2$ for $|\zeta| > 1$. Therefore an A-function is analytic and different from zero for $|\zeta| > 1$ and for $\zeta = \infty$, because the variation of $\arg\phi(\zeta)$ on a small circle around a zero or a pole (a large circle in the case $\zeta = \infty$) is a multiple of 2π. There may however be poles (and zeros) on the boundary of the unit circle. Definition (4.3.7) is thus also equivalent to the statement that $|\zeta| \geq 1$ implies $\mathrm{Re}(\phi(\zeta)) \geq 0$ or $\phi(\zeta) = \infty$. For instance, the function $\phi = \rho/\sigma$ associated with the trapezoidal rule has the values $\phi(\zeta) = 2(\zeta - 1)/(\zeta + 1)$. Every pole ζ_0 with $|\zeta_0| = 1$ must be simple and $\phi(\zeta)$ must behave like $\gamma(\zeta + \zeta_0)/(\zeta - \zeta_0)$ with $\gamma > 0$ in a neighborhood of ζ_0; cf. Dahlquist [78b].

(4.3.8) Lemma. *A linear multistep method* (4.1.4) *is A-stable iff* $\phi = \rho/\sigma$ *is an A-function.*

Proof. Observe that $\eta = \phi(\zeta) = \rho(\zeta)/\sigma(\zeta)$ follows from $\pi(\zeta,\eta) = \rho(\zeta) - \eta\sigma(\zeta) = 0$. If the method is A-stable then $\mathrm{Re}(\phi(\zeta)) \leq 0$ implies $|\zeta| \leq 1$ by (4.3.4) hence $|\zeta| > 1$ implies $\mathrm{Re}(\phi(\zeta)) > 0$. On the other side, if $|\zeta| > 1$ implies $\mathrm{Re}(\phi(\zeta)) > 0$ then $\mathrm{Re}(\phi(\zeta)) \leq 0$ implies $|\zeta| \leq 1$ which is (4.3.4).

Now the following result shows together with Theorem (4.3.6) that A-stability and G-stability are equivalent for linear multistep methods.

(4.3.9) Theorem. (Dahlquist [78b].) *Let* ρ *and* σ *be two polynomials of degree* k *such that* ρ/σ *is an A-function. Then there exists a hermitean positive definite* (k,k)*-matrix* G *such that for arbitrary* $w_i \in \mathbb{C}$, $i = 0,\dots,k$, *and* $W_n = (w_{n-k+1},\dots,w_n)^T$

$$W_k^H G W_k - W_{k-1}^H G W_{k-1} \leq 2\mathrm{Re}(\overline{\sigma(T)w_0}\,\rho(T)w_0).$$

If $w_i \in \mathbb{R}$, $i = 0,\dots,k$, *or if* ρ *and* σ *are real then* G *can be replaced by the real symmetric matrix* $\mathrm{Re}(G) = (G + G^H)/2$ *which is also positive definite.*

The proof of this theorem yields at the same time a general device for the computation of the matrix G. Because of its length we however must refer to Dahlquist [78b] for the details.

The next result is written as a lemma only for convenience; cf. also the error equation (1.2.13).

(4.3.10) Lemma. *Let* $-1 < \Gamma_0 \le \Gamma$ *and let*

$$\|E_n\|^2 \le (1+\Gamma)\|E_{n-1}\|^2 + (\|E_n\| + \|E_{n-1}\|)\|D_n\|, \qquad n = k,k+1,\ldots,$$

then

$$\|E_n\| \le (1+\tilde{\Gamma})^{n-k+1}\|E_{k-1}\| + \kappa\sum_{\nu=k}^{n}(1+\tilde{\Gamma})^{n-\nu}\|D_\nu\|, \qquad n = k,k+1,\ldots,$$

where $\tilde{\Gamma} = \Gamma$ *if* $\Gamma \ge 0$ *and* $\tilde{\Gamma} = \Gamma/2$ *if* $\Gamma < 0$, *and* $\kappa \le 2(1+\Gamma_0)^{-1/2}$.

Proof. By induction we obtain for $\mu = k-1,k,\ldots,$

$$(4.3.11)\quad \|E_\mu\|^2 \le (1+\Gamma)^{\mu-k+1}\|E_{k-1}\|^2 + \sum_{\nu=k}^{\mu}(1+\Gamma)^{\mu-\nu}(\|E_\nu\| + \|E_{\nu-1}\|)\|D_\nu\|.$$

Hence, if $\Gamma \ge 0$ then

$$(\max_{k-1\le\mu\le n}\|E_\mu\|)^2 \le (1+\Gamma)^{n-k+1}\|E_{k-1}\|^2 + \sum_{\nu=k}^{n}(1+\Gamma)^{n-\nu}(\|E_\nu\| + \|E_{\nu-1}\|)\|D_\nu\|$$

$$\le (\max_{k-1\le\mu\le n}\|E_\mu\|)((1+\Gamma)^{n-k+1}\|E_{k-1}\| + 2\sum_{\nu=k}^{n}(1+\Gamma)^{n-\nu}\|D_\nu\|),\ n = k,k+1,\ldots,$$

which proves the assertion in this case.

If $\Gamma < 0$ then we observe that by (4.3.11)

$$(1+\Gamma)^{-\mu}\|E_\mu\|^2$$

$$\le (1+\Gamma)^{-(k-1)}\|E_{k-1}\|^2 + \sum_{\nu=k}^{\mu}((1+\Gamma)^{-\nu/2}\|E_\nu\| + (1+\Gamma)^{-(\nu-1)/2}\|E_{\nu-1}\|)\tilde{\kappa}(1+\Gamma)^{-\nu/2}\|D_\nu\|.$$

Hence we obtain for $\mu = k,k+1,\ldots,$

$$\|\hat{E}_\mu\|^2 \le \|\hat{E}_{k-1}\|^2 + \sum_{\nu=k}^{\mu}(\|\hat{E}_\nu\| + \|\hat{E}_{\nu-1}\|)\|\hat{D}_\nu\|$$

writing $\hat{E}_\mu = (1+\Gamma)^{-\mu/2}E_\mu$ and $\hat{D}_\nu = \tilde{\kappa}(1+\Gamma)^{-\nu/2}D_\nu$. Now the first part of the proof yields

$$\|E_n\| \le (1+\Gamma)^{(n-k+1)/2}\|E_{k-1}\| + \sum_{\nu=k}^{n}(1+\Gamma)^{(n-\nu)/2}\kappa\|D_\nu\|$$

which proves the assertion in the second case because $(1+\Gamma)^{1/2} \le (1+\Gamma/2)$ if $\Gamma > -1$.

With these aids we now are able to derive error bounds for A-stable multistep methods and stable nonlinear differential equations. The first one concerns one-leg methods and the second one ordinary linear multistep methods which allow an estimation of the discretization error by means of Lemma (1.1.8). However, both types coincide for the test equation $y' = \lambda y$ and a repeatedly mentioned result of Dahlquist [63] says that the order of an A-stable linear multistep method cannot exceed two. Therefore both subsequent theorems concern only methods up to order two.

(4.3.11) Theorem. (Dahlquist [75 , 78b].) *Let the initial value problem* (4.1.3) *satisfy* (4.2.3) *and let the one-leg method* (4.3.1) *be A-stable. Then for* $n = k,k+1,\dots,$

$$|y_n - v_n| \le \kappa[|Y_{k-1} - V_{k-1}| + \textstyle\sum_{\nu=k}^{n} |d(\Delta t,y)_{\nu-k}|]$$

where $d(\Delta t,y)(t)$ *denotes the discretization error.*

Proof. The error $e_n = y_n - v_n$ satisfies for $n = 0,1,\dots,$

$$\rho(T)e_n - \Delta t[f(\sigma(T)t_n,\sigma(T)y_n) - f(\sigma(T)t_n,\sigma(T)v_n)] = d(\Delta t,y)_n.$$

We multiply from left by $\sigma(T)e_n$ and obtain using the monotonicity condition (4.2.3)

$$(4.3.12)\quad (\sigma(T)e_n)^T\rho(T)e_n \le (\sigma(T)e_n)^T d(\Delta t,y)_n.$$

Now, by Theorem (4.3.9) there exists a fixed real symmetric and positive definite matrix G such that

$$E_n^T G E_n - E_{n-1}^T G E_{n-1} \le 2(\sigma(T)e_{n-k})^T d(\Delta t,y)_{n-k} \le \tilde{\kappa}(|E_n| + |E_{n-1}|)|d(\Delta t,y)_{n-k}|$$

where $\tilde{\kappa} = 2(\sum_{i=0}^{k}|\beta_i|^2)^{1/2}$ and $E_n = (e_{n-k+1},\dots,e_n)^T$. But

$$|E_n| \le |G^{-1/2}||G^{1/2}E_n| \equiv |G^{-1/2}|\|E_n\|_G$$

accordingly

$$\|E_n\|_G^2 \le \|E_{n-1}\|_G^2 + (\|E_n\|_G + \|E_{n-1}\|_G)\kappa|G^{-1/2}||d(\Delta t,y)_{n-k}|,\qquad n = k,k+1,\dots,$$

and an application of Lemma (4.3.10) proves the result.

This result is only stated to show the simple way of error bounding in A-stable linear one-leg methods. Naturally, it cannot bear a comparison with the special result for the modified midpoint rule of Theorem (4.2.5) but it can be improved substantially for two-step methods using the version of Lemma (4.3.10) for $\Gamma < 0$. In the error estimation of linear multistep methods additional difficulties arise in spite of the equivalence theorem (4.3.9). So, for instance, we have to introduce here the condition that $\infty \in \mathring{S}$ which excludes the trapezoidal rule.

(4.3.13) Theorem. (Nevanlinna and Odeh [81].) *(i) Let the initial value problem* (4.1.3) *satisfy* (4.2.3) *and let the solution* y *be (p+1)-times continuously differentiable.*

(ii) Let the linear multistep method (4.1.4) *be consistent of order* p *and A-stable.*
(iii) Let $\infty \in \mathring{S}$, *i.e., let* $\sigma(\zeta)$ *have only roots of modulus less than one.*
Then for $n = k,k+1,\ldots$,

$$|y_n - v_n| \le \kappa[\max_{0\le i\le k-1}\{|y_i - v_i| + \Delta t|f_i(y_i) - f_i(v_i)|\} + \Delta t^p\int_0^{n\Delta t}|y^{(p+1)}(\tau)|d\tau].$$

Proof. The error $e_n = y_n - v_n$ of the method (4.1.4) satisfies

$$\rho(T)e_n - \Delta t\sigma(T)(f_n(y_n) - f_n(v_n)) = d(\Delta t,y)_n, \qquad n = 0,1,\ldots \quad .$$

Recall that

$$\rho(\zeta) = \sum_{i=0}^k \alpha_i\zeta^i, \ \sigma(\zeta) = \sum_{i=0}^k \beta_i\zeta^i,$$

and write

$$\text{(4.3.14)} \quad \begin{aligned} \tilde{d}(\Delta t,y)_\nu &= \sum_{i=0}^\nu \alpha_i e_i - \Delta t\sum_{i=0}^\nu \beta_i[f_i(y_i) - f_i(v_i)], & \nu &= 0,\ldots,k-1,\\ \tilde{d}(\Delta t,y)_\nu &= d(\Delta t,y)_\nu, & \nu &= k,k+1,\ldots \ . \end{aligned}$$

Then the sequence $\{e_n\}_{n=-\infty}^\infty$ satisfies

$$\text{(4.3.15)} \quad \rho(T)e_n - \Delta t\sigma(T)(f_n(y_n) - f_n(v_n)) = \tilde{d}(\Delta t,y)_n, \qquad n = 0,1,\ldots \quad .$$

with $e_n = \tilde{d}(\Delta t,y)_n = 0$ for $-n \in \mathbb{N}$. We now consider (4.3.15). As $\sigma(\zeta)$ has only roots of modulus less than one we can write

$$\rho(\zeta)/\sigma(\zeta) = \sum_{j=0}^\infty \chi_j\zeta^{-j}, \qquad |\zeta| \ge 1,$$

and then obtain from (4.3.15)

$$e_n^T\sigma(T)^{-1}\rho(T)e_n - \Delta t e_n^T(f_n(y_n) - f_n(v_n)) = e_n^T\sigma(T)^{-1}\tilde{d}(\Delta t,y)_n, \quad n = 0,1,\ldots$$

We apply the monotonicity condition (4.2.3) and write $\sigma(T)^{-1}e_n = \tilde{e}_n$. By this way we obtain

$$(\sigma(T)\tilde{e}_n)^T\rho(T)\tilde{e}_n \le (\sigma(T)\tilde{e}_n)^T\sigma(T)^{-1}\tilde{d}(\Delta t,y)_n, \qquad n = 0,1,\ldots,$$

which is now the same situation as in (4.3.12). Therefore Theorem (4.3.9) and Lemma (4.3.10) yield in the same way as in Theorem (4.3.11)

$$|\tilde{E}_{n+k}| \leq \kappa\sum_{\nu=k}^{n+k}|\sigma(T)^{-1}\tilde{d}(\Delta t,y)_{\nu-k}|, \qquad n = k,k+1,\ldots .$$

But

$$(4.3.16)\quad \begin{aligned}|e_n| &= |\sigma(T)\sigma(T)^{-1}e_n| = |\sigma(T)\tilde{e}_n| \leq (\textstyle\sum_{i=0}^{k}|\beta_i|^2)^{1/2}(\sum_{i=0}^{k}|\tilde{e}_{n+i}|^2)^{1/2}\\ &\leq \tilde{\kappa}(|\tilde{E}_{n+k}| + |\tilde{E}_{n+k-1}|) \leq 2\tilde{\kappa}\kappa\textstyle\sum_{\nu=k}^{n+k}|\sigma(T)^{-1}\tilde{d}(\Delta t,y)_{\nu-k}|.\end{aligned}$$

Now observe that

$$\sigma(\zeta)^{-1} = \beta_k^{-1}\prod_{i=1}^{k}(\zeta-\zeta_i)^{-1} = \beta_k^{-1}\prod_{i=1}^{k}\zeta^{-1}(1-(\zeta_i/\zeta))^{-1} = \sum_{j=k}^{\infty}\sigma_j\zeta^{-j}$$

where $\sum_{j=k}^{\infty}|\sigma_j| < \infty$ because $\sigma(\zeta)^{-1}$ is analytic in the exterior of a ball $\{\eta \in \mathbb{C}, |\zeta| \leq r\}$ with $r < 1$. Therefore we obtain

$$|\sigma(T)^{-1}\tilde{d}(\Delta t,y)_\nu| = |\sum_{j=k}^{\infty}\sigma_j\tilde{d}(\Delta t,y)_{\nu-j}| = |\sum_{j=k}^{\nu}\sigma_j\tilde{d}(\Delta t,y)_{\nu-j}|, \quad \nu = k,k+1,\ldots,$$

and accordingly

$$\begin{aligned}\sum_{\nu=2k}^{n+k}|\sigma(T)^{-1}\tilde{d}(\Delta t,y)_{\nu-k}| &\leq \sum_{\nu=k}^{n}\sum_{j=k}^{\nu}|\sigma_j||\tilde{d}(\Delta t,y)_{\nu-j}|\\ &= \sum_{\nu=k}^{n}(\sum_{\mu=k}^{n-\nu}|\sigma_\mu|)|d(\Delta t,y)_{\nu-k}| \leq \kappa\sum_{\nu=k}^{n}|d(\Delta t,y)_{\nu-k}|.\end{aligned}$$

A similar estimation yields by (4.3.14)

$$\sum_{\nu=k}^{2k-1}|\sigma(T)^{-1}\tilde{d}(\Delta t,y)_{\nu-k}| \leq \kappa\max_{0\leq i\leq k-1}\{|e_i| + \Delta t|f_i(y_i) - f_i(v_i)|\}.$$

A substitution of these bounds into (4.3.16), finally, and Lemma (1.1.8) prove the assertion.

As an example we consider the modified trapezoidal rule (4.2.1),

$$(4.3.17)\quad v_{n+1} = v_n + \Delta t[\omega f_{n+1}(v_{n+1}) + (1-\omega)f_n(v_n)], \qquad n = 0,1,\ldots,$$

with the parameter proposed by Kreth [81],

$$(4.3.18)\quad \omega = [-1 + \Delta t\alpha + (1 + \Delta t^2\alpha^2)^{1/2}]/2\Delta t\alpha = 1/2 + \mathcal{O}(\Delta t) \geq 1/2.$$

Multiplying the modified error equation (4.3.15) from left by $\sigma(T)^{-1}$ and applying the monotonicity condition (4.2.3) we obtain at first for $\tilde{e}_n = \sigma(T)^{-1}e_n$

$$(\sigma(T)\tilde{e}_n)^T\rho(T)\tilde{e}_n + \Delta t\alpha(\sigma(T)\tilde{e}_n)^T\rho(T)\tilde{e}_n \leq e_n^T\sigma(T)^{-1}\tilde{d}(\Delta t,y)_n.$$

By Dahlquist [75] the only possible 'matrix' G is here G = 1 and unfortunately there exists no positive constant δ such that

$$(\sigma(T)\tilde{e}_n)^T\sigma(T)\tilde{e}_n \geqq \delta(|\tilde{e}_{n+1}|^2 + |\tilde{e}_n|^2)$$

therefore we have only

$$|\tilde{e}_{n+1}|^2 - |\tilde{e}_n|^2 \leq 2e_n^T\sigma(T)^{-1}\tilde{d}(\Delta t,y)_n, \qquad n = 0,1,\dots \quad .$$

Moreover, we deduce easily that in the present case

$$\sigma(\zeta)^{-1} = \sum_{j=1}^{\infty}\sigma_j\zeta^{-j} = \frac{1}{\omega}\sum_{j=1}^{\infty}\left(\frac{\omega-1}{\omega}\right)^j\zeta^{-j}, \ \sum_{j=1}^{\infty}|\sigma_j| = (2\omega - 1)^{-1},$$

and, for $\nu = 1,2,\dots$,

$$|\tilde{d}(\Delta t,y)(t)| = |d(\Delta t,y)(t)| \leq |\tfrac{1}{2} - \omega|\Delta t^2|y''(t)| + \Gamma\Delta t^2\int_t^{t+\Delta t}|y^{(3)}(\tau)|d\tau.$$

Therefore, if $\omega = \frac{1}{2} + \mathcal{O}(\Delta t) \geqq \frac{1}{2}$ then for $\nu = 1,2,\dots$,

$$|\sigma(T)^{-1}\tilde{d}(\Delta t,y)_\nu| = |\sum_{j=1}^{\nu}\sigma_j d(\Delta t,y)_{\nu-j}| \leq \kappa(\Delta t^2|y''| + \Delta t\int_{\nu\Delta t}^{(\nu+1)\Delta t}|y^{(3)}(\tau)|d\tau),$$

and we obtain the following error bound for the method (4.3.17) with the parameter (4.3.18) in the case of stable nonlinear problems:

$$|y_n - v_n| \leq \kappa[|y_0 - v_0| + \Delta t|f_0(y_0) - f_0(v_0)| + n\Delta t^2\max_{0\leq\nu\leq n-1}|y_\nu''| + \Delta t\int_0^{n\Delta t}|y^{(3)}(\tau)|d\tau].$$

For the classical trapezoidal rule and nonlinear problems, i.e. (4.3.17) with $\omega = 1/2$, no error bounds are known to the author besides the classical one (cf. Henrici [62 , Theorem 5.10]) even of this form where the order of convergence is smaller than the order of consistence. But, in concluding this section we quote a result concerning the linear case. Recall that $A \geqq 0$ means for a real (m,m)-matrix A that $x^TAx \geqq 0 \ \forall x \in \mathbb{R}^m$, i.e., that $\mathrm{Re}(A) = (A + A^T)/2$ is positive semidefinite.

<u>(4.3.19) Theorem</u>. *(i) Let the real (m,m)-matrix* A(t) *in* (4.1.1) *be symmetric, negative semidefinite, and two times continuously differentiable; let the solution* y *be three times continuously differentiable.*
(ii) Let $A''(t)$ *be positive semidefinite and* $\Delta t^2|||A'|||_n/4 < \delta < 1$ *or let* $(\Delta t^2|||A'|||_n + n\Delta t^3|||A''|||_n)/4 < \delta < 1$.
Then the method (4.3.17) *with* $\omega = 1/2$ *satisfies for* $n = 1,2,\dots$,

$$|y_n - v_n| \leq (1 - \delta)^{-1}\kappa[|y_0 - v_0| + \Delta t|A_0(y_0 - v_0)| + \Delta t^2\int_0^{n\Delta t}|y^{(3)}(\tau)|d\tau].$$

If $A'(t)$ *is negative semidefinite and* $A''(t)$ *is positive semidefinite then this error bound holds with* $\delta = 0$ *and assumption (ii) cancelled.*

Proof. We change temporarily the notations and write $E_n = (e_1,\ldots,e_n)^T$, $e_n = y_n - v_n$, and

$$D_n = (e_0 + (\Delta t/2)A_0e_0 + d(\Delta t,y)_0,\ d(\Delta t,y)_1,\ldots,\ d(\Delta t,y)_{n-1})^T.$$

$\underline{A}_n$ denotes the block diagonal matrix with the diagonal block elements $A_1,\ldots,A_n$, T is the 'translation' matrix,

$$T = [t_{ij}]_{i,j=1}^{n},\quad t_{i+1,i} = 1,\ t_{ij} = 0 \text{ else},$$

and I is the identity matrix of dimension both m and m·n. Then, writing (4.3.17) for $\nu = 1,\ldots,n$ as a large system we obtain for the error E_n the following equation,

$$(I - T)E_n - \Delta t\tfrac{1}{2}(I + T)\underline{A}_nE_n = D_n, \qquad n = 1,2,\ldots,$$

and a multiplication from left by $(I + T)E_n$ yields

$$(4.3.20)\quad E_n^T(I + T)^T(I - T)E_n - \Delta t\,\frac{1}{2}\,E_n^T(I + T)^T(I + T)\underline{A}_nE_n = E_n^T(I + T)^TD_n.$$

Writing briefly $B_\nu = A_\nu + A_{\nu+1}$ we obtain for the real part of the matrix $G = (I + T)^T(I + T)\underline{A}_n$ the following representation because of the symmetry of A(t),

$$2\mathrm{Re}(G) = G + G^T = H + \begin{bmatrix} B_1 & B_1 & & & & & \\ B_1 & B_1 + B_2 & B_2 & & & \bigcirc & \\ & B_2 & B_2 + B_3 & B_3 & & & \\ & & \ddots & \ddots & \ddots & & \\ & \bigcirc & & B_{n-2} & B_{n-2} + B_{n-1} & B_{n-1} \\ & & & & B_{n-1} & B_{n-1} \end{bmatrix}$$

where H is a block diagonal matrix of the form

$$(4.3.21)\quad H = ((2A_1 - A_0 - A_2) + (A_1 + A_0), 2A_2 - A_1 - A_3,\ldots, 2A_{n-1} - A_{n-2} - A_n, A_n - A_{n-1}).$$

But the matrix 2Re(G) - H is negative semidefinite because the marked submatrices have this property and

$$E_n^T(I + T)^T(I - T)E_n = |e_n|^2$$

therefore we have by (4.3.20)

(4.3.22) $|e_n|^2 - \frac{\Delta t}{4} E_n^T H E_n \leq E_n^T(I + T)D_n \leq 2(\max_{1\leq\nu\leq n}|e_\nu|)\|D_n\|_1.$

Now observe that the classical Mean Value Theorem holds for the real scalar function $t \mapsto x^T A(t)x$ where $x \in \mathbb{R}^m$ is fixed. Therefore we have

$$e_n^T(A_n - A_{n-1})e_n = \Delta t e_n^T A'(\xi_n)e_n$$

and

$$e_\nu^T(A_{\nu+1} - 2A_\nu + A_{\nu-1})e_\nu = \Delta t\delta_\nu e_\nu^T A''(\xi_\nu)e_\nu, \qquad 0 < \delta_\nu < \Delta t.$$

Accordingly, as $-e_1^T(A_0 + A_1)e_1 \geq 0$, (4.3.21) yields

$$-\frac{\Delta t}{4} E_n^T H E_n \geq \frac{\Delta t^2}{4}[\textstyle\sum_{\nu=1}^{n-1}\delta_\nu e_\nu^T A''(\xi_\nu)e_\nu - e_n^T A'(\xi_n)e_n]$$

$$\geq -\frac{\Delta t^2}{4}[n\Delta t|||A''|||_n + |||A'|||_n]\max_{1\leq\nu\leq n}|e_\nu|^2$$

or, if $A''(t) \geq 0$,

$$-\frac{\Delta t}{4} E_n^T H E_n \geq -\frac{\Delta t^2}{4}|||A'|||_n \max_{1\leq\nu\leq n}|e_\nu|^2$$

or, if $A''(t) \geq 0$ and $A'(t) \leq 0$,

(4.3.23) $-\frac{\Delta t}{4} E_n^T H E_n \geq 0.$

By these three different estimations, (4.3.22) yields under assumption (ii)

$$(1 - \delta)\max_{1\leq\nu\leq n}|e_\nu| \leq 2\|D_n\|_1 \leq 2[|e_0 + (\Delta t/2)A_0 e_0| + \textstyle\sum_{\nu=0}^{n-1}|d(\Delta t,y)_\nu|]$$

where $\delta = 0$ in the third case, (4.3.23). This proves the assertion after an application of Lemma (1.1.8).

This error bound contains as a peculiarity an assumption on the second derivative of the matrix $A(t)$, $A''(t) \geq 0$, and carries no restriction of the time step Δt only if $A''(t) \geq 0$ and $A'(t) \leq 0$. This is a somewhat surprising result because in the next section error bounds of the above form with $\delta = 0$ are derived for strongly $A(\alpha)$-stable methods under the assumption that $A'(t) \geq 0$ and $A(t) \leq 0$.

The symmetry of $A(t)$ seems to be an unsuitable assumption in the trapezoidal rule but the author was unable to cancel it.

4.4. Uniform Stability under Stronger Assumptions on the Differential System

Thanks to Dahlquist's Theorem (4.3.9) we were able to establish a satisfactory stability analysis for A-stable linear multistep methods and stable nonlinear problems without any data-dependent exponential growing factor. But these results concern only methods up to order two because methods of higher consistence order have no longer this property. This leads immediately to the question whether the error bounds of LeRoux, cf. Theorem (4.1.2), can be improved under stronger assumptions on the differential system. In the search for a generalization of the theory developed in the preceding section to the class of $A(\alpha)$-stable methods, a very promising tool is the multiplier technique proposed by Nevanlinna and Odeh [81]:

(4.4.1) Definition. *A rational function* $\mu = \nu/\chi$ *is a multiplier for the linear multistep method with the characteristic polynomial* $\pi(\zeta,\eta) = \rho(\zeta) - \eta\sigma(\zeta)$ *if*
(i) $\mu(\zeta) = \sum_{j=0}^{\infty}\mu_j\zeta^{-j}$, $|\zeta| > 1$, $\mu_j \in \mathbb{R}$, $j = 0,1,\ldots,$
(ii) $\sum_{j=0}^{\infty}|\mu_j| < \infty$,
(iii) $(\chi\rho)/(\nu\sigma)$ *is an A-function (cf.* Definition (4.3.7)*).*
A multiplier is finitely supported if there exists a $N \in \mathbb{N}$ *such that* $\mu_j = 0$ *for* $j > N$.

Recall that an A-function is analytic and nonzero in $\zeta = \infty$ therefore $\chi\rho$ and $\nu\sigma$ must have the same degree.

(4.4.2) Definition. (Nevanlinna and Odeh [81].) *Let* μ *be a rational function satisfying* (4.4.1)(i)*. Then the differential system* $y' = f(t,y)$ *satisfies the angle-bounded monotonicity condition with respect to* μ *if*

$$\sum_{n=0}^{N}[\mu(T)(v_n - w_n)]^T[f_n(v_n) - f_n(w_n)] \le 0 \;\forall\; v_n, w_n \in \mathbb{R}^m, \qquad N = 0,1,\ldots,$$

where $Tv_n = v_{n+1}$, $n = 0,1,\ldots,$ *and* $v_n = w_n = 0$ *for* $-n \in \mathbb{N}$.

For instance, an A-stable linear multistep method has the multiplier $\mu(\zeta) \equiv 1$.

Before we discuss these notions in some detail we first prove the following general result:

(4.4.3) Theorem. (Nevanlinna and Odeh [81].) *(i) Let the solution* y *of the initial value problem* (4.1.3) *be (p+1)-times continuously differentiable.*
(ii) Let the linear multistep method (4.1.4) *be implicit and consistent of order* p, *let* $\mu = \nu/\chi$ *be a multiplier for this method, and let* $\nu(\zeta)\sigma(\zeta)$ *have only roots of modulus less than one.*
(iii) Let the differential system in (4.1.3) *satisfy the angle-bounded monotonicity condition with respect to* μ.

Then the assertion of Theorem (4.3.13) *holds for* n = k,k+1,...:

$$|y_n - v_n| \le \kappa[\max_{0\le i\le k-1} \{|y_i - v_i| + \Delta t|f_i(y_i) - f_i(v_i)|\} + \Delta t^p \int_0^{n\Delta t} |y^{(p+1)}(\tau)|d\tau].$$

Proof. We consider again the modified error equation (4.3.15) with $e_n = \tilde{d}(\Delta t,y)_n = 0$ for $-n \in \mathbb{N}$. As $\zeta \to \sigma(\zeta)^{-1}$ and $\zeta \to \chi(\zeta)^{-1}$ are analytic in the exterior of a ball $\{\zeta \in \mathbb{C}, |\zeta| \le r\}$ with $r < 1$ we are allowed to multiply (4.3.15) by $\sigma(T)^{-1}$ and, after this operation, scalarly by $\mu(T)e_n$. The result is

$$(\chi(T)^{-1}\nu(T)e_n)^T\sigma(T)^{-1}\rho(T)e_n - \Delta t(\mu(T)e_n)^T(f_n(y_n) - f_n(v_n))$$

$$= (\mu(T)e_n)^T\sigma(T)^{-1}\tilde{d}(\Delta t,y)_n, \qquad n = 0,1,\ldots \quad .$$

Writing $\tilde{e}_n = \chi(T)^{-1}\sigma(T)^{-1}e_n$ we obtain by assumption (iii)

$$(\nu(T)\sigma(T)\tilde{e}_n)^T\chi(T)\rho(T)\tilde{e}_n \le (\nu(T)\sigma(T)\tilde{e}_n)^T\sigma(T)^{-1}\tilde{d}(\Delta t,y)_n, \qquad n = 0,1,\ldots \quad .$$

Now $\phi = (\chi\rho)/(\nu\sigma)$ is an A-function by assumption (ii). Let $\chi\rho$ and $\nu\sigma$ have the degree $r \ge k$, then there exists by Theorem (4.3.9) a real symmetric and positive definite (r,r,)-matrix G such that for n = r,r+1,...,

$$\tilde{E}_n^T G\tilde{E}_n - \tilde{E}_{n-1}^T G\tilde{E}_{n-1} - \Delta t(\mu(T)e_n)^T(f_n(y_n) - f_n(v_n)) \le (\nu(T)\sigma(T)\tilde{e}_{n-r})^T\sigma(T)^{-1}\tilde{d}(\Delta t,y)_{n-r},$$

where $\tilde{E}_n = (\tilde{e}_{n-r+1},\ldots,\tilde{e}_n)^T$ and $\tilde{E}_n = 0$ for $n < r$. But

$$|\nu(T)\sigma(T)\tilde{e}_{n-r}| \le \tilde{\kappa}(\textstyle\sum_{i=0}^r |\tilde{e}_{n+r-i}|^2)^{1/2} \le \kappa(\|\tilde{E}_{n-1}\|_G + \|\tilde{E}_n\|_G)$$

with $\|\tilde{E}_n\|_G = |G^{-1/2}\tilde{E}_n|$ hence assumption (iii) yields after a recursive computation

$$\|\tilde{E}_{n+r}\|_G^2 \le \kappa(\max_{r\le\nu\le n+r}\|\tilde{E}_\nu\|_G)\textstyle\sum_{\nu=r}^{n+r}|\sigma(T)^{-1}\tilde{d}(\Delta t,y)_{\nu-r}|.$$

This inequality remains true if on the left side n is replaced by ν, $r \le \nu \le n+r$, and we obtain after simplifying

$$\|\tilde{E}_{n+r}\|_G \le \max_{r\le\nu\le n+r}\|\tilde{E}_\nu\|_G \le \kappa\textstyle\sum_{\nu=r}^{n+r}|\sigma(T)^{-1}\tilde{d}(\Delta t,y)_{\nu-r}|.$$

The remaining part of the proof is the same as in Theorem (4.3.13).

Let us first turn to the monotonicity condition (4.4.2). Nevanlinna and Odeh verify assumption (4.4.3)(iii) under several different assumptions on the differential system and derive also conditions for the multiplier in the nonlinear case (4.1.3). However, we shall restrict ourselves here to one of the most interesting results for linear initial value problems (4.1.1). The following auxiliary result is e.g. found in Ortega and Rheinboldt [70 , Lemma 3.4.4].

<u>(4.4.4) Lemma</u>. *Let* $\Phi: D \subset \mathbb{R}^m \to \mathbb{R}$ *be continuously differentiable on a convex subset* $D_0 \subset D$. *Then* Φ *is convex on* D_0, *i.e.,*

$$\Phi(\omega v + (1-\omega)w) \le \omega\Phi(v) + (1-\omega)\Phi(w) \qquad \forall\ v,w \in D_0,\ 0 \le \omega \le 1,$$

iff

$$(v - w)^T \mathrm{grad}(\Phi(v)) \ge \Phi(v) - \Phi(w).$$

Notice that this lemma yields $v^T\mathrm{grad}(\Phi(v)) \ge \Phi(v)$ if moreover $\Phi(0) = 0$.

<u>(4.4.5) Lemma</u>. (Nevanlinna and Odeh [81].) *(i) Let* $\Phi: \mathbb{R}^{m+1} \ni (t,v) \to \Phi(t,v) \in \mathbb{R}$ *be nonnegative with* $\Phi(t,0) = 0$ *and let* $\Phi(t,\cdot)$ *be convex and continuously differentiable. (ii) Let* $\mu(\zeta) = \sum_{j=0}^{\infty}\mu_j\zeta^{-j}$ *be a multiplier with* $\mu_j \le 0$ *and* $\sum_{\nu=0}^{j}\mu_j \ge 0$ *for* $j = 1,2,\ldots$. *Then*

$$(\mu(T)v_n)^T \mathrm{grad}_v(\Phi_n(v_n)) \ge \mu(T)\Phi_n(v_n) + \sum_{j=1}^{n}\mu_j(\Phi_n(v_{n-j}) - \Phi_{n-j}(v_{n-j}))$$

for all $\{v_n\}_{n=-\infty}^{\infty}$ *with* $v_n = 0$ *for* $-n \in \mathbb{N}$.

Proof. We have

$$(\mu(T)v_n)^T \mathrm{grad}_v(\Phi_n(v_n)) = \sum_{j=0}^{n}\mu_j v_{n-j}^T \mathrm{grad}_v(\Phi_n(v_n))$$

$$= \sum_{j=1}^{n}(-\mu_j)(v_n - v_{n-j})^T \mathrm{grad}_v(\Phi_n(v_n)) + (\mu_0 + \sum_{j=1}^{n}\mu_j)v_n^T \mathrm{grad}_v(\Phi_n(v_n))$$

$$\ge \sum_{j=1}^{n}(-\mu_j)(\Phi_n(v_n) - \Phi_n(v_{n-j})) + (\mu_0 + \sum_{j=1}^{n}\mu_j)\Phi_n(v_n)$$

$$= \mu(T)\Phi_n(v_n) + \sum_{j=1}^{n}\mu_j(\Phi_n(v_{n-j}) - \Phi_{n-j}(v_{n-j})).$$

Now we reconsider the linear initial value problem (4.1.1) and prove the following modification of Theorem (4.4.3); cf. Nevanlinna and Odeh [81 , Example 3.16]. In order to avoid too many back-references all assumptions are once more collected:

<u>(4.4.6) Theorem</u>. *(i) Let the (m,m)-matrix* $A(t)$ *in* (4.1.1) *be real symmetric and continuously differentiable, let* $A(t) \le 0$ *and* $A'(t) \ge 0$; *and let the solution* y *be (p+1)-times continuously differentiable.*

(ii) Let the linear multistep method (4.1.4) *be implicit, consistent of order* p, *and let* $\sigma(\zeta)$ *have only roots of modulus less than one.*
(iii) Let $\mu = \nu/\chi = \sum_{j=0}^{\infty}\mu_j\zeta^{-j}$ *be a multiplier for the method* (4.1.4) *with* $\mu_j \leq 0$ *and* $\sum_{\nu=0}^{j}\mu_\nu \geq 0$, $j = 1,2,\ldots$, *and let* ν *have only roots of modulus less than one.*
Then the assertion of Theorem (4.4.3) *holds.*

Proof. We have only to verify assumption (iii) of Theorem (4.4.3), i.e., the angle-bounded monotonicity condition (4.4.2) with respect to $f(t,v) = A(t)v$. Writing $\Phi(t,v) = -v^TA(t)v/2$ we obtain by Lemma (4.4.5)

$$(4.4.7)\qquad \begin{aligned} -\sum_{n=0}^{N}(\mu(T)e_n)^TA_ne_n &= \sum_{n=0}^{N}(\mu(T)e_n)^T\mathrm{grad}_v(\Phi_n(e_n)) \\ &\geq \sum_{n=0}^{N}\mu(T)\Phi_n(e_n) + \sum_{n=0}^{N}\sum_{j=1}^{n}\mu_j(\Phi_n(v_{n-j}) - \Phi_{n-j}(v_{n-j})). \end{aligned}$$

But

$$\Phi_n(e_{n-j}) - \Phi_{n-j}(e_{n-j}) = e_{n-j}^TA'(\xi)e_{n-j} \geq 0$$

where ξ depends on j, n, and e_{n-j}, and

$$2\sum_{n=0}^{N}\mu(T)\Phi_n(e_n) = -\sum_{n=0}^{N}\sum_{j=0}^{n}\mu_je_{n-j}^TA_{n-j}e_{n-j} = -\sum_{n=0}^{N}\left(\sum_{j=0}^{N}\mu_j\right)e_{N-n}^TA_{N-n}e_{N-n} \geq 0.$$

Hence both terms on the right side of the inequality in (4.4.7) are nonnegative which proves the result.

Assumption (4.4.6)(iii) is rather restrictive but allows nevertheless the construction of multipliers in many important cases as shall now be explained. Recall that a linear multistep method (4.1.4) is $A(\alpha)$-stable for $0 \leq \alpha \leq \pi/2$ iff

$$A(\alpha) \equiv \{\eta \in \mathbb{C}, |\arg\eta - \pi| \leq \alpha\} \cup \{\infty\} \subset S.$$

For $0 < \alpha \leq \pi/2$ this definition is equivalent to the classical one, $\mathring{A}(\alpha) \subset \mathring{S}$, introduced by Widlund [67] because for $\alpha > 0$ no points of $\bar{S} \setminus S$ can lie on the straight lines $\{\eta \in \mathbb{C}, \pm\arg\eta = \pi - \alpha\}$ or in $\eta = \infty$ by Appendix A.1. The first part of the following lemma includes for $\alpha = \pi/2$ Lemma (4.3.8) where we have proved that a method (4.1.4) is A-stable iff $\phi = \rho/\sigma$ is an A-function or, in other words, iff $|\zeta| \geq 1$ implies $|\arg\phi(\zeta)| \leq \pi/2$ or $\phi(\zeta) = \infty$.

<u>(4.4.8) Lemma</u>. *A linear multistep method* (4.1.4) *is A(α)-stable,* $0 < \alpha \leq \pi/2$,
(i) iff $\phi = \rho/\sigma$ *satisfies* $|\arg\phi(\zeta)| \leq \pi - \alpha$ *or* $\phi(\zeta) = \infty$ *for* $|\zeta| \geq 1$;
(ii) iff $A(\alpha) \cap \mathring{S} \neq \emptyset$ *and* $|\zeta| = 1$ *implies* $|\arg\phi(\zeta)| \leq \pi - \alpha$ *or* $\phi(\zeta) = \infty$.

Proof. The first assertion follows in the same way as Lemma (4.3.8). For the proof of the second assertion it suffices to show the A(α)-stability. Recall that the root locus curve $\eta = \rho(e^{i\theta})/\sigma(e^{i\theta})$, $0 \le \theta \le 2\pi$, divides the complex η-plane into several open and connected components $\mathfrak{C}_\nu$, $\nu = 1,\dots,r$, and $\mathfrak{C}_\nu \subset \mathring{S}$ iff $\mathfrak{C}_\nu \cap \mathring{S} \ne \emptyset$. Therefore we obtain $\mathring{A}(\alpha) \subset \mathfrak{C}_\nu \subset \mathring{S}$ for some ν if $|\arg\phi(e^{i\theta})| \le \pi - \alpha$ or $\phi(\zeta) = \infty$ for $0 \le \theta \le 2\pi$, and if $\mathring{A}(\alpha) \cap \mathring{S} \ne \emptyset$.

Especially we obtain from (4.4.8)(ii) that $\phi = \rho/\sigma$ is an A-function - and the method (4.1.4) is A-stable - iff

(4.4.9) $\{\eta \in \mathbb{C},\ \mathrm{Re}(\eta) < 0\} \cap \mathring{S} \ne \emptyset$

and

(4.4.10) $\mathrm{Re}(\phi(e^{i\theta})) \ge 0$ or $\phi(e^{i\theta}) = \infty$, $\quad \forall\ \theta \in \mathbb{R}$.

If the method (4.1.4) is convergent and strongly D-stable in $\eta = 0$, i.e., if $\rho'(1) = \sigma(1) \ne 0$ and $\rho(\zeta)/(\zeta - 1)$ has only roots of modulus less than one then condition (4.4.9) is fulfilled, and condition (4.4.10) is fulfilled with $\phi(e^{i\theta}) \ne \infty$ iff $\infty \in \mathring{S}$, i.e., iff $\sigma(\zeta)$ has only roots of modulus less than one. Hence, with respect to the function $(\chi\rho)/(\nu\sigma)$ we obtain the following corollary to Lemma (4.4.8).

<u>(4.4.11) Corollary</u>. *(i) Let the method* (4.1.4) *be convergent, strongly D-stable in* $\eta = 0$, *and let* $\infty \in \mathring{S}$.
(ii) Let the real polynomials ν *and* χ *have only roots of modulus less than one. Then* $\mu = \nu/\chi$ *is a multiplier for the method* (4.1.4) *iff*

(4.4.12) $\mathrm{Re}((\chi\rho)/(\nu\sigma)(e^{i\theta})) \ge 0 \quad \forall\ \theta \in \mathbb{R}$.

Notice that $\phi = (\chi\rho)/(\nu\sigma)$ has real coefficients hence $\mathrm{Re}(\phi(e^{i\theta})) \ge 0\ \forall\ \theta \in \mathbb{R}$ if this is true for $0 \le \theta \le \pi$. Moreover, as

$$\mathrm{Re}((\rho/\sigma\mu)(e^{i\theta})) = |\mu(e^{i\theta})|^{-2}\mathrm{Re}(\mu(e^{-i\theta})(\rho/\sigma)(e^{i\theta}))$$

and $\mu(\zeta) = \sum_{j=0}^{\infty}\mu_j\zeta^{-j}$, condition (4.4.12) is equivalent to

(4.4.13) $\mathrm{Re}[(\sum_{j=0}\mu_j e^{ij\theta})\frac{\rho}{\sigma}(e^{i\theta})] \ge 0 \quad 0 \le \theta \le \pi$.

<u>(4.4.14) Lemma</u>. *Let the method* (4.1.4) *and the multiplier* μ *satisfy the assumption of Corollary* (4.4.11) *then the method is A(α)-stable if*

(4.4.15) $|\arg\mu(e^{-i\theta})| \le \frac{\pi}{2} - \alpha \quad \forall\ \theta \in \mathbb{R}$.

Proof. Observe that (4.4.12) is equivalent to

$$|\arg((\rho/\sigma\mu)(e^{i\theta}))| \leq \pi/2 \qquad \forall\ \theta \in \mathbb{R}.$$

because $\arg((\rho/\sigma\mu)(e^{i\theta})) = \arg(\mu(e^{-i\theta})(\rho/\sigma)(e^{i\theta}))$ we obtain

$$-\frac{\pi}{2} - |\arg(\mu(e^{-i\theta}))| \leq \arg(\frac{\rho}{\sigma}(e^{i\theta})) \leq \frac{\pi}{2} + |\arg(\mu(e^{-i\theta}))|$$

or

$$-\pi + \alpha \leq \arg(\frac{\rho}{\sigma}(e^{i\theta})) \leq \pi - \alpha \qquad \forall\ \theta \in \mathbb{R}$$

and the assertion follows by Lemma (4.4.8)(ii).

This result shows that a method (4.1.4) is $A(\alpha)$-stable with large α if it has a multiplier μ with small $\arg(\mu(e^{-i\theta}))$. A converse of this fact is stated in the following lemma without proof.

<u>(4.4.16) Lemma</u>. (Nevanlinna and Odeh [81].) *Let the method* (4.1.4) *be $A(\alpha')$-stable and let it satisfy assumption* (4.4.11)(i). *Then there is a finitely supported multiplier μ satisfying assumption* (4.4.11)(ii) *and* (4.4.15) *for a* $0 < \alpha < \alpha'$.

The next result gives some more insight in the form of multipliers and is also stated without proof.

<u>(4.4.17) Lemma</u>. (Nevanlinna and Odeh [81].) *(i) There exists for every* k *a k-step method of order* p = k *with the multiplier* $\mu(\zeta) = 1 - \eta\zeta^{-1}$, $0 < \eta < 1$.
(ii) If $\mu(\zeta) = (\zeta - \omega)/(\zeta - \lambda)$ *is a multiplier for a method* (4.1.4) *of order* $p \geq 2$ *then* $\lambda \leq \omega$.
(iii) If the method (4.1.4) *satisfies assumption* (4.4.11)(i) *and*

$$\operatorname{Im}(\frac{\rho}{\sigma}(e^{i\theta})) > 0 \qquad 0 < \theta < \pi$$

then there exists a multiplier $\mu(\zeta) = (\zeta - \omega)/(\zeta - \lambda)$ *with* $-1 < \lambda \leq \omega < 1$.

Naturally, the error constant of the method can be large in (i) and the assumption in (iii) cannot replace assumption (iii) of Theorem (4.4.6) because it is not claimed here that λ is nonnegative.

By (4.4.13) a method (4.1.4) has the multiplier

$$(4.4.18)\qquad \mu(\zeta) = 1 - \omega\zeta^{-1}, \qquad 0 \leq \omega < 1,$$

if

$$(4.4.19)\quad \mathrm{Re}((1-\omega e^{i\theta})\tfrac{\rho}{\sigma}(e^{i\theta})) \geq 0, \qquad 0 \leq \theta \leq \pi.$$

If we introduce the *modified root locus curve*

$$(4.4.20)\quad \eta(\theta) = \mathrm{Re}(\tfrac{\rho}{\sigma}(e^{i\theta})) + i[\sin(\theta)\mathrm{Im}(\tfrac{\rho}{\sigma}(e^{i\theta})) - \cos(\theta)\mathrm{Re}(\tfrac{\rho}{\sigma}(e^{i\theta}))], \qquad 0 \leq \theta \leq \pi,$$

then this condition is fulfilled if

$$\mathrm{Re}(\eta(\theta)) + \omega\mathrm{Im}(\eta(\theta)) \geq 0, \qquad 0 \leq \theta \leq \pi,$$

or, in other words, if the line $\mathrm{Re}\eta = -\omega\mathrm{Im}\eta$ stays below the curve (4.4.20). By this way it can be easily checked whether a method (4.1.4) has a multiplier (4.4.18). Besides, as

$$|\arg(1-\omega e^{i\theta})| \leq \arctan(\omega/(1-\omega^2)^{1/2}) = \frac{\pi}{2} - \arccos\omega,$$

Lemma (4.4.14) says that a method (4.1.4) with the multiplier (4.4.18) is at least $A(\beta)$-stable with $\beta = \arccos\omega$. For the backward differentiation methods up to order 6 (cf. Appendix A.4) the following table is found in Nevanlinna and Odeh [81].

Table I: Backward Differentiation Methods

k	ω	arccosω	α
2	0	π/2	π/2
3	0.0836	85°13'	88°02'
4	0.2878	73°16'	73°21'
5	0.8160	35°19'	51°50'
6	5		17°50' .

The methods of Cryer [73] (cf. Appendix A.4) are k-step methods of order k defined uniquely by $\sigma(\zeta) = (\zeta + d)^k$. For $d = -1 + 2/(1 + 2^{k+1})$ we obtain the following table.

Table II: Cryer's Methods

k	ω	arccosω	α
2	0	π/2	π/2
3	0.3000	72°33'	88.8°
4	0.6046	52°48'	86.3°
5	0.7952	37°20'	83.6°
6	0.8979	26°07'	81.0°
7	0.9498	18°14'	78.5° .

The α-values in both tables are optimum and were computed by Norsett [69] and Jeltsch [76] respectively. The modified root locus curves are plotted in Appendix A.4.

Finally, by means of these data we can state the following consequence of Theorems (4.4.3) and (4.4.6):

(4.4.21) Corollary. *Let the (m,m)-matrix* $A(t)$ *in* (4.1.1) *be real symmetric and continuously differentiable, let* $A(t) \leq 0$ *and* $A'(t) \geq 0$; *and let the solution* y *be (k+1)-times continuously differentiable. Then the error bound of Theorem* (4.4.3) *holds*
(i) for the k-step backward differentiation methods of order k *up to order* k = 5,
(ii) for the k-step methods of order k *defined by* $\sigma(\zeta) = (\zeta + \beta)^k$ *and* $\beta = -1 + 2/(1 + 2^{k+1})$ *(Cryer's methods) up to order* k = 7.

For the backward differentiation methods up to order k = 6 a very similar result was proved by Gekeler [82a] in a different way.

V. Runge-Kutta Methods for Differential Systems of First Order

5.1. General Multistage Methods and Runge-Kutta Methods

Up to now we have considered linear multistep methods and multistep methods with higher derivatives in this volume. These methods suffer from two drawbacks if the step number k is greater than one: They need a special procedure for the computation of the initial values $v_1,\dots,v_{k-1}$ and a change of the step length during the calculation is complicated and deteriorates the otherwise favorable relation between exactness and computational effort per time step. Moreover, the computation of derivatives of f can be difficult for instance if f is given in tabular form. These disadvantages are avoided to some extent by the introduction of intermediate time steps in which the numerical approximation is then to be computed by additional recurrence equations and which on the other side also augments the computational amount of work.

Let us briefly recall that a multistep multiderivative method for the initial value problem

$$(5.1.1)\qquad y' = f(t,y),\ t > 0,\ y(0) = y_0,$$

is a device in which *one* unknown, the approximation v_{n+k} of $y((n+k)\Delta t)$, is computed in each time step by *one* recurrence equation,

$$(5.1.2)\qquad \textstyle\sum_{i=0}^{k}\sigma_i(\Delta t\Theta)T^i f_n^{(-1)}(v_n) = 0,\ \Theta = \partial/\partial t, \qquad n = 0,1,\dots\ .$$

If in this equation some of the translation operators

$$T^i\colon v_n \to v_{n+i} = v((n+i)\Delta t), \qquad i = 1,\dots,k,$$

say for $i = k-r+1,\dots,k$, $1 \le r < k$, are formally replaced by arbitrary translation operators,

$$T_j\colon v_n \to v_{n+\tau_j} = v((n+\tau_j)\Delta t),\ \tau_j \notin \mathbb{N}, \qquad j = 1,\dots,r,$$

then we obtain a (k-r)-step device of the form

$$(5.1.3)\qquad \textstyle\sum_{i=0}^{k-r}\sigma_i(\Delta t\Theta)T^i f_n^{(-1)}(v_n) + \sum_{j=1}^{r}\sigma_j^*(\Delta t\Theta)T_j f_n^{(-1)}(v_n) = 0, \qquad n = 0,1,\dots\ .$$

(Naturally, this is an entire new formula and the polynomials $\sigma_i(\eta)$ do no longer agree with those of (5.1.2).) Here we have to compute the unknown vectors

(5.1.4) v_{n+i}, $i = 1,\dots,k-r$, $T_j v_n = v_{n+\tau_j}$, $j = 1,\dots,r$,

in each time step and we must therefore add r further equations of the same type as (5.1.3) to the recurrence equation (5.1.3). In this context, r is called the stage number, and the complete k-step ℓ-derivative r-stage method can be written as

(5.1.5) $\sum_{i=0}^{k}\sigma_{i\nu}(\Delta t\Theta)T^i f_n^{(-1)}(v_n) + \sum_{j=0}^{r}\sigma_{j\nu}^*(\Delta t\Theta)T_j f_n^{(-1)}(v_n) = 0, \qquad \nu = 0,\dots,r;$

see e.g. Lambert [73 , Chap. 5] and Stetter [73 , Section 5.3]. The data of this method, i.e., the off-step points τ_j and the coefficients of the polynomials $\sigma_{i\nu}(n)$ and $\sigma_{j\nu}^*(n)$ are to be chosen in such a way that the method has the desired order of consistence and stability region S and, above all, that the vectors (5.1.4) are determined in a unique way.

Let now

$$t_{n,i} = (n + \tau_i)\Delta t,\ 0 \le \tau_i \le 1,\ v_{n,i} = v(t_{n,i}),\ f_{n,i}(v) = f(t_{n,i},v),\ i = 1,\dots,r,$$

then Runge-Kutta methods are single step single derivative multistage methods of the form

(5.1.6) $v_{n,i} = v_n + \Delta t\sum_{j=1}^{r}\alpha_{ij}f_{n,j}(v_{n,j}), \qquad i = 1,\dots,r,$

(5.1.7) $v_{n+1} = v_n + \Delta t\sum_{j=1}^{r}\beta_j f_{n,j}(v_{n,j}), \qquad n = 0,1,\dots ,$

with real coefficients α_{ij} and β_j. Obviously, in each time step the unknown vectors

$$k_{n,j} = f_{n,j}(v_{n,j}), \qquad j = 1,\dots,r,$$

are to be computed hence the scheme (5.1.6) and (5.1.7) is frequently written as

$$k_{n,i} = f(t_{n,i}, v_n + \Delta t\sum_{j=1}^{r}\alpha_{ij}k_{n,j}), \qquad i = 1,\dots,r,$$

$$y_{n+1} = y_n + \Delta t\sum_{j=1}^{r}\beta_j k_{n,j}, \qquad n = 0,1,\dots .$$

In this chapter we consider mainly the linear initial value problem

(5.1.8) $y' = A(t)y + c(t),\ t > 0,\ y(0) = y_0,$

with a (m,m)-matrix A(t) and introduce the following notations:

$\underline{A}_n = [A_{n,1},\dots, A_{n,r}]$ block diagonal matrix of block dimension r,

$\underline{c}_n = (c_{n,1},\ldots, c_{n,r})^T$, $\underline{v}_n = (v_{n,1},\ldots, v_{n,r})^T$ etc., and

$\underline{\tilde{v}}_n = (v_n,\ldots, v_n)^T$ block vectors of block dimension r,

$P = [\alpha_{ij}]_{i,j=1}^r$ (r,r)-matrix, $T = [\tau_1,\ldots, \tau_r]$ (r,r,)-diagonal matrix

$q = (\beta_1,\ldots, \beta_r)^T$, $z = (1,\ldots, 1)^T$ r-vectors,

$\underline{P} = [\alpha_{ij}I]_{i,j=1}^r$, $q^T\underline{v}_n = \sum_{j=1}^r \beta_j v_{n,j}$, $q^T\underline{P}z = \sum_{i,j=1}^r \beta_i \alpha_{ij} I z_j$ (m,m)-matrix.

A Runge-Kutta method is called *semi-implicit* or *explicit* if - possibly after a suitable permutation of rows and corresponding columns - the matrix P is lower triangular or strictly lower triangular the latter meaning that $\alpha_{ij} = 0$ for $i \leq j$. In the other cases the method is called *implicit*.

For the problem (5.1.8) the computational device (5.1.6), (5.1.7), i.e.,

$$v_{n,i} = v_n + \Delta t \sum_{j=1}^r \alpha_{ij}(A_{n,j} v_{n,j} + c_{n,j}), \qquad i = 1,\ldots,r,$$

$$v_{n+1} = v_n + \Delta t \sum_{j=1}^r \beta_j (A_{n,j} v_{n,j} + c_{n,j}), \qquad n = 0,1,\ldots,$$

can now be written in the following form,

$$(\underline{I} - \Delta t \underline{P}\underline{A}_n)\underline{v}_n = \underline{\tilde{v}}_n + \Delta t \underline{P}\underline{c}_n,$$

$$v_{n+1} = v_n + \Delta t q^T(\underline{A}_n \underline{v}_n + \underline{c}_n), \qquad n = 0,1,\ldots,$$

or, if $(\underline{I} - \Delta t \underline{P}\underline{A}_n)$ is invertible,

$$\text{(5.1.9)} \quad v_{n+1} = G(\Delta t A)_n v_n + r_n, \qquad n = 0,1,\ldots,$$

with the notations

$$\text{(5.1.10)} \quad G(\Delta t A)_n = I + \Delta t q^T \underline{A}_n (\underline{I} - \Delta t \underline{P}\underline{A}_n)^{-1} z$$

and

$$\text{(5.1.11)} \quad r_n = \Delta t q^T(\underline{I} + \Delta t \underline{A}_n (\underline{I} - \Delta t \underline{P}\underline{A}_n)^{-1}\underline{P})\underline{c}_n = \Delta t q^T (\underline{I} - \Delta t \underline{P}\underline{A}_n)^{-1}\underline{c}_n.$$

In particular, we obtain for the test equation $y' = \lambda y$

$$\text{(5.1.12)} \quad v_{n+1} = G(\Delta t \lambda) v_n, \qquad n = 0,1,\ldots,$$

where

$$\text{(5.1.13)} \quad G(\eta) = 1 + \eta q^T (I - \eta P)^{-1} z = -\sigma_0(\eta)/\sigma_1(\eta).$$

The real polynomials $\sigma_0(\eta)$ and $\sigma_1(\eta) = \det(I - \eta P)$ have degree not greater than r and $\sigma_1(\eta) = 1$ if the method is explicit.

The error estimation of Runge-Kutta methods differs entirely from that of multistep methods. Whereas in linear problems (5.1.8) with constant matrix A the Uniform Boundedness Theorem does not come to application here, suitable bounds for the discretization error are rather cumbersome to derive in the general case. Over a long period the discretization error was only estimated using the Landau symbolic and neglecting the influence of an ill-conditioned leading matrix A. However, due to Crouzeix [75] who closed this gap in the theory of numerical methods by his doctoral thesis we have today error bounds for Runge-Kutta methods at our disposal which are of the same efficiency as those for multistep methods at least as concerns linear initial value problems.

5.2. Consistence

Let y be again the solution of the initial value problem (5.1.1) and let $w_{n,i}$, $i = 1,\ldots, r$, be the solution of the system

(5.2.1) $$w_{n,i} = y_n + \Delta t\sum_{j=1}^{r}\alpha_{ij}f_{n,j}(w_{n,j}), \qquad i = 1,\ldots,r,$$

then the *discretization error* of the method (5.1.6), (5.1.7) is

$$d(\Delta t,y)_n = y_{n+1} - y_n - \Delta t\sum_{j=1}^{r}\beta_j f(t_{n,j},w_{n,j}), \qquad n = 0,1,\ldots \quad .$$

Because of the strong nonlinearity of the scheme (5.1.6), (5.1.7) the discretization error is defined here only for the exact solution y in opposition to multistep methods.

(5.2.2) Definition. *A Runge-Kutta method* (5.1.6), (5.1.7) *is consistent with a differential system* $y' = f(t,y)$, $t > 0$, *if there exist a positive integer* p *and a* $\Gamma > 0$ *not depending on* Δt *such that*

$$\|d(\Delta t,y)(t)\| \le \Gamma\Delta t^{p+1}$$

for every solution $y \in C^{p+1}(\mathbb{R}^+;\mathbb{R}^m)$ *of* $y' = f(t,y)$. *The maximum* p *is the order of the method.*

Every r-stage method (5.1.6), (5.1.7) can be associated with r+1 numerical integration formulae, namely

(5.2.3) $$\int_0^{\tau_i} f(t)dt \cong \sum_{j=1}^{r}\alpha_{ij}f(\tau_j), \qquad i = 1,\ldots,r,$$

(5.2.4) $\int_0^1 f(t)dt \cong \sum_{j=1}^r \beta_j f(\tau_j)$,

and it is convenient to introduce the following notations of which the first one is well-known and the second plays a particular role in the subsequent error estimation of ill-conditioned differential systems.

(5.2.5) Definition. *(i) A numerical integration formula* (5.2.4) *has order* ℓ *if it is exact for all polynomials of degree less than or equal* ℓ.
(ii) For $i = 1,\dots,r$ *let* ℓ_i *be the maximum order of* (5.2.3) *then*

$$\ell = \min_{1\le i\le r}\{\ell_i\}$$

is the degree of the Runge-Kutta method (5.1.6), (5.1.7).

The following lemma shows that consistence order p implies order p-1 of the formula (5.2.4) in the case of the trivial differential equation $y' = c(t)$ and so explains why in (5.2.5)(ii) the formula (5.2.4) is not included.

(5.2.6) Lemma. *If a method* (5.1.6), (5.1.7) *has order* p *for every differential equation* $y' = c(t)$, $c \in C^p(\mathbb{R}^+;\mathbb{R}^m)$, *then* (5.2.4) *has order* p-1,

(5.2.7) $q^T T^k z = \frac{1}{k+1}$, $\qquad k = 0,\dots,p-1.$

Proof. (Cf. Crouzeix and Raviart [80].) By assumption we have

(5.2.8) $d(\Delta t,y)_n = y_{n+1} - y_n - \Delta t\sum_{j=1}^r \beta_j c(\tau_j) = \mathcal{O}(\Delta t^{p+1})$ $\qquad c \in C^p(\mathbb{R}^+;\mathbb{R}^m)$.

On the other side, a Taylor expansion of $y' = c(t)$ provides

$$\int_{n\Delta t}^{(n+1)\Delta t} c(t) = y_{n+1} - y_n = \sum_{k=0}^{p-1} \frac{\Delta t^{k+1}}{(k+1)!} c_n^{(k)} + \mathcal{O}(\Delta t^{p+1})$$

and

$$\sum_{j=1}^r \beta_j c_{n,j} = \sum_{k=0}^{p-1}\sum_{j=1}^r \beta_j \frac{c_n^{(k)}}{k!}(t_{n,j} - n\Delta t)^k + \mathcal{O}(\Delta t^p) = \sum_{k=0}^{p-1}\Big(\sum_{j=1}^r \beta_j \tau_j^k\Big)\frac{\Delta t^k}{k!} c_n^{(k)} + \mathcal{O}(\Delta t^p).$$

A substitution of these representations into (5.2.8) yields

$$d(\Delta t,y)_n = \sum_{k=0}^{p-1}\Big(\frac{1}{k+1} - \sum_{j=1}^r \beta_j \tau_j^k\Big)\frac{\Delta t^{k+1}}{k!} + \mathcal{O}(\Delta t^{p+1}) = \mathcal{O}(\Delta t^{p+1})$$

which proves the assertion.

Because of this result *we henceforth assume that every Runge-Kutta method of order* p *satisfies* (5.2.7). In particular we obtain for k = 0

$$\sum_{j=1}^{r} \beta_j = 1$$

which is the well-known necessary and sufficient condition for the consistence if $f(t,y) \not\equiv 0$ in (5.1.1). Recall that consistence implies convergence here in the classical meaning without further conditions if the basic problem (5.1.1) is well-conditioned; see e.g. Henrici [62 , Theorem 2.1] and Grigorieff [72].

The next result provides *sufficient* conditions for order p with respect to the general problem (5.1.1). It has been proved by Butcher [64] in an algebraic way and by Crouzeix [75 , Theorem 1.2] in a more analytic way.

(5.2.9) Theorem. *If the following three conditions are fulfilled:*
(i) The integration formula (5.2.4) *has order* p-1,
(ii) the integration formulae (5.2.3) *have order* k-1,

$$\sum_{j=1}^{r} \alpha_{ij}\tau_j^{\kappa} = \frac{\tau_i^{\kappa+1}}{\kappa+1}, \qquad \kappa = 0,\dots,k-1,\ i = 1,\dots,r,$$

(iii)

$$\sum_{j=1}^{r} \alpha_{ij}\beta_j\tau_j^{\lambda} = \frac{1}{\lambda+1}\beta_j(1-\tau_i^{\lambda+1}), \qquad \lambda = 0,\dots,\ell-1,\ i = 1,\dots,r,$$

then the method (5.1.6), (5.1.7) *has order*

$$p = \min\{k+\ell+1,\ 2k+2\}.$$

for the differential system (5.1.1) *with* $y \in C^{p+1}(\mathbb{R}^+;\mathbb{R}^m)$.

In particular, we obtain for $\ell = 0$ and $k = p-1$:

(5.2.10) Corollary. *If the integration formula* (5.2.4) *has order* p-1 *and the formulae* (5.2.3) *have order* p-2 *then the method* (5.1.6), (5.1.7) *has order* p *for the general problem* (5.1.1).

A further consequence of Theorem (5.2.9) is due to Crouzeix [75 , Corollary 1.2]:

(5.2.11) Corollary. *Let* r* *be the number of the different intermediate time steps* τ_i, $i = 1,\dots,r$, *let the formulae* (5.2.3) *be of order* r*-1, *and let the formula* (5.2.4) *be of order p-1. Then the method* (5.1.6), (5.1.7) *has order* p *for the general problem* (5.1.1).

With respect to the general *linear* problem (5.1.8) Crouzeix [75 , Theorem 1.3] has deduced *necessary and sufficient* conditions for consistence order p; cf. also Crouzeix and Raviart [80 , Theorem 3.1]. In order to present this result we introduce some further notations. Let

(5.2.12) $d_0(\Delta t,y)_n = y_{n+1} - y_n - \Delta t\sum_{j=1}^{r}\beta_j y'_{n,j}$,

(5.2.13) $d_i(\Delta t,y)_n = y_{n,i} - y_n - \Delta t\sum_{j=1}^{r}\alpha_{ij} y'_{n,j}$, $\quad i = 1,\dots,r$,

and

$$\underline{d}(\Delta t,y)_n = (d_1(\Delta t,y)_n,\dots, d_r(\Delta t,y)_n)^T.$$

For the linear problem (5.1.8) we can write

$$y_{n,i} = y_n + \Delta t\sum_{j=1}^{r}\alpha_{ij}(A_{n,j}y_{n,j} + c_{n,j}) + y_{n,i} - y_n - \Delta t\sum_{j=1}^{r}\alpha_{ij}(A_{n,j}y_{n,j} + c_{n,j})$$

then, by (5.2.1),

$$y_{n,i} - w_{n,i} = \Delta t\sum_{j=1}^{r}\alpha_{ij}A_{n,j}(y_{n,j} - w_{n,j}) + d_i(\Delta t,y)_n$$

and accordingly, if $\underline{I} - \Delta t\underline{\underline{PA}}_n$ is regular,

$$\underline{y}_n - \underline{w}_n = (\underline{I} - \Delta t\underline{\underline{PA}}_n)^{-1}\underline{d}_n.$$

For the discretization error $d(\Delta t,y)_n$ we thus obtain

$$\begin{aligned} d(\Delta t,y)_n &= y_{n+1} - y_n - \Delta t\sum_{j=1}^{r}\beta_j(A_{n,j}w_{n,j} + c_{n,j}) \\ &= d_0(\Delta t,y)_n + \Delta t\sum_{j=1}^{r}\beta_j A_{n,j}(y_{n,j} - w_{n,j}) \qquad (5.2.14) \\ &= d_0(\Delta t,y)_n + \Delta t q^T\underline{A}_n(\underline{I} - \Delta t\underline{\underline{PA}}_n)^{-1}\underline{d}(\Delta t,y)_n. \end{aligned}$$

A substitution of the Taylor expansion

$$y'(t) = \sum_{k=0}^{p-1}\frac{(t - n\Delta t)^k}{k!}y_n^{(k+1)} + \int_{n\Delta t}^{t}\frac{(\tau - n\Delta t)^{p-1}}{(p-1)!}y^{(p+1)}(\tau)d\tau$$

into (5.2.12) yields

$$d_0(\Delta t,y)_n = \sum_{k=0}^{p-1}\left[\int_{n\Delta t}^{(n+1)\Delta t}\frac{(\tau - n\Delta t)^k}{k!}d\tau - \Delta t\sum_{j=1}^{r}\beta_j\frac{(t_{n,j} - n\Delta t)^k}{k!}\right]y_n^{(k+1)}$$
$$+ \int_{n\Delta t}^{(n+1)\Delta t}\int_{n\Delta t}^{\sigma}\frac{(\tau - n\Delta t)^{p-1}}{(p-1)!}y^{(p+1)}(\tau)d\tau d\sigma - \Delta t\sum_{j=1}^{r}\beta_j\int_{n\Delta t}^{t_{n,j}}\frac{(\tau - n\Delta t)^{p-1}}{(p-1)!}y^{(p+1)}(\tau)d\tau$$

or

$$(5.2.15)\quad d_0(\Delta t,y)_n = \sum_{k=0}^{p-1}\left[\frac{1}{k+1} - \sum_{j=1}^{r}\beta_j\tau_j^k\right]\frac{\Delta t^{k+1}}{k!}y_n^{(k+1)} + \delta_{0,n}$$

where

$$|\delta_{0,n}| = \frac{\Delta t^p}{(p-1)!}\left|\int_0^1\int_0^\sigma \tau^{p-1}y^{(p+1)}(n\Delta t+\tau)d\tau d\sigma - \sum_{j=1}^r\beta_j\int_0^{\tau_j}\tau^{p-1}y^{(p+1)}(n\Delta t+\tau)d\tau\right|$$

$$\le \frac{\Delta t^p}{(p-1)!}(1+\sum_{j=1}^r|\beta_j|)\int_0^1|y^{(p+1)}(n\Delta t+\tau)|d\tau \le \Gamma\Delta t^p\int_{n\Delta t}^{(n+1)\Delta t}|y^{(p+1)}(\tau)|d\tau.$$

The stipulated condition (5.2.7) implies that the bracketed terms in the sum of (5.2.15) are zero hence

$$(5.2.16)\quad |d_0(\Delta t,y)_n| \le |\delta_{0,n}|.$$

In the same way as (5.2.15) we find

$$(5.2.17)\quad d_i(\Delta t,y)_n = \sum_{k=0}^{p-1}\left[\frac{\tau_i^{k+1}}{k+1} - \sum_{j=1}^r\alpha_{ij}\tau_j^k\right]\frac{\Delta t^{k+1}}{k!}y_n^{(k+1)} + \delta_{i,n}, \qquad i = 1,\dots,r,$$

where the $|\delta_{i,n}|$ have the same bound as $|\delta_{0,n}|$, i.e.,

$$(5.2.18)\quad \max_{0\le i\le r}|\delta_{i,n}| \le \Gamma\Delta t^p\int_{n\Delta t}^{(n+1)\Delta t}|y^{(p+1)}(\tau)|d\tau.$$

We substitute (5.2.17) into (5.2.14) and obtain

$$(5.2.19)\quad \begin{aligned} d(\Delta t,y)_n &= d_0(\Delta t,y)_n + \sum_{k=1}^p q^T\underline{A}_n(\underline{I} - \Delta t\underline{PA}_n)^{-1}(\tfrac{1}{k}\underline{T}^k - \underline{PT}^{k-1})z\frac{\Delta t^{k+1}}{(k-1)!}y_n^{(k)} \\ &+ \Delta t q^T\underline{A}_n(\underline{I} - \Delta t\underline{PA}_n)^{-1}\underline{\delta}_n. \end{aligned}$$

But

$$\underline{A}_n(\underline{I} - \Delta t\underline{PA}_n)^{-1} = (\underline{I} - \Delta t\underline{A}_n\underline{P})^{-1}\underline{A}_n$$

and if the method has degree p^*-2 then the numerical integration formulas (5.2.3) are exact for the functions $x \to x^k$, $k = 0,\dots,p^*-2$ which yields

$$\frac{1}{k}T^k - PT^{k-1} = 0, \qquad k = 1,\dots,p^*-1.$$

Therefore a substitution of the bounds (5.2.16) and (5.2.18) into (5.2.19) leads to

$$(5.2.20)\quad \begin{aligned} |d(\Delta t,y)_n| &\le \Gamma(1 + |(\underline{I} - \Delta t\underline{A}_n\underline{P})^{-1}\Delta t\underline{A}_n|)\Delta t^p\int_{n\Delta t}^{(n+1)\Delta t}|y^{(p+1)}(\tau)|d\tau \\ &+ \sum_{k=p^*}^p\frac{\Delta t^{k+1}}{(k-1)!}|u_{k,n}| \end{aligned}$$

with the abbreviating notation

(5.2.21) $u_{k,n} \equiv q^T(\underline{I} - \Delta t \underline{A}_n \underline{P})^{-1} \underline{A}_n (\frac{1}{k}\underline{T}^k - \underline{PT}^{k-1}) z y_n^{(k)}$, $\qquad k = p^*,\dots,p.$

(5.2.22) Theorem. (Crouzeix [75].) *In* (5.1.8) *let* $A \in C^p(\mathbb{R}^+;\mathbb{R}^{m,m})$, $c \in C^p(\mathbb{R}^+;\mathbb{R}^m)$ *and let* $\underline{I} - \Delta t \underline{PA}_n$ *be regular for* $t > 0$ *and* $0 < \Delta t \le \Delta t_0$. *Then the method* (5.1.6), (5.1.7) *has order* p *for the linear problem* (5.1.8) *and* $0 < \Delta t \le \Delta t_0$ *iff*

(5.2.23)
$$\forall \ell \in \mathbb{N}\ \forall \lambda_j \in \mathbb{N} \cup \{0\}\ \sum_{j=1}^{\ell} \lambda_j \le p - \ell \implies$$
$$\Phi(\lambda;\ell) \equiv q^T T^{\lambda_1} P T^{\lambda_2} P \dots P T^{\lambda_\ell} z = \prod_{i=1}^{\ell} [(\ell - i + 1) + \sum_{j=i}^{\ell} \lambda_j]^{-1}$$

where $\Phi(\lambda;1) = q^T T^{\lambda_1} z$ *and* $\Phi(\lambda;2) = q^T T^{\lambda_1} P T^{\lambda_2} z$. *For* $m \ge 2p$ *this condition is also necessary.*

By (5.2.20) and (5.2.21) a Runge-Kutta method has order p iff (5.2.7) holds and in (5.2.21)

(5.2.24) $|q^T(\underline{I} - \Delta t \underline{A}_n \underline{P})^{-1} \underline{A}_n (\frac{1}{k}\underline{T}^k - \underline{PT}^{k-1}) z| \le \Gamma \Delta t^{p-k}$

where Γ does not depend on Δt. (5.2.7) is (5.2.23) for $\ell = 1$. For the proof of the equivalence of (5.2.23) for $\ell > 1$ and (5.2.24) we refer the reader to Crouzeix and Raviart [80 , Theorem 3.1]. Instead we prove below Theorem (5.2.22) for linear problems with constant matrix A (Corollary (5.2.27)). But let us quote before a further result of Crouzeix [75] without proof concerning the general nonlinear problem (5.1.1):

(5.2.25) Theorem. (Crouzeix [75].) *Let the solution* y *of the problem* (5.1.1) *be (p+1)-times continuously differentiable then the method* (5.1.6), (5.1.7) *has order* p *for this problem and sufficiently small* Δt *if* (5.2.23) *is fulfilled and*

(5.2.26) $PT^{\ell} z = \frac{1}{\ell+1} T^{\ell+1} z$, $\qquad \ell = 0,\dots,[\frac{p-3}{2}].$

The conditions (5.2.23) *are necessary. If* $\beta_j > 0$, $j = 1,\dots,r$, *in* (5.1.7) *then the conditions* (5.2.26) *are necessary, too.*

(5.2.27) Corollary. *Let the assumptions of Theorem* (5.2.22) *be fulfilled but let the matrix* A *be* constant. *Then the method* (5.1.6), (5.1.7) *has order* p *for the linear problem* (5.1.8) *and* $0 < \Delta t \le \Delta t_0$ *iff*

(5.2.28) $\forall k,\ell \in \mathbb{N} \cup \{0\}\quad \ell \le k \le p-1 \implies q^T P^{k-\ell} T^{\ell} z = \prod_{i=\ell}^{k} [i + 1]^{-1}.$

Proof. After the above preliminaries we have only to verify the equivalence of (5.2.7) and (5.2.24) with (5.2.28) for $\underline{A}_n = A\underline{I}$. A substitution of

$$(\underline{I} - \Delta t A\underline{P})^{-1} = \sum_{\lambda=0}^{\ell}(\Delta t A\underline{P})^{\lambda} + \mathcal{O}(\Delta t^{\ell+1})$$

into (5.2.24) yields the condition

$$\sum_{\ell=0}^{p-k-1} \Delta t^{\ell} q^T A(A\underline{P})^{\ell}(\tfrac{1}{k}\underline{T}^k - \underline{PT}^{k-1})z = \mathcal{O}(\Delta t^{p-k}), \qquad k = 1,\dots,p-1,$$

or

$$(5.2.29)\quad q^T P^k(\tfrac{1}{\ell}T^{\ell} - PT^{\ell-1})z = 0, \qquad k \in \mathbb{N} \cup \{0\},\ \ell \in \mathbb{N},\ k+\ell \le p-1,$$

or

$$(5.2.30)\quad q^T P^{k-\ell}(\tfrac{1}{\ell}T^{\ell} - PT^{\ell-1})z = 0, \qquad 1 \le \ell \le k \le p-1.$$

(5.2.28) implies (5.2.30) and, for $k = \ell$, (5.2.7). On the other side, (5.2.7) and (5.2.29) define recursively $q^T P^{k-\ell} T^{\ell} z$ for $k,\ell \in \mathbb{N} \cup \{0\}$, $0 \le \ell \le k \le p-1$, in a unique way. This proves the assertion.

In order to derive bounds for the vectors $u_{k,n}$ defined by (5.2.21) we substitute

$$(\underline{I} - \Delta t\underline{A}_n\underline{P})^{-1} = \sum_{j=0}^{p-k-1}(\Delta t\underline{A}_n\underline{P})^j + (\underline{I} - \Delta t\underline{A}_n\underline{P})^{-1}(\Delta t\underline{A}_n\underline{P})^{p-k}$$

and obtain

$$(5.2.31)\quad \begin{aligned} u_{k,n} &= \sum_{j=0}^{p-k-1} \Delta t^j q^T(\underline{A}_n\underline{P})^j \underline{A}_n(\tfrac{1}{k}\underline{T}^k - \underline{PT}^{k-1})zy_n^{(k)} \\ &+ \Delta t^{p-k} q^T(\underline{I} - \Delta t\underline{A}_n\underline{P})^{-1}(\underline{A}_n\underline{P})^{p-k}\underline{A}_n(\tfrac{1}{k}\underline{T}^k - \underline{PT}^{k-1})zy_n^{(k)}. \end{aligned}$$

Recall that $\underline{A}_n$ is a block diagonal matrix with the diagonal elements $A_{n,j} = A((n+\tau_j)\Delta t)$ and that in this context $\underline{P}$ is a block matrix with the elements $\alpha_{ij}I$ and $\underline{T}$ is a block diagonal matrix with the diagonal elements $\tau_i I$. We consider two cases:

(i) If $A(t) = A$ is a constant matrix then $\underline{A}_n = A\underline{I}$ commutes with $\underline{P}$ and $\underline{T}$ and we obtain

$$\begin{aligned} u_{k,n} &= \sum_{j=0}^{p-k-1} \Delta t^j q^T \underline{P}^j(\tfrac{1}{k}\underline{T}^k - \underline{PT}^{k-1})z\, A^{j+1}y_n^{(k)} \\ &+ \Delta t^{p-k} q^T(\underline{I} - \Delta tA\underline{P})^{-1}\underline{P}^{p-k}(\tfrac{1}{k}\underline{T}^k - \underline{PT}^{k-1})zA^{p-k+1}y_n^{(k)}. \end{aligned}$$

As the method is supposed to have order p the sum on the right side disappears by (5.2.26) hence we have in this case

$$(5.2.32)\quad |u_{k,n}| \le |(\underline{I} - \Delta tA\underline{P})^{-1}|\Delta t^{p-k}|A^{p+1-k}y_n^{(k)}|.$$

(ii) If A(t) varies with time then we can write $(\underline{A}_n\underline{P})^j\underline{A}_n$ as a function of the step length Δt,

$$(\underline{A}_n\underline{P})^j\underline{A}_n = [(\underline{A}_n\underline{P})^j\underline{A}_n](\Delta t).$$

<u>(5.2.33) Corollary</u>. *If the method* (5.1.6), (5.1.7) *has order* p *then*

$$q^T(\underline{A}_n\underline{P})^j\underline{A}_n(\tfrac{1}{k}\underline{T}^k - \underline{PT}^{k-1})zy_n^{(k)}$$
$$= q^T\int_0^{\Delta t}\frac{(\Delta t-\tau)^{p-k-j-1}}{(p-k-j-1)!}[(\underline{A}_n\underline{P})^j\underline{A}_n]^{(p-k-j)}(\tau)d\tau(\tfrac{1}{k}\underline{T}^k - \underline{PT}^{k-1})zy_n^{(k)}, \quad j+k \le p-1.$$

Proof. We have

$$[(\underline{A}_n\underline{P})^j\underline{A}_n]^{(i)}(\Delta t) = \sum_{\substack{\lambda_1+\cdots+\lambda_{j+1}=i\\ \lambda_\nu\in\mathbb{N}\cup\{0\}}}\frac{i!}{\lambda_1!\cdots\lambda_{j+1}!}\underline{T}^{\lambda_1}\underline{A}_n^{(\lambda_1)}\underline{PT}^{\lambda_2}\underline{A}_n^{(\lambda_2)}\cdots\underline{PT}^{\lambda_{j+1}}\underline{A}_n^{(\lambda_{j+1})}$$

and

$$[(\underline{A}_n\underline{P})^j\underline{A}_n]^{(i)}(0) = \sum_{\substack{\lambda_1+\cdots+\lambda_{j+1}=i\\ \lambda_\nu\in\mathbb{N}\cup\{0\}}}\frac{i!}{\lambda_1!\cdots\lambda_{j+1}!}\underline{A}_n^{(\lambda_1)}\cdots\underline{A}_n^{(\lambda_{j+1})}\underline{T}^{\lambda_1}\underline{PT}^{\lambda_2}\underline{P}\cdots\underline{PT}^{\lambda_{j+1}}.$$

As the method is supposed to be of order p, (5.2.23) yields for $i+k \le p-j-1$

$$q^T T^{\lambda_1}PT^{\lambda_2}\cdots PT^{\lambda_{j+1}}(\tfrac{1}{k}T^k - PT^{k-1})z = 0$$

and thus we have

$$q^T[(\underline{A}_n\underline{P})^j\underline{A}_n]^{(i)}(0)(\tfrac{1}{k}\underline{T}^k - \underline{PT}^{k-1})z = 0, \qquad i+k \le p-j-1,$$

hence a Taylor expansion,

$$(\underline{A}_n\underline{P})^j\underline{A}_n(\Delta t) = \sum_{i=0}^{p-k-j-1}\frac{\Delta t^i}{i!}[(\underline{A}_n\underline{P})^j\underline{A}_n]^{(i)}(0) + \int_0^{\Delta t}\frac{(\Delta t-\tau)^{p-k-j-1}}{(p-k-j-1)!}[(\underline{A}_n\underline{P})^j\underline{A}_n]^{(p-k-j)}(\tau)d\tau,$$

proves the assertion.

Now, (5.2.31) yields by Corollary (5.2.33)

$$|u_{k,n}| \le \Gamma\Big[\Delta t^{p-k}\sum_{j=0}^{p-k-1}\max_{0\le\tau\le\Delta t}|[(\underline{A}_n\underline{P})^j\underline{A}_n]^{(p-k-j)}(\tau)(\tfrac{1}{k}\underline{T}^k - \underline{PT}^{k-1})zy_n^{(k)}$$
$$+ \Delta t^{p-k}|(\underline{I} - \Delta t\underline{A}_n\underline{P})^{-1}||(\underline{A}_n\underline{P})^{p-k}\underline{A}_n(\tfrac{1}{k}\underline{T}^k - \underline{PT}^{k-1})zy_n^{(k)}|\Big]$$

or

(5.2.34)
$$|u_{k,n}| \le \Gamma\max\{1, |(\underline{I} - \Delta t\underline{A_n}\underline{P})^{-1}|\}\Delta t^{p-k}$$
$$\times \sum_{j=0}^{p-k}\max_{0\le\tau\le\Delta t}|[(\underline{A_n}\underline{P})^j\underline{A_n}]^{(p-k-j)}(\tau)(\tfrac{1}{k}\underline{T}^k - \underline{PT}^{k-1})zy_n^{(k)}|.$$

If $A(t) = a(t)A$ in (5.1.8) with a scalar-valued function a and $\underline{a_n} = (a_{n,1}I,\dots,a_{n,r}I)$ denotes the diagonal matrix with the elements $a_{n,j}I = a((n+\tau_j)\Delta t)I$ then

$$[(\underline{A_n}\underline{P})^j\underline{A_n}]^{(p-k-j)}(\tau)(\tfrac{1}{k}\underline{T}^k - \underline{PT}^{k-1})zy_n^{(k)} = [(\underline{a_n}\underline{P})^j\underline{a_n}]^{(p-k-j)}(\tau)(\tfrac{1}{k}\underline{T}^k - \underline{PT}^{k-1})zA^{j+1}y_n^{(k)}$$

and thus we obtain in this case

(5.2.35)
$$|u_{k,n}| \le \Gamma\max\{1, |(\underline{I} - \Delta t\underline{A_n}\underline{P})^{-1}|\}\Delta t^{p-k}$$
$$\times \max_{0\le i\le p-k+1}\max_{0\le j\le p-k}|||(a^i)^{(j)}|||_{n+1}\sum_{j=0}^{p-k}|A^{j+1}y_n^{(k)}|.$$

We substitute the bounds (5.2.34), (5.2.35), and (5.2.32) successively into (5.2.20) and assemble the result in the following lemma.

<u>(5.2.36) Lemma</u>. (Crouzeix [75].) *(i) Let the (m,m)-matrix* $A(t)$ *in* (5.1.8) *and the solution* y *be respectively p-times and (p+1)-times continuously differentiable.*
(ii) Let the method (5.1.6), (5.1.7) *be of order* p *for this problem and of degree* $p^* - 2$.
(iii) Let $\underline{I} - \Delta t\underline{A_\nu}\underline{P}$ *be regular for* $0 < \Delta t \le \Delta t_0$ *and* $\nu = 0,\dots,n$, *and let*

(5.2.37) $$\Gamma_{\Delta tA,n} \equiv \Gamma\max\{1 + |(\underline{I} - \Delta t\underline{A_n}\underline{P})^{-1}\Delta t\underline{A_n}|, |(\underline{I} - \Delta t\underline{A_n}\underline{P})^{-1}|\}$$

where Γ *is a generic positive constant depending only on the data of the method. Then, for* $0 < \Delta t \le \Delta t_0$ *and* $n = 0,1,\dots,$

$$|d(\Delta t,y)_n| \le \Gamma_{\Delta tA,n}\Delta t^p[\int_{n\Delta t}^{(n+1)\Delta t}|y^{(p+1)}(\tau)|d\tau + \Delta t\sum_{k=p^*}^{p}w_k(\Delta t,y)_n]$$

where

(5.2.38) $$w_k(\Delta t,y)_n = \sum_{j=0}^{p-k}\max_{0\le\tau\le\Delta t}|[(\underline{A_n}\underline{P})^j\underline{A_n}]^{(p-k-j)}(\tau)(\tfrac{1}{k}\underline{T}^k - \underline{PT}^{k-1})zy_n^{(k)}|,$$

$$w_k(\Delta t,y)_n = \max_{0\le i\le p-k+1}\ \max_{0\le j\le p-k}|||(a^i)^{(j)}|||_{n+1}\sum_{j=0}^{p-k}|A^{j+1}y_n^{(k)}|$$

if $A(t) = a(t)A$, *and*

$$w_k(\Delta t,y)_n = |A^{p+1-k}y_n^{(k)}|$$

if $A(t) = A$ *is a constant matrix.*

5.3. Error Bounds for Stable Linear Differential Systems

A general Runge-Kutta method of stage r and a general single step r-derivative method do not differ from each other for the homogeneous differential system $y' = Ay$ with constant matrix A, they both have the form

$$\sigma_1(\Delta tA)v_{n+1} + \sigma_0(\Delta tA)v_n = 0, \qquad n = 0,1,\ldots,$$

$\sigma_0(\eta)$ and $\sigma_1(\eta)$ being polynomials of degree not greater than r; cf. (5.2.12) and (5.2.13). The region of absolute stability S is therefore defined for Runge-Kutta methods in the same way as in Chapter I, Definition (1.2.7). However, the characteristic polynomial is now always linear in ζ,

$$\pi(\zeta,\eta) = \sigma_1(\eta)\zeta + \sigma_0(\eta),$$

hence the stability region S is always closed in $\overline{\mathbb{C}}$.

In this section we consider the linear initial value problem

$$(5.3.1)\quad y' = A(t)y + c(t) + h(t),\ t > 0,\ y(0) = y_0,$$

with the perturbation h(t) which is again omitted in the numerical approximation, (5.1.6), (5.1.7), as in the former chapters. Writing the Runge-Kutta method in the form (5.1.9) we then obtain the following equation for the error $e_n = y_n - v_n$ by definition of the discretization error $d(\Delta t,y)_n$ and by (5.1.11),

$$(5.3.2)\quad e_n = G(\Delta tA)_{n-1}e_{n-1} + d(\Delta t,y)_{n-1} + \Delta tq^T(\underline{I} - \Delta t\underline{PA}_n)^{-1}\underline{h}_{n-1}, \qquad n = 1,2,\ldots .$$

If the 'constant' $\Gamma_{\Delta tA,n}$ defined by (5.2.37) is modified slightly,

$$(5.3.3)\quad \Gamma^*_{\Delta tA,n} = \max\{\Gamma_{\Delta tA,n}, |(\underline{I} - \Delta t\underline{PA}_n)^{-1}|\}$$

then Lemma (5.2.36) yields immediately

$$(5.3.4)\quad \begin{aligned} |e_n| &\le |G(\Delta tA)_{n-1}||e_{n-1}| \\ &+ \Gamma^*_{\Delta tA,n-1}\Big(\Delta t^p \int_{(n-1)\Delta t}^{n\Delta t} |y^{(p+1)}(\tau)|d\tau + \Delta t^{p+1}\textstyle\sum_{k=p^*}^{p} w_k(\Delta t,y)_{n-1} + \Delta t|||h|||_n\Big). \end{aligned}$$

For a constant and diagonable matrix $A(t) = A = X\Lambda X^{-1}$ we can write instead of (5.3.2)

$$(5.3.5)\quad X^{-1}e_n = G(\Delta t\Lambda)X^{-1}e_{n-1} + X^{-1}d(\Delta t,y)_{n-1} + \Delta tq^T(\underline{I} - \Delta t\underline{P\Lambda})^{-1}X^{-1}\underline{h}_{n-1},\ n = 1,2,\ldots,$$

and a slight modification of Lemma (5.2.36) yields

$$|X^{-1}d(\Delta t,y)_n + \Delta t q^T(\underline{I} - \Delta t\underline{P}\Lambda)^{-1}X^{-1}\underline{h}_n|$$

$$\leq \Gamma^*_{\Delta t\Lambda}[|X^{-1}|(\Delta t^p \int_{n\Delta t}^{(n+1)\Delta t} |y^{(p+1)}(\tau)|d\tau + \Delta t|||h|||_n) + \Delta t^{p+1}\textstyle\sum_{k=p^*}^{p}|X^{-1}A^{p+1-k}y_n^{(k)}|].$$

Recall that $G(\eta) = -\sigma_0(\eta)/\sigma_1(\eta)$ and that $\Lambda = [\lambda_1,\ldots,\lambda_m]$ is the diagonal matrix of the eigenvalues of A. Therefore we have for $Sp(\Delta tA) \subset S$

$$|G(\Delta t\Lambda)| = \max_{1\leq\mu\leq m}|G(\Delta t\lambda_\mu)| \leq \sup_{\eta\in S}|G(\eta)| \leq 1,$$

and (5.3.5) yields

$$|X^{-1}e_n| \leq |X^{-1}e_0| + \Gamma^*_{\Delta t\Lambda}\Big[|X^{-1}|(\Delta t^p\int_0^{n\Delta t}|y^{(p+1)}(\tau)|d\tau + n\Delta t|||h|||_n) + n\Delta t^{p+1}\textstyle\sum_{k=p^*}^{p}|||X^{-1}A^{p+1-k}y_n^{(k)}|||\Big]. \tag{5.3.6}$$

So we are faced with two problems here: an estimation of $\Gamma^*_{\Delta tA,n}$ and, in the general case (5.3.2), with an estimation of $|G(\Delta tA)_n|$.

We begin with a collection of estimations of $\Gamma^*_{\Delta tA,n}$ under different assumptions. The first one makes no assumptions on the matrix P.

<u>(5.3.7) Lemma</u>. *Let* $P^* = [|\alpha_{ij}|]_{i,j=1}^r$ *and let* $|A(t)| \leq \Gamma$ *for* $t > 0$ *such that* $\Gamma spr(P^*) < 1$. *Then* $\Gamma^*_{A,n} \leq \kappa_\Gamma$.

Proof. Cf. also Crouzeix [75 , Proposition 3.3]. As $|A(t)|$ is bounded it suffices to consider $(\underline{I} - \underline{A}(t)\underline{P})^{-1}$. Let $U = (u_1,\ldots,u_r)^T$, $u_j \in \mathbb{C}^m$, and let

$$W = \underline{A}(t)\underline{P}U.$$

Then we have $w_i = \sum_{j=1}^r A_{t,j}\alpha_{ij}u_j$, $i = 1,\ldots,r$, and hence

$$|w_i| \leq \textstyle\sum_{j=1}^r |A_{t,j}||\alpha_{ij}||u_j| \leq \Gamma\sum_{j=1}^r|\alpha_{ij}||u_j|. \tag{5.3.8}$$

If $|U|_\infty = \max_{1\leq i\leq r}|u_i|$ and $|\underline{Q}|_\infty$ is the associated matrix norm then we obtain by this way recursively

$$|(\underline{A}(t)\underline{P})^n|_\infty \leq \Gamma^n\|(P^*)^n\|_\infty.$$

By e.g. Stoer and Bulirsch [80 , Theorem 6.9.2] there exists for P^* and $\varepsilon > 0$ a norm $\|\cdot\|_*$ such that $\|P^*\|_* \leq spr(P^*) + \varepsilon$ and $\|Q\|_\infty \leq \kappa\|Q\|_*$ for all (r,r)-matrices Q. Accordingly,

$$|(\underline{A}(t)\underline{P})^n|_\infty \le \Gamma^n \|(P^*)^n\|_\infty \le \kappa\Gamma^n\|(P^*)^n\|_* \le \kappa\Gamma^n\|P^*\|_*^n = \kappa\Gamma^n(\mathrm{spr}(P^*) + \varepsilon)^n.$$

Choosing ε sufficiently small such that $\chi = \Gamma(\mathrm{spr}(P^*) + \varepsilon) < 1$ we thus find

$$|(\underline{I} - \underline{A}(t)\underline{P})^{-1}|_\infty \le \lim_{n\to\infty}\sum_{\nu=0}^n |(\underline{A}(t)\underline{P})^n|_\infty \le \kappa\lim_{n\to\infty}\sum_{\nu=0}^n \chi^\nu \le \kappa(1 - \Gamma(\mathrm{spr}(P^*) + \varepsilon))^{-1}.$$

Let now $|\underline{Q}|_1$ be the matrix norm associated with the vector norm $|U|_1 = \sum_{i=1}^r |u_i|$ then we derive in the same way a bound for $|(\underline{I} - \underline{A}(t)\underline{P})^{-1}|_1$ depending only on P^*, ε, and Γ. With these two bounds finally

$$|(\underline{I} - \underline{A}(t)\underline{P})^{-1}|^2 = \mathrm{spr}((\underline{I} - \underline{A}(t)\underline{P})^{-H}(\underline{I} - \underline{A}(t)\underline{P})^{-1})$$

$$\le |(\underline{I} - \underline{A}(t)\underline{P})^{-H}(\underline{I} - \underline{A}(t)\underline{P})^{-1}|_\infty \le |(\underline{I} - \underline{A}(t)\underline{P}^{-1}|_1 |(\underline{I} - \underline{A}(t)\underline{P})^{-1}|_\infty \le \kappa_\Gamma$$

proves the assertion for $|(\underline{I} - \underline{A}(t)\underline{P})^{-1}|$. For $|(\underline{I} - \underline{PA}(t))^{-1}|$ the proof follows in an analogeous way.

Let now $P = WJW^{-1}$ be the Jordan canonical decomposition of P, let $\mathrm{diag}(J)$ be the diagonal of J, i.e., the diagonal matrix of the eigenvalues of P, and let $Q = J - \mathrm{diag}(J)$. Then Q is a nilpotent matrix with $Q^r = 0$. The next two results concern the case of a constant diagonable matrix $A(t) = A$. Here we have $A\underline{P} = \underline{P}A$ and hence in (5.3.6)

$$\Gamma^*_{\Delta t\Lambda} = \Gamma_{\Delta t\Lambda} \le \Gamma(1 + \max_{\nu=0,1}|(\underline{I} - \Delta t\Lambda\underline{P})^{-1}(\Delta t\Lambda)^\nu|) \le \Gamma(1 + \sup_{\eta\in \mathrm{Sp}(\Delta tA)}|(I - \eta P)^{-1}\eta^\nu|).$$

<u>(5.3.9) Lemma.</u> *(i) Let* $\mathrm{Re}\,\eta \le 0$ *and let all nonzero eigenvalues of* P *have positive real part or let* $\mathrm{Re}\,\eta < 0$ *and let all eigenvalues of* P *have nonnegative real part.*
(ii) Let P *be regular or let the dimension of the kernel of* P *be equal to the multiplicity of the eigenvalue* 0 *of* P.
Then $I - \eta P$ *is regular and*

$$\max_{\nu=0,1}|\eta^\nu(I - \eta P)^{-1}| \le \kappa$$

where κ *depends only on* P.

Proof. If P is regular then

$$|\eta^\nu(I - \eta P)^{-1}| = |\eta^\nu W(I - \eta J)^{-1})W^{-1}| \le \kappa|\eta^\nu(I - \eta J)^{-1}|$$

where J is regular. If P is singular then J can be chosen of the form

$$J = \begin{bmatrix} 0 & 0 \\ 0 & \tilde{J} \end{bmatrix}$$

where $\tilde{J}$ is regular, and we have

$$|\eta^\nu(I - \eta P)^{-1}| = |W\begin{bmatrix} I & 0 \\ 0 & \eta^\nu(I - \eta\tilde{J})^{-1} \end{bmatrix}W^{-1}| \leq \kappa(1 + |\eta^\nu(I - \eta\tilde{J})^{-1}|)$$

hence it suffices to consider the case where P is regular. Then, as J is regular,

$$|(I - \eta J)^{-1}| = |[(I - \eta J)^{-1} - I]J^{-1}| \leq \kappa(1 + |(I - \eta J)^{-1}|)$$

and it suffices to find a bound of $|(I - \eta J)^{-1}|$ for $\eta \neq 0$. But if $\mathrm{Re}\,\eta \leq 0$ (< 0) then $\mathrm{Re}(\eta^{-1}) \leq 0$ (< 0) and therefore

$$I - \eta J = \eta(\eta^{-1}I - J)$$

is regular because $\mathrm{Re}(\eta^{-1}I - \mathrm{diag}(J)) < 0$ by assumption. Moreover, $|(I - \eta\,\mathrm{diag}(J))^{-1}|$ is bounded in a neighborhood of $\eta = 0$ hence bounded and

$$(5.3.10)\quad |(I - \eta J)^{-1}| \leq |(I - \eta\,\mathrm{diag}(J))^{-1}||(I - (I - \eta\,\mathrm{diag}(J))^{-1}Q)^{-1}| \leq \kappa$$

proves the assertion.

(5.3.11) Lemma. *(i) Let* $\mathrm{Re}\,\eta \leq 0$ *and let all nonzero eigenvalues of* P *have positive real part or let* $\mathrm{Re}\,\eta < 0$ *and let all eigenvalues of* P *have nonnegative real part.*
(ii) Let $\eta \in R$ *where* R *is bounded in* $\mathbb{C}$.
Then $I - \eta P$ *is regular and*

$$\max_{\nu=0,1}|\eta^\nu(I - \eta P)^{-1}| \leq \kappa_R.$$

Proof. Obviously it suffices to consider $(I - \eta P)^{-1}$ and the regularity of this matrix follows in the same way as above. Then $I - \eta\,\mathrm{diag}(J)$ is regular and the assertion follows from (5.3.10).

The results received hitherto for linear problems with constant diagonable matrix can be assembled in the following theorem:

(5.3.12) Theorem. (Crouzeix [75].) *(i) Let the (m,m)-matrix* A *in* (5.3.1) *be constant and diagonable,* $A = X\Lambda X^{-1}$, *and let the solution* y *of* (5.3.1) *be (p+1)-times continu-*

ously differentiable.
(ii) Let the method (5.1.6), (5.1.7) *be consistent of order* p *and of degree* $p^* - 2$ *with the stability region* S.
(iii) Let $Sp(\Delta tA) \subset S \cap \{\eta \in \overline{\mathbb{C}},\ \mathrm{Re}\,\eta < 0\}$ *and let all eigenvalues of* P *have nonnegative part or let* $Sp(\Delta tA) \subset S \cap \{\eta \in \overline{\mathbb{C}},\ \mathrm{Re}\,\eta \leq 0\}$ *and let all nonzero eigenvalues of* P *have positive real part.*
(iv) Let P *be regular or let the dimension of the kernel of* P *be equal to the multiplicity of the eigenvalue* 0 *of* P.
Then for $n = 1,2,\ldots,$

$$|X^{-1}(y_n - v_n)| \leq \kappa\Big[|X^{-1}|(|y_0 - v_0| + \Delta t^p \int_0^{n\Delta t} |y^{(p+1)}(\tau)|d\tau + n\Delta t|||h|||_n$$
$$+ \Delta t^p n\Delta t \textstyle\sum_{k=p^*}^{p} |||X^{-1}A^{p+1-k}y^{(k)}|||_n\Big].$$

If assumption (iv) is not fulfilled then κ *depends on* $\Delta t|A|$. *Without any assumption on* P *the estimation holds for* $\Delta t|A| < \mathrm{spr}(P^*)^{-1}$ *with* $P^* = [|\alpha_{ij}|]_{i,j=1}^{r}$ *and* κ *depending on* $\Delta t|A|$.

In the sequel, $A > 0$ means again that the hermitean (m,m)-matrix A is positive definite and $\mathrm{Re}(A) = (A + A^H)/2$ denotes the hermitean part of A. If the constant matrix A in (5.3.1) is not diagonable then we have to modify slightly Lemma (5.3.9) in order to obtain uniform bounds. (In Lemma (5.3.11), $|A| \leq \kappa$ follows uniformly from $\mathrm{spr}(A) \leq \tilde{\kappa}$ only if A is normal hence diagonable.)

<u>(5.3.13) Lemma</u>. *(i) Let* A *be regular with* $\mathrm{Re}(A) \leq 0$ *and let all nonzero eigenvalues of* P *be positive or let* $\mathrm{Re}(A) < 0$ *and let all eigenvalues of* P *be (real and) nonnegative.*
(ii) Let P *be regular or let the dimension of the kernel of* P *be equal to the dimension of the eigenvalue* 0 *of* P.
Then $\underline{I} - A\underline{P}$ *is regular and*

$$\max_{\nu=0,1}|(\underline{I} - A\underline{P})^{-1}A^{\nu}| \leq \kappa_*.$$

Proof. As A commutes with $\underline{P}$ we can follow Lemma (5.3.9) and have to find a bound for $|(\underline{I} - A\underline{J})^{-1}|$ only. If A is regular with $\mathrm{Re}(A) \leq 0$ or if $\mathrm{Re}(A) < 0$ then $\mathrm{Re}(A^{-1}) \leq 0$ or $\mathrm{Re}(A^{-1}) < 0$ hence $\underline{I} - A\underline{J} = A(A^{-1}\underline{I} - \underline{J})$ is regular because $\mathrm{Re}(A^{-1}\underline{I} - \mathrm{diag}(\underline{J})) < 0$ by assumption. Moreover we have $|A^{-T}| \leq \alpha^{-1}$ if $\mathrm{Re}(A) \geq \alpha I > 0$ and $\mathrm{Re}(\underline{I} - A\mathrm{diag}(\underline{J})) \geq I$ because diag(J) is real and nonnegative therefore the assertion follows from (5.3.10) with η replaced by A.

If A(t) varies with time then $\underline{A}(t)$ does no longer commute with $\underline{P}$ and we have to impose still more conditions on P:

(5.3.14) Lemma. *Let* Re(A(t)) ≤ 0 *for* t > 0 *and let there exist a regular diagonal matrix* D *such that* $\mathrm{Re}(DPD^{-1}) > 0$. *Then* $\Gamma^*_{\Delta tA,n} \le \kappa$ *where* κ *depends only on* P *(and* D*)*.

Proof. As $\underline{A}(t)$ and $\underline{D}$ are block diagonal matrices they commute with each other and we have

$$\underline{I} - \underline{A}(t)\underline{P} = \underline{D}^{-1}(\underline{I} - \underline{A}(t)\underline{DPD}^{-1})\underline{D}$$

hence it suffices to prove the assertion for Re(P) > 0. Then $\mathrm{Re}(P^{-1}) > 0$ and, as P is regular,

$$\underline{I} - \underline{A}(t)\underline{P} = (\underline{P}^{-1} - \underline{A}(t))\underline{P}$$

with

$$\mathrm{Re}(\underline{P}^{-1} - \underline{A}(t)) \ge \mathrm{Re}(P^{-1}) \ge \tilde{\kappa} I > 0$$

whence

$$|(\underline{I} - \underline{A}(t)\underline{P})^{-1}| \le |P^{-1}||(\underline{P}^{-1} - \underline{A}(t))^{-1}| \le |P^{-1}|\tilde{\kappa}^{-1} \le \kappa.$$

Now the assertion follows from

$$(\underline{I} - \underline{A}(t)\underline{P})^{-1}\underline{A}(t) = [(\underline{I} - \underline{A}(t)\underline{P})^{-1} - \underline{I}]\underline{P}^{-1}$$

and

$$\underline{I} - \underline{A}(t)\underline{P} = \underline{P}^{-1}(\underline{I} - \underline{PA}(t))\underline{P}.$$

If P is a lower triangular matrix then, choosing $D = [\varepsilon, \varepsilon^2, \ldots, \varepsilon^n]$ with sufficiently small ε > 0, the assumption on P in this lemma is revealed to be equivalent to the condition diag(P) > 0.

Now we turn to the estimation of the iteration operator $G(\Delta tA)_n$ under the assumption that the matrix A is not necessarily diagonable. The main tool is here the following result due to J. von Neumann:

(5.3.15) Theorem. *Let* ϕ *be regular in a neighborhood of the unit disk* $\{\eta \in \mathbb{C}, |\eta| \le 1\}$ *and let* B *be a (m,m)-matrix with* |B| ≤ 1. *Then*

$$|\phi(B)| \le \sup_{\eta\in\mathbb{C}, |\eta|\le 1} |\phi(\eta)|.$$

Proof. See e.g. Riesz and Nagy [52].

(5.3.16) Lemma. (Crouzeix [75].) *Let* Re(A) ≥ αI, α ∈ ℝ, *and let* G *be a rational function which is bounded in the half-plane* {η ∈ ℂ, Re η ≥ α}. *Then*

$$|G(A)| \leq \sup_{\eta \in \mathbb{C}, \mathrm{Re}\,\eta \geq \alpha} |G(\eta)|.$$

Proof. The matrix

$$B = (I - (A - \alpha I))(I + (A - \alpha I))^{-1}$$

satisfies |B| ≤ 1 by Lemma (1.5.5). Writing

$$\phi(\eta) = G(\alpha + \frac{1 - \eta}{1 + \eta})$$

we have

$$G(\eta) = \phi(\frac{1 - (\eta - \alpha)}{1 + (\eta - \alpha)})$$

and $\eta \to \alpha + (1 - \eta)/(1 + \eta)$ is a bijective mapping of the unit disk onto the half-plane {η ∈ ℂ, Re η ≥ α} ∪ {∞}. Thus we have

$$\sup_{|\eta| \leq 1} |\phi(\eta)| = \sup_{\mathrm{Re}\,\eta \geq \alpha} |G(\eta)|$$

and the rational function ϕ has no poles in the unit disk. ϕ is therefore regular in a neighborhood of the unit disk and because G(A) = ϕ(B) Theorem (5.3.10) proves the assertion.

Before we prove the next result let us note once more that Runge-Kutta methods and single step multiderivative methods coincide for the test equation y' = λy. In particular, a Runge-Kutta method which is consistent with y' = λy must also be consitent in the sense of Definition (1.1.7) (and Lemma (1.1.12)). Accordingly, the principal root of a consistent method (being here the only root at all) has the same form as in (1.3.6),

$$\zeta(\eta) = 1 + \eta + \mathcal{O}(\eta^2) \qquad \eta \to 0,$$

and Lemma (A.1.41) yields that every consistent Runge-Kutta method has a stability region S containing a disk {η ∈ ℂ, |η + ρ| ≤ ρ}, ρ > 0.

(5.3.17) Lemma. (Crouzeix [75].) *Let* Re(A) > 0, *let* G *be a rational function which is bounded in the disk* D = {η ∈ ℂ, |η - ρ| ≤ ρ}, ρ > 0, *and let*

(5.3.18) $0 \leq \Delta t\ \mathrm{spr}[(\frac{A + A^H}{2})^{-1} A^H A] \leq 2\rho.$

Then

$$|G(\Delta tA)| \le \sup_{\eta \in D} |G(\eta)|.$$

Proof. By straightforward computation we verify that $|\Delta tA - \rho| \le \rho$ is equivalent to the condition

$$\sup_{w \ne 0} \frac{(Aw)^H Aw}{\mathrm{Re}(w^H Aw)} \le 2\rho,$$

and this condition is equivalent to (5.3.18) because

$$\mathrm{spr}[(\frac{A + A^H}{2})^{-1} A^H A] = \mathrm{spr}[(\frac{A + A^H}{2})^{-1/2} A^H A (\frac{A + A^H}{2})^{-1/2}] = |A(\frac{A + A^H}{2})^{-1/2}|.$$

Writing now $B = (\Delta tA - \rho I)/\rho$,

$$\phi(\eta) = G(\rho(\eta + 1)), \text{ and } G(\eta) = \phi((\eta - \rho)/\rho),$$

we have $\phi(B) = G(\Delta tA)$, ϕ is regular in a neighborhood of the unit disk, and hence Theorem (5.3.10) proves the result.

With respect to the linear problem (5.3.1) with constant but not necessarily diagonable matrix A we can now estimate $\Gamma^*_{\Delta tA}$ in (5.3.4) by Lemma (5.3.7) or Lemma (5.3.13) and find that $|G(\Delta tA)| \le 1$ under the assumptions of Lemma (5.3.16) or Lemma (5.3.17). The result can be assembled in the following theorem:

<u>(5.3.19) Theorem.</u> *(i) Let the (m,m)-matrix* A *in* (5.3.1) *be regular and constant, and let the solution* y *be (p+1)-times continuously differentiable.*
(ii) Let the Runge-Kutta method be consistent of order p *with the problem* (5.3.1), *of degree* p* - 2, *and A-stable.*
(iii) Let Re(A) < 0 *and let all eigenvalues of* P *be real and nonnegative or let* Re(A) ≤ 0 *and let all nonzero eigenvalues of* P *be positive.*
(iv) Let P *be regular or let the dimension of the kernel of* P *be equal to the multiplicity of the eigenvalue* 0 *of* P.
Then for n = 1,2,...,

$$|y_n - v_n| \le \kappa\Big[|y_0 - v_0| + \Delta t^p \int_0^{n\Delta t} |y^{(p+1)}(\tau)|d\tau + n\Delta t|||h|||_n + n\Delta t^{p+1}\textstyle\sum_{k=p^*}^{p} |||A^{p+1-k}y^{(k)}|||_n\Big].$$

There exists a ρ, $0 < \rho < \mathrm{spr}(P^*)^{-1}/2$, $P^* = [|\alpha_{ij}|]^r_{i,j=1}$, *such that A-stability, assumption (iii) and (iv) can be replaced by* $\mathrm{Re}(A) < 0$ *and* $|\Delta tA + \rho I| \le \rho$ *but then* κ *depends on* ρ.

With respect to the general linear problem (5.3.1) we can estimate $\Gamma^*_{\Delta tA,n}$ by Lemma (5.3.7) or Lemma (5.3.14). In order to find a bound for $|G(\Delta tA)_n|$ we suppose that P is regular and obtain

$$G(\Delta tA)_n \equiv I + \Delta t q^T \underline{A}_n(\underline{I} - \Delta t \underline{PA}_n)^{-1} z$$

$$= I + q^T \underline{P}^{-1}[(\underline{I} - \Delta t\underline{PA}_n)^{-1} - \underline{I}]z = I - q^T\underline{P}^{-1}z + q^T\underline{P}^{-1}(\underline{I} - \Delta t\underline{PA}_n)^{-1}z$$

(5.3.20)

$$= G(\Delta tA_n) + q^T\underline{P}^{-1}[(\underline{I} - \Delta t\underline{PA}_n)^{-1} - (\underline{I} - \Delta t\underline{P}A_n)^{-1}]z$$

$$= G(\Delta tA_n) + \Delta t q^T\underline{P}^{-1}(\underline{I} - \Delta t\underline{PA}_n)^{-1}\underline{P}[(\underline{A}_n - A_n\underline{I})A_n^{-1}]A_n(\underline{I} - \Delta t\underline{P}A_n)^{-1}z.$$

Here, $G(\Delta tA_n) = -\sigma_1(\Delta tA_n)^{-1}\sigma_0(\Delta tA_n)$ is a rational function with the argument ΔtA_n. Hence Lemma (5.3.14) and Lemma (5.3.16) yield for an A-stable method and $\mathrm{Re}(A(t)) \le 0$

$$|G(\Delta tA)_n| \le 1 + \kappa^*\Theta_{n+1}\Delta t$$

with the notation

(5.3.21) $$\Theta_n = \max_{0\le t\le(n-1)\Delta t,\ t\le\tau\le t+\Delta t}\{|A(t)^{-1}(A(\tau) - A(t))|,\ |(A(\tau) - A(t))A(t)^{-1}|\}\Delta t^{-1}$$

and

$$|q||\Delta t\underline{A}_n(\underline{I} - \Delta t\underline{PA}_n)^{-1}||(\underline{I} - \Delta t\underline{P}A_n)^{-1}||z| \le \kappa^*.$$

The complete result reads as follows:

<u>(5.3.22) Theorem</u>. (Crouzeix [75].) *(i) Let the (m,m)-matrix* $A(t)$ *in* (5.3.1) *and the solution* y *be respectively p-times and (p+1)-times continuously differentiable; let* $A(t)$ *be regular and* $\mathrm{Re}(A(t)) \le 0$ *for* $t > 0$.
(ii) Let the Runge-Kutta method be of order p *for the problem* (5.3.1), *of degree* $p^* - 2$, *and A-stable,* $S = \{\eta \in \mathbb{C},\ \mathrm{Re}\,\eta \le 0\} \cup \{\infty\}$.
(iii) Let there exist a regular diagonal matrix D *such that* $\mathrm{Re}(DPD^{-1}) > 0$.
Then for $n = 1,2,\ldots,$

$$|y_n - v_n| \le e^{\kappa^*\Theta_n n\Delta t}\Big[|y_0 - v_0| + \kappa[\Delta t^p\int_0^{n\Delta t}|y^{(p+1)}(\tau)|d\tau + n\Delta t^{p+1}\textstyle\sum_{k=p^*}^{p}\|w_k(\Delta t,y)\|_n + n\Delta t|||h|||_n]\Big]$$

where $\|w_k(\Delta t,y)\|_n = \max_{0\le\nu\le n} w_k(\Delta t,y)_\nu$ *and* $w_k(\Delta t,y)_n$ *is defined in* (5.2.38). *For regular* P *exists a* ρ, $0 < \rho < \mathrm{spr}(P^*)^{-1}/2$, *such that A-stability and ass. (iii) can be replaced by* $\mathrm{Re}(A(t)) < 0$ *and* $|\Delta tA(t) + \rho I| \le \rho$ *for* $t > 0$ *but then* κ *depends on* ρ.

The assumption of regular P is somewhat unsatisfactory here because not only explicit methods but also some implicit methods have a singular matrix P. However, if there exists a vector g such that

$$q = Pg$$

then we can write instead of (5.3.20)

$$(5.3.23)\quad G(\Delta tA)_n = G(\Delta tA_n) + g^T(\underline{I} - \Delta t\underline{P}\underline{A}_n)^{-1}P[(\underline{A}_n - A_n\underline{I})A_n^{-1}]\Delta tA_n(\underline{I} - \Delta t\underline{P}\underline{A}_n)^{-1}z.$$

Here, $|G(\Delta tA_n)| \le 1$ follows as above, $|\Delta tA_n(\underline{I} - \Delta t\underline{P}\underline{A}_n)^{-1}| \le \kappa$ follows from Lemma (5.3.13) but for $\Gamma^*_{\Delta tA,n}$ (cf. (5.2.37) and (5.3.3)) we have to find a new bound under the assumption of Lemma (5.3.13):

<u>(5.3.24) Lemma</u>. *Let the assumption of Lemma* (5.3.13) *be fulfilled, let* κ_* *be the constant of Lemma* (5.3.13), *and let*

$$(5.3.25)\quad \kappa_*|P|\Theta_n\Delta t_0 \le \Gamma_0 < 1, \qquad n = 0,1,\dots .$$

Then for $0 < \Delta t \le \Delta t_0$

$$\max\{|(\underline{I} - \Delta t\underline{P}\underline{A}_n)^{-1}|, |(\underline{I} - \Delta t\underline{A}_n\underline{P})^{-1}|\} \le \kappa(1 - \Gamma_0)^{-1},$$

$$|(\underline{I} - \Delta t\underline{A}_n\underline{P})^{-1}\Delta t\underline{A}_n| \le \kappa(1 - \Gamma_0)^{-1} \max_{n\Delta t\le\tau\le(n+1)\Delta t} |A_n^{-1}A(\tau)|.$$

Proof. We write

$$(\underline{I} - \Delta t\underline{P}\underline{A}_n) = (\underline{I} - \Delta t\underline{P}A_n)(\underline{I} - \underline{B}_n)$$

where

$$\underline{B}_n = \Delta t(\underline{I} - \Delta t\underline{P}A_n)^{-1}P(\underline{A}_n - A_n\underline{I}) = (\underline{I} - \Delta t\underline{P}A_n)^{-1}\Delta tA_nP[A_n^{-1}(\underline{A}_n - A_n\underline{I})]$$

hence by Lemma (5.3.13), as $A_n\underline{I}$ commutes with $\underline{P}$,

$$|\underline{B}_n| \le \kappa_*|P|\Theta_n\Delta t_0 \le \Gamma_0 < 1.$$

Accordingly we obtain again by Lemma (5.3.13)

$$|(\underline{I} - \Delta t\underline{P}\underline{A}_n)^{-1}| \le |(\underline{I} - \Delta t\underline{P}A_n)^{-1}||(\underline{I} - \underline{B}_n)^{-1}| \le \kappa_*(1 - \Gamma_0)^{-1}.$$

This bound holds also for $|(\underline{I} - \Delta t\underline{A}_n\underline{P})^{-1}|$, furthermore

$$(\underline{I} - \Delta t\underline{A}_n P)^{-1}\Delta t\underline{A}_n = (\underline{I} - \underline{B}_n)^{-1}[(\underline{I} - \Delta t\underline{PA}_n)^{-1}\Delta tA_n](A_n^{-1}\underline{A}_n)$$

which proves the second result after a further application of Lemma (5.3.13).

By means of this lemma, Theorem (5.3.22) can be modified as follows, see also Crouzeix [75].

(5.3.26) Theorem. *Let ass. (i) and (ii) of Theorem* (5.3.22) *be fulfilled, and let the assumption of Lemma* (5.3.13) *be fulfilled for* $A(t)$ *and* $t > 0$. *Let* $\Delta t_0 > 0$ *be defined by* (5.3.25) *and let there exist a* g *such that* $q = Pg$. *Then the error bound of Theorem* (5.3.22) *holds for* $0 < \Delta t < \Delta t_0$ *with* κ *and* κ^* *multiplied by* $(1 - \Gamma_0)^{-1}$ *and* κ *depending on* $\max_{0 \leq t \leq (n-1)\Delta t, t \leq \tau \leq t+\Delta t}|A(t)^{-1}A(\tau)|$.

There exists a $\rho > 0$ *such that A-stability and the assumption of Lemma* (5.3.13) *can be replaced by* $Re(A(t)) < 0$ *and* $|\Delta tA(t) + \rho I| \leq \rho$ *for* $t > 0$ *but then* κ *depends on* ρ.

In concluding this section we note that instead of (5.3.20) we also can write without any assumption on P

$$G(\Delta tA)_n = G(\Delta tA_n) + \Delta tq^T[\underline{A}_n - A_n\underline{I}](\underline{I} - \Delta t\underline{PA}_n)^{-1}z$$
$$+ \Delta t^2 q^T A_n(\underline{I} - \Delta t\underline{PA}_n)^{-1}\underline{P}[\underline{A}_n - A_n\underline{I}](\underline{I} - \Delta t\underline{PA}_n)^{-1}z.$$

This leads to further error bounds in the case where

$$|\Delta tA_n(\underline{I} - \Delta t\underline{PA}_n)^{-1}| \quad \text{and} \quad |[\underline{A}_n - A_n\underline{I}](\underline{I} - \Delta t\underline{PA}_n)^{-1}|$$

is bounded, e.g., if $A(t) = a(t)A$ with a scalar function a.

5.4. Examples and Remarks

The characteristic polynomial $\pi(\zeta,\eta) = \sigma_1(\eta)\zeta + \sigma_0(\eta)$ of a single step multistage or multiderivative method has only one root $\zeta(\eta)$ which is the amplification factor $G(\eta)$ with respect to the test equation $y' = \lambda y$,

$$v_{n+1} = G(\eta)v_n, \qquad n = 0,1,\ldots, \; \eta = \Delta t\lambda,$$

cf. (5.1.12). Hence the stability region S is closed in $\overline{\mathbb{C}}$ as we have already mentioned above. In particular, we have $0 \in S$ if the method is consistent and

(5.4.1) $\zeta(\eta) = G(\eta) = 1 + \eta + \mathcal{O}(\eta^2) = e^\eta + \mathcal{O}(\eta^2), \quad \eta \to 0.$

Therefore, an explicit consistent Runge-Kutta method is convergent in the classical meaning, i.e., for well-conditioned problems with Lipschitz-continuous f in (5.1.1) and finite time intervals. Because of (5.4.1), Lemma (A.1.41) or a simple direct verification shows:

(5.4.2) Corollary. *If the general Runge-Kutta method* (5.1.6), (5.1.7) *is consistent then there exists a* $\rho > 0$ *such that*

$$\mathfrak{D}_\rho = \{\eta \in \mathbb{C},\ |\eta + \rho| \leq \rho\} \subseteq S.$$

Obviously we have

$$G(\eta) = 1 + \eta + \textstyle\sum_{j=2}^{r} \alpha_j \eta^j$$

in explicit consistent methods of stage r. Jeltsch and Nevanlinnna [78] have proved the following result:

(5.4.3) Lemma. *The stability region* S *of an explicit r-stage method satisfies* $\mathfrak{D}_r \subseteq S$ *iff*

$$G(\eta) = (1 + (\eta/r))^r.$$

A generalization of this result to nonlinear multistep methods is found in Jeltsch and Nevanlinna [81 , Theorem 3.1].

For explicit r-stage Runge-Kutta methods the maximum attainable order $p^\dagger$ has been derived by Butcher [65 et al.] in a tedious work. The result can be represented in the following table (cf. also Lambert [73 , p. 122]):

(5.4.4) Table.

r	1	2	3	4	5	6	7	8	9	$r \geq 10$
$p^\dagger$	1	2	3	4	4	5	6	6	7	$p^* \leq r - 2$

It can be shown that there exist explicit Runge-Kutta methods of arbitrary order. For a given stage r = 1,...,4 the amplification factor $G(\eta)$ of an explicit r-stage method of (maximum) order p = r has the form

$$G(\eta) = \sum_{j=0}^{r} \frac{1}{j!}\eta^j, \qquad p = r = 1,\dots,4.$$

Accordingly, all these methods have the same stability region for fixed $p = r$. Plots are given e.g. in Grigorieff [72 , p. 109], Lambert [73 , p. 227], and Stetter [73 , p. 176], and we refer also to these books for further details and special examples.

After these brief remarks on explicit methods let us now turn to implicit methods which deserve more interest in the solution of stiff problems.

The general consistent single-stage method is the method (4.2.2) with $\omega \in \mathbb{R}$. We consider the linear problem (5.3.1),

$$(5.4.5) \quad y' = A(t)y + c(t) + h(t),\ t > 0,\ y(0) = y_0,$$

and obtain the computational device

$$(5.4.6) \quad v_{n+1} = (I - \omega\Delta tA_{n+\omega})^{-1}[(I + (1-\omega)\Delta tA_{n+\omega})v_n + \Delta tc_{n+\omega}], \qquad n = 0,1,\dots,$$

where $c_{n+\omega} = c((n+\omega)\Delta t)$. Let $0 < \omega < 1$ and let $A(t)$ be regular with $\mathrm{Re}(A(t)) \leq 0$. Then Lemma (5.3.13) yields

$$|(I - \omega\Delta tA_{n+\omega})^{-1}| \leq \kappa,\ |(I - \omega\Delta tA_{n+\omega})^{-1}(I + (1-\omega)\Delta tA_{n+\omega})| \leq \kappa.$$

The degree of the method is $p^* - 2 = 0$ for all $\omega \in \mathbb{R}$ hence (5.2.19) and Lemma (5.3.14) yield for the discretization error

$$|d(\Delta t,y)(t)| \leq |\tfrac{1}{2} - \omega|\Delta t^2|y^{(2)}(t)| + \Gamma[\Delta t^2 \int_t^{t+\Delta t} |y^{(3)}(\tau)|d\tau$$
$$+ \Delta t^3|(I - \omega\Delta tA(t+\omega\Delta t))^{-1}A(t+\omega\Delta t)y^{(2)}(t)|]$$

where the third term on the right side cannot be cancelled. We state the result as follows:

<u>(5.4.7) Corollary</u>. *Let the real (m,m)-matrix* $A(t)$ *in* (5.4.5) *be regular with* $\mathrm{Re}(A(t)) \leq 0$ *for* $t > 0$ *and let the solution* y *be three times continuously differentiable. Then the error of the method* (5.4.6) *satisfies for* $0 < \omega < 1$ *and* $n = 1,2,\dots,$

$$|y_n - v_n| \leq |y_0 - v_0| + \Gamma\Big[\Delta t^2 \int_0^{n\Delta t} |y^{(3)}(\tau)|d\tau + n\Delta t|||h|||_n$$
$$+ n(|\tfrac{1}{2} - \omega| + |||(I - \omega\Delta tA(\cdot))^{-1}\Delta tA(\cdot)|||_n)\Delta t^2|||y^{(2)}|||_n\Big].$$

With exception of the second row this error bound is rather simple in comparison with the corresponding result for the trapezoidal rule, Theorem (4.3.19). For $\omega = 1/2$ the method has convergence order two for well-conditioned problems. However, if $A(t)$ is ill-conditioned then

$$|(I - \omega\Delta tA(t))^{-1}\Delta tA(t)| \le \kappa$$

by Lemma (5.3.14) and the uniform convergence order with respect to (5.4.5) is only *one*. A similar remark holds obviously if ω is modified slightly following the proposition of Kreth, (4.2.8).

In Section 1.5 the stability of diagonal Padé approximants has been proved in a direct way which however seems difficult to apply to other methods. But Lemma (5.3.16) generalizes the second part of Lemma (1.5.5) to every A-stable single step method and, by Corollary (5.4.2), Lemma (5.3.17) applies to every consistent single step method. Recall now that $\sigma_0(\eta)$ and $\sigma_1(\eta)$ are always supposed to have no common factor and write

$$G_{\mu,\nu}(\eta) = -\sigma_0(\eta)/\sigma_1(\eta)$$

if $\deg(\sigma_0(\eta)) \le \mu$, $\deg(\sigma_1(\eta)) \le \nu$. Then we have by Lemma (1.3.5)

$$G_{\mu,\nu}(\eta) = e^{\eta} + \mathcal{O}(\eta^{p+1}), \qquad \eta \to 0,\ p \ge 1,$$

in the case of consistence, and $p = \mu + \nu$ is the maximum attainable order of the corresponding single step method for the test equation $y' = \lambda y$.

<u>(5.4.8) Definition</u>. $G_{\mu,\nu}(\eta)$ *is a (μ,ν)-Padé approximant (of the exponential function near $\eta = 0$) iff*

$$G_{\mu,\nu}(\eta) = e^{\eta} + \mathcal{O}(\eta^{\mu+\nu+1}), \qquad \eta \to 0.$$

A Padé approximant determines uniquely a single step multiderivative method but not a Runge-Kutta method for the general problem (5.1.1). Nevertheless we say briefly that a method is a Padé approximant if the corresponding $G_{\mu,\nu}(\eta)$ is a Padé approximant. By definition these methods have maximum order for the test equation which justifies their popularity.

(μ,ν)-Padé approximants of an arbitrary function are determined uniquely if they exist and can be computed explicitely, see Hummel and Seebeck [49]. For the exponential function we obtain the following result; cf. also Grigorieff [72].

<u>(5.4.9) Lemma</u>. *Let*

$$\sigma_{\mu,\nu}(\eta) = \sum_{j=0}^{\mu} \frac{\mu!}{(\mu-j)!}\,\frac{(\mu+\nu-j)!}{(\mu+\nu)!}\,\frac{\eta^j}{j!}$$

and

$$G_{\mu,\nu}(\eta) = \sigma_{\mu,\nu}(\eta)/\sigma_{\nu,\mu}(-\eta)$$

then

$$e^{\eta} = G_{\mu,\nu}(\eta) + (-1)^{\nu}\kappa\eta^{\mu+\nu+1} + \mathcal{O}(\eta^{\mu+\nu+2}), \qquad \eta \to 0,$$

where $\kappa > 0$, *and* $\sigma_{\mu,\nu}(\eta)$, $\sigma_{\nu,\mu}(-\eta)$ *have no common factor.*

Crouzeix and Raviart [80 , Theorem 2.4.3] have shown in a simple way that every (μ,ν)-Padé approximant is A(0)-stable for $\mu \leq \nu$. The following result has been proved by Wanner, Hairer, and Norsett [78a] by a study of the order star:

(5.4.10) Lemma. *(μ,ν)-Padé approximants are A-stable iff* $\mu \leq \nu \leq \mu + 2$.

For $\nu = \mu$, $\mu + 1$ this result is also found in Grigorieff [72]. If $\nu = \mu + 1$ or $\nu = \mu + 2$ then we have $\infty \in \mathring{S}$ which is favorable for the error propagation in ill-conditionend problems, cf. e.g. Section 4.2.

Lemma (5.4.10) implies that no poles of $G_{\mu,\nu}(\eta)$ lie in the left half-plane or, in other words, that all roots of the denominator of $G_{\mu,\nu}(\eta)$ have positive real part. Therefore the first part of Lemma (1.5.5) is valid here, too,

$$(5.4.11) \quad |\sigma_{\nu,\mu}(-\Delta tA)^{-1}| \leq \kappa, \qquad \mathrm{Re}(A) \leq 0,\ \mu \leq \nu \leq \mu + 2.$$

Furthermore, Lemma (5.3.16) yields

$$(5.4.12) \quad |G_{\mu,\nu}(\Delta tA)| \leq 1, \qquad \mathrm{Re}(A) \leq 0,\ \mu \leq \nu \leq \mu + 2.$$

(5.4.11) and (5.4.12) together provide the following result:

(5.4.13) Corollary. *Let the (m,m)-matrix* A *in* (5.4.5) *be constant, let* $\mathrm{Re}(A) \leq 0$, *and let the solution* y *be $(\mu+\nu+1)$-times continuously differentiable. Then the error of a (μ,ν)-Padé approximant satisfies for* $\mu \leq \nu \leq \mu + 2$

$$|y_n - v_n| \leq |y_0 - v_0| + \kappa[\Delta t^{\mu+\nu} \int_0^{n\Delta t} |y^{(\mu+\nu+1)}(\tau)|d\tau + n\Delta t \max_{0\leq i\leq \max\{\mu,\nu\}-1} |||h^{(i)}|||_n].$$

An application of this result to the first order transformation (1.4.2),

$$(5.4.14) \quad z' = A^*z + c^*(t) + h^*(t),\ t > 0,\ z(0) = z_0,$$

of the second order problem

(5.4.15) $y'' = Ay + By' + c(t) + h(t),\ t > 0,\ y(0) = y_0,\ y'(0) = y_0^*,$

yields after the transformation (1.4.10) immediately the following generalization of Theorem (1.5.9):

(5.4.16) Theorem. *In the initial value problem* (5.4.15) *let* A, B *be real symmetric*, $A \leq -\gamma I < 0$, $B \leq 0$, *and let the solution* y *be* $(\mu+\nu+2)$*-times continuously differentiable. Let* $v_n^* = (v_n, w_n)^T$, $n = 1,2,\ldots$, *be obtained by a* (μ,ν)*-Padé approximant with* $\mu \leq \nu \leq \mu+2$ *applied to the transformed problem* (5.4.14). *Then*

$$|y_n - v_n| + |y_n' - w_n| \leq \kappa(1+\gamma^{-1})^{1/2}\Big[|(-A)^{1/2}(y_0 - v_0)| + |y_0' - w_0|$$
$$+ \Delta t^{\mu+\nu}\int_0^{n\Delta t}(|(-A)^{1/2}y^{(\mu+\nu+1)}(\tau)| + |y^{(\mu+\nu+2)}(\tau)|)d\tau + n\Delta t \max_{0\leq i\leq \max\{\mu,\nu\}-1} |||h^{(i)}|||_n\Big].$$

Up today Runge-Kutta methods haven't lost anything from their attraction for numerical analysis and application. On the contrary, methods in which the matrix P has only one eigenvalue are an essential subject of current research. A thorough presentation of the results available here in the meanwhile would go far beyond the scope of this volume. For a concise treatment and some interesting existence and uniqueness statements we refer to the forthcoming book of Crouzeix and Raviart [80].

VI. Approximation of Initial Boundary Value Problems

6.1. Initial Boundary Value Problems and Galerkin Procedures

Unlike elliptic and parabolic problems there are in hyperbolic problems basically two different ways of numerical approximation in dependence of the underlying form of the differential equation and the given initial and boundary conditions: The method of characteristics and the *method of lines*. In the former method the solution is computed along the characteristic curves which implies a strong connection between time and space discretization whereas in the latter method time and space are discretized in a separated way. On each time level $t = n\Delta t$ an 'elliptic' problem is solved here by a finite difference method or a Galerkin procedure. The connection between time and space discretization consists if at all in a Courant-Friedrichs-Lewy condition which guarantees that the spectral radius of the iteration operator with respect to the time direction is not greater than one. The method of lines has the advantage that numerical methods for elliptic problems and methods for initial value problems with ordinary differential equations can be applied in space and time direction respectively without much preliminary work.

In this chapter we study the numerical approximation of linear parabolic problems and hyperbolic problems of second order by the method of lines choosing Galerkin procedures for the discretization in the space direction. Some specific assumptions are then made for the error estimations which are fulfilled by a large class of problems and finite element methods.

The description of the analytic problems to be considered needs some further notations which are listed up for shortness:

$\Omega \subset \mathbb{R}^r$ bounded and open domain;

$(f,g) = \int_\Omega f(x)g(x)dx$, $|f|^2 = (f,f)$;

$\|f\|_s = (\sum_{|\sigma|\le s}|D^\sigma f|^2)^{1/2}$, $s \in \mathbb{N}$, Sobolev norm with the standard multi-index notation, $\sigma = (\sigma_1,\dots,\sigma_r)$, $\sigma_i \in \mathbb{N}$, $D^\sigma f = \partial^{|\sigma|}f/\partial x_1^{\sigma_1}\cdots\partial x_r^{\sigma_r}$, $|\sigma| = \sigma_1 + \dots + \sigma_r$;

$|||f|||_{s,n} = \max_{0\le t\le n\Delta t}\|f(\cdot,t)\|_s$;

$W^s(\Omega) = \{f \in L^2(\Omega),\ \partial^\sigma f \in L^2(\Omega),\ \forall\ \sigma,\ |\sigma| \le s\}$ Sobolev space, $W_0^s(\Omega) = \{f \in W^s(\Omega),\ f(x) = 0\ \forall\ x \in \partial\Omega\}$, $H \subset W^s(\Omega)$ Hilbert space with $W_0^s(\Omega) \subseteq H$;

a: $H\times H \ni (u,v) \mapsto a(u,v) \in \mathbb{R}$ symmetric bilinear form such that $a(v,v)^{1/2}$ defines a norm which is equivalent to $\|\cdot\|_s$ over H, $0 < \gamma\|v\|_s^2 \le a(v,v) \le \Gamma\|v\|_s^2 \; \forall\, 0 \ne v \in H$;

b: $\mathbb{R}_+ \times H \times H \ni (t,u,v) \mapsto b(t;u,v) \in \mathbb{R}$ symmetric bilinear form in u and v such that

$$0 < b(t;v,v) \le \beta\|v\|_0^2 \;\forall\, 0 \ne v \in H \text{ or } b \equiv 0.$$

$\phi_{\Delta x,1}, \ldots, \phi_{\Delta x,m(\Delta x)}$ denote linear independent functions which span the Galerkin subspace $G_{\Delta x} \subset H$ and define the Galerkin procedure. In finite element analysis Δx denotes the maximum diameter of all 'patches' by which the domain Ω is partitioned. We omit the index Δx if it is not necessary and use the following further notations:

$$\Phi = (\phi_1,\ldots,\phi_m)^T,\; M = [(\phi_\mu,\phi_\nu)]_{\mu,\nu=1}^m,\; K = [a(\phi_\mu,\phi_\nu)]_{\mu,\nu=1}^m;$$

$\tilde{w} \in G$ Ritz projection of $w \in H$ defined by $a(w - \tilde{w}, v) = 0 \;\forall\, v \in G$;

$\hat{w} \in G$ L^2-projection of $w \in L^2(\Omega) = W^0(\Omega)$ defined by $(w - \hat{w}, v) = 0 \;\forall\, v \in G$.

In order to make a difference between scalar- and vector-valued functions we denote in the sequel vectors of dimension m by capitals and write for instance

$$\tilde{w}(x) = \tilde{W}^T\Phi(x) \equiv \sum_{\mu=1}^m \tilde{w}_\mu \phi_\mu(x),\; \hat{w}(x) = \hat{W}^T\Phi(x).$$

After these preliminaries we can define an elliptic model problem by

$$(6.1.1)\quad a(w,v) = (c,v) \;\forall\, v \in H, \qquad c \in L^2(\Omega).$$

The Ritz projection $\tilde{w} \in G$ of the exact solution $w \in H$ is the finite element approximation of w by the main theorem of the Rayleigh-Ritz-Galerkin theory, cf. e.g. Strang and Fix [73 , Theorem 1.1]. (Naturally, we can speak of a finite element approximation only if G is a finite element subspace and not a general Galerkin subspace.) A-priori estimations of $w - \tilde{w}$ take a large place in modern finite element analysis. In the present work we however do not treat error estimations for elliptic problems but refer the reader e.g. to the book of Ciarlet [79] for a detailled analysis of finite element methods. Instead we *suppose* here that the elliptic problem (6.1.1) and the Galerkin procedure, i.e., the subspace $G \subset H$, have the following property in which Δx denotes again the small parameter introduced in the definition of G.

<u>(6.1.2) Assumption</u>. *Let* $w \in H$ *be the solution of* (6.1.1) *and let* $\tilde{w} \in G$ *be the Ritz projection of* w *then there exist positive integers* q *and* q_* *such that for all* $c \in L^2(\Omega)$

$$|w - \tilde{w}| \le \kappa_G \Delta x^q \|w\|_{q_*}.$$

The announced parabolic model problem now reads

(6.1.3) $(u_t(\cdot,t),v) + a(u(\cdot,t),v) = (c(\cdot,t),v) \ \forall\, v \in H,\ t > 0,\ u(\cdot,0) = u_0,$

where $u_t = \partial u/\partial t$, $u_0 \in H$, and $c(\cdot,t) \in L^2(\Omega)$ for $t > 0$. We assume that all data are sufficiently smooth such that Assumption (6.1.2) holds with respect to the underlying bilinear form a and that there exists a unique solution u with $u(\cdot,t) \in H$ and $u_t(\cdot,t) \in L^2(\Omega)$ for $t > 0$.

The Galerkin approximation $u_G(\cdot,t) \in G$ of $u(\cdot,t)$ is the solution of (6.1.3) for all $v \in G$. We substitute

(6.1.4) $$u_G(x,t) = U_G(t)^T\phi(x) = \textstyle\sum_{\mu=1}^m u_{G,\mu}(t)\phi_\mu(x),$$

$$\hat{c}(x,t) = \hat{C}(t)^T\phi(x),\ (c(\cdot,t) - \hat{c}(\cdot,t),v) = 0 \ \forall\, v \in G,$$

$$\hat{u}_0(x) = \hat{U}_0^T\phi(x),\ (u_0 - \hat{u}_0,v) = 0 \ \forall\, v \in G,$$

into (6.1.3) and set $v = \phi_\mu$, $\mu = 1,\ldots,m$, successively. Then we obtain an initial value problem with an ordinary differential system for the unknown function $U_G\colon [0,\infty] \to \mathbb{R}^m$:

(6.1.5) $$MU_G' + KU_G = M\hat{C}(t),\ t > 0,\ U_G(0) = \hat{U}_0.$$

The hyperbolic model problem we consider reads

(6.1.6) $$(u_{tt}(\cdot,t),v) + b(t;u_t(\cdot,t),v) + \chi a(u_t(\cdot,t),v) + a(u(\cdot,t),v) = (c(\cdot,t),v)$$
$$\forall\, v \in H,\ t > 0,\ u(\cdot,0) = u_0,\ u_t(\cdot,0) = u_0^*.$$

Here we assume that $\chi \geq 0$ and that in addition to the above assumptions $u_0^* \in H$ and the exact solution u satisfies $u_{tt}(\cdot,t) \in L^2(\Omega)$ and $u_t(\cdot,t) \in H$ if $\chi > 0$. The special form of the damping term is chosen here to enclose all cases which are considered below. Let

(6.1.7) $$N(t) = [b(t;\phi_\mu,\phi_\nu)]_{\mu,\nu=1}^m$$

then the semi-discrete problem associated with (6.1.6) reads

(6.1.8) $$MU_G'' + (N(t) + \chi K)U_G' + KU_G = M\hat{C}(t),\ t > 0,\ U_G(0) = \hat{U}_0,\ U_G'(0) = \hat{U}_0^*.$$

Note that M, N, and K are real symmetric and positive definite matrices.

In engineering mechanics this system is called the equilibrium equations of dynamic finite element analysis and plays a fundamental role. The basic partial differential equation is however not available in matrix structural analysis. Instead the original body is partitioned into more or less small cells of which the equations of motion can be approximated in a simpler way. These interdependent equations are then assembled to a large system which has the form (6.1.8). M, $N + \chi K$, and K are then the mass, damping, and stiffness matrix, and C(t) is the external load vector. See e.g. Bathe and Wilson [76], Fried [79], and Przemienicki [68]. In the meanwhile, these notations have also become customary in numerical analysis. If damping is not disregarded then it is frequently of the above form; cf. e.g. Przemienicki [68 , ch. 13], Clough [71], and Cook [74 , p. 303].

After having discretized the problem in the space direction it remains to solve the semi-discrete problem (6.1.5) or (6.1.8) numerically. For this we always write the differential system in explicit form, e.g. instead of (6.1.5)

(6.1.9) $U_G' = -M^{-1}KU_G + \hat{C}(t)$,

and then try to avoid the explicit computation of M^{-1} as a rule. For instance, the multistep multiderivative method (1.1.3) has for (6.1.9) the form (1.2.2) with $A = -M^{-1}K$ and $c = \hat{C}$,

(6.1.10) $\sum_{i=0}^{k}\sigma_i(-\Delta tM^{-1}K)T^iV_n = -\sum_{i=0}^{k}\sum_{j=1}^{\ell}\sigma_{ij}(-\Delta tM^{-1}K)\Delta t^jT^i\hat{C}_n^{(j-1)}$, $\quad n = 0,1,\dots$.

Of course, this scheme is multiplied by M again. As

$$M\hat{C}(t) = C^*(t),\ C^*(t) = ((c(\cdot,t),\phi_1),\dots,(c(\cdot,t),\phi_m))^T,$$

we get for linear multistep methods the computational device

(6.1.11) $M\rho_0(T)V_n - \Delta tK\rho_1(T)V_n = \Delta t\rho_1(T)C_n^*$, $\quad n = 0,1,\dots$.

In the general case (6.1.10) a linear system of the form

(6.1.12) $M\sigma_k(-\Delta tM^{-1}K)V_{n+k} = R_n$

is to be solved in every time step. The computation of R_n requires for $\ell > 1$ some matrix-vector multiplications and the solution of linear systems with the mass matrix M. The matrix on the left side of (6.1.12) is regular if $Sp(-\Delta tM^{-1/2}KM^{-1/2}) \subset S$ because

$$M\sigma_k(-\Delta tM^{-1}K) = M^{1/2}\sigma_k(-\Delta tM^{-1/2}KM^{-1/2})M^{1/2}$$

and the definition of the stability region S, (1.2.7). If the polynomial $\sigma_k(\eta)$ is nonlinear,

$$\sigma_k(\eta) = \alpha_{hk}\prod_{j=1}^{h}(\eta - \eta_j),\ \eta_j \in \mathbb{C},\ 1 < h \leq \ell,$$

then (6.1.12) can be written as

$$\alpha_{hk}(\Delta tK + M\eta_1)M^{-1}(\Delta tK + M\eta_2)\cdots M^{-1}(\Delta tK + M\eta_h)V_{n+k} = (-1)^h R_n,$$

and the solution V_{n+k} is computed by solving successively the h linear systems

$$\text{(6.1.13)}\quad \begin{aligned} &\alpha_{hk}(\Delta tK + M\eta_1)Z_1 = (-1)^h R_n,\ (\Delta tK + M\eta_j)Z_j = MZ_{j-1},\ j = 2,\ldots,h-1, \\ &(\Delta tK + M\eta_h)V_{n+k} = MZ_{h-1}. \end{aligned}$$

Here it is advantageous for an application of the Cholesky decomposition to use methods in which the leading polynomial $\sigma_k(\eta)$ has only roots η_j with positive real part. This requirement is e.g. fulfilled by the diagonal and subdiagonal Padé approximants presented in Section 1.5, and by the methods of Enright given in table (A.4.5). Obviously, if the nonlinear $\sigma_k(\eta)$ has the form

$$\sigma_k(\eta) = \alpha_{hk}(\eta - \eta_1)^h,\ 1 < h \leq \ell,$$

then (6.1.13) leads to the solution of h linear systems with the *same* matrix $\Delta tK + M\eta_1$. Methods with this property (and $\mathrm{Re}\,\eta_1 > 0$) are e.g. Calahan's method, cf. A.4.(iib), Enright's methods II given in table (A.4.6), and the restricted Padé approximants presented e.g. in the forthcoming monograph of Crouzeix and Raviart [80].

Naturally, the same arguments concerning the computational amount of work hold also for the numerical solution of second order initial value problems (6.1.8).

6.2. Error Estimates for Galerkin-Multistep Procedures and Parabolic Problems

In this section we use Theorem (1.2.12) and (1.2.18) to derive a-priori error bounds for the parabolic model problem (6.1.3). Instead of (6.1.9) and the numerical approximation (6.1.10) we write

$$\text{(6.2.1)}\quad M^{1/2}U_G' = AM^{1/2}U_G + M^{1/2}\hat{C}(t)$$

and

$$\text{(6.2.2)}\quad \textstyle\sum_{i=0}^{k}\sigma_i(\Delta tA)T^iM^{1/2}V_n = -\sum_{i=0}^{k}\sum_{j=1}^{\ell}\sigma_{ij}(\Delta tA)\Delta t^jT^iM^{1/2}\hat{C}_n^{(j-1)},\qquad n = 0,1,\ldots,$$

with the leading matrix

$$A = -M^{-1/2}KM^{-1/2}$$

being real symmetric and negative definite. The full-discrete scheme (6.2.2) yields an approximation $u_\Delta(\cdot,t) \in G$ of the exact solution $u(\cdot,t)$ of the form

$$u_\Delta(x,t) = V(t)^T\Phi(x),\ t = n\Delta t, \qquad n = k,k+1,\ldots .$$

By the fundamental relation

$$(6.2.3)\qquad |M^{1/2}W| = |W^T\Phi(\cdot)| \equiv |w|\ \forall\ w = W^T\Phi(\cdot) \in G$$

we then obtain immediately an error bound for the Galerkin approximation u_G defined by (6.1.4), (6.1.5), i.e., a bound of

$$|M^{1/2}(U_{G,n} - V_n)| = |(u_G - u_\Delta)(\cdot,n\Delta t)|, \qquad n = k,k+1,\ldots,$$

when an estimation of Section 1.2 is applied to the pair (6.2.1), (6.2.2). However, an error estimation via the decomposition

$$(6.2.4)\qquad u - u_\Delta = (u - \tilde{u}) + (\tilde{u} - u_\Delta)$$

$\tilde{u}$ denoting the Ritz projection of u again distinguishes more exactly between space and time discretization and moreover an estimation of $u - u_G$ needs also the approximation properties of $\tilde{u}$; cf. e.g. Fairweather [78]. Therefore we use the decomposition (6.2.4) in this chapter. Assumption (6.1.2) yields immediately

$$(6.2.5)\qquad |(u - u_\Delta)(\cdot,t)| \le \kappa_G\Delta x^q\|u(\cdot,t)\|_{q_*} + |(\tilde{u} - u_\Delta)(\cdot,t)|$$

hence it suffices to deduce error bounds with respect to the Ritz projection $\tilde{u}$ in the sequel.

If the data are sufficiently smooth then the parabolic problem (6.1.3) yields

$$a(u^{(\mu)}(\cdot,t),v) = (c^{(\mu)}(\cdot,t) - u^{(\mu+1)}(\cdot,t),v)\ \forall\ v \in H$$

writing shortly $u^{(\mu)} = \partial^\mu u/\partial t^\mu$ and Assumption (6.1.2) yields again

$$(6.2.6)\qquad |u^{(\mu)}(\cdot,t) - \widetilde{u^{(\mu)}}(\cdot,t)| \le \kappa_G\Delta x^q\|u^{(\mu)}(\cdot,t)\|_{q_*}.$$

As $\widetilde{u^{(\mu)}} = \tilde{u}^{(\mu)}$ in the present case of a time-independent bilinear form a, time derivatives of $\tilde{u}$ in the error bounds can be replaced by corresponding time-derivatives of u using the triangle inequality and (6.2.6) or, in a more direct way, by

$$|\tilde{u}^{(\mu)}(\cdot,t)| \le (\Gamma/\gamma)^{1/2}\|u^{(\mu)}(\cdot,t)\|_s$$

which results from $|w| \le \|w\|_s$ and

$$\gamma\|\tilde{w}\|_s^2 \le a(\tilde{w},\tilde{w}) \le a(w,w) \le \Gamma\|w\|_s^2 \;\forall\, w \in H.$$

Now, the Ritz projection $\tilde{u}$ satisfies

$$(\tilde{u}_t(\cdot,t),v) + a(\tilde{u}(\cdot,t),v) = (c(\cdot,t),v) - ([u_t - \tilde{u}_t](\cdot,t),v) \;\forall\, v \in G,\; t > 0.$$

We substitute $\tilde{u}(x,t) = \tilde{U}(t)^T\Phi(x)$ and obtain in the same way as above the following differential system for the unknown function $\tilde{U}$: $[0,\infty] \to \mathbb{R}^m$,

$$(6.2.7)\quad M^{1/2}\tilde{U}' = AM^{1/2}\tilde{U} + M^{1/2}\hat{C}(t) - M^{1/2}\hat{H}(t),$$

where $\hat{h}(\cdot,t)$ denotes the L^2-projection of $h(\cdot,t) = (u_t - \tilde{u}_t)(\cdot,t)$ and $\hat{h}(x,t) = \hat{H}(t)^T\Phi(x)$. For instance, Theorem (1.2.12) then yields immediately the following error bound:

$$|M^{1/2}(\tilde{U}_n - V_n)| \le \kappa_R\Big[\textstyle\sum_{i=0}^{k-1}|M^{1/2}(\tilde{U}_i - V_i)| + \Delta t^p \displaystyle\int_0^{n\Delta t} |M^{1/2}\tilde{U}^{(p+1)}(\tau)|d\tau$$

$$+ n\Delta t \max_{0\le i\le \ell-1} |||M^{1/2}\hat{H}^{(i)}|||_n\Big].$$

But, by (6.2.3),

$$|M^{1/2}(\tilde{U}_n - V_n)| = |(\tilde{u} - u_\Delta)(\cdot)_n|,\; |M^{1/2}\tilde{U}^{(p+1)}(\tau)| = |\tilde{u}^{(p+1)}(\cdot,\tau)|,$$

and the Projection Theorem together with Assumption (6.1.2) yields

$$(6.2.8)\quad \begin{aligned} |M^{1/2}\hat{H}^{(i)}(t)| &= |\hat{h}^{(i)}(\cdot,t)| \le |[u_t - \tilde{u}_t]^{(i)}(\cdot,t)| \\ &= |[u^{(i+1)} - u^{(i+1)\sim}](\cdot,t)| \le \kappa_G\Delta x^q\|u^{(i+1)}(\cdot,t)\|_{q_*}. \end{aligned}$$

For an application of Theorem (1.2.18) and Corollary (1.3.16) we have to verify that the (m,m)-matrix $A = -M^{-1/2}KM^{-1/2}$ in (6.2.1) is negative definite. For this let again $0 \ne w \in G$, $w(x) = W^T\Phi(x)$, then the ellipticity condition $0 < \gamma\|v\|_s^2 \le a(v,v) \;\forall\, v \in H$ implies

$$0 < \gamma|M^{1/2}W|^2 = \gamma|w|^2 \le a(w,w) = W^TKW.$$

Thus we have

$$0 < \gamma W^T M W \le W^T K W \qquad \forall\, 0 \neq W \in \mathbb{R}^m$$

and a substitution of $W = M^{-1/2}Z$ proves the desired result, $A \le -\gamma I$.

We summarize the results of this section in the following two theorems:

(6.2.9) Theorem. *(i) Let the parabolic problem* (6.1.3) *and the Galerkin subspace* $G \subset H$ *satisfy Assumption* (6.1.2); *let the exact solution* u *satisfy* $u^{(i)}(\cdot,t) \in H$, $i = 0,\dots,\ell$, $u^{(\ell+1)}(\cdot,t) \in L^2(\Omega)$ *for* $t > 0$, *and let the Ritz projection* $\tilde{u}$ *be (p+1)-times continuously differentiable with respect to* t.
(ii) Let the method (6.1.10) *be consistent of order* $p \ge \ell$ *with stability region* S.
(iii) Let $Sp(-\Delta t M^{-1/2} K M^{-1/2}) \subseteq R \subseteq S$ *where* R *is closed in* $\bar{\mathbb{C}}$.
Then for $n = k, k+1, \dots,$

$$|(u - u_\Delta)(\cdot)_n| \le \kappa_G \Delta x^q |||u|||_{q_*,n}$$
$$+ \kappa_R\Big[\textstyle\sum_{i=0}^{k-1}|(\tilde{u} - u_\Delta)(\cdot)_i| + \Delta t^p \int_0^{n\Delta t} |\tilde{u}^{(p+1)}(\cdot,\tau)|\,d\tau + \kappa_G n\Delta t \Delta x^q \max_{1\le i\le \ell} |||u^{(i)}|||_{q_*,n}\Big].$$

(6.2.10) Theorem. *Let the assumptions of Theorem* (6.2.9) *be fulfilled and let the method* (6.1.10) *fulfil Assumption (ii) of Corollary* (1.3.16). *Then for* $n = k, k+1, \dots,$

$$|(u - u_\Delta)(\cdot)_n| \le \kappa_G \Delta x^q |||u|||_{q_*,n} + \kappa_S\Big[e^{-\kappa_S^*\gamma(n-k)\Delta t}\textstyle\sum_{i=0}^{k-1}|(\tilde{u} - u_\Delta)(\cdot)_i|$$
$$+ \Delta t^p \int_0^{n\Delta t} e^{-\kappa_S^*\gamma((n-k)\Delta t-\tau)} |\tilde{u}^{(p+1)}(\cdot,\tau)|\,d\tau$$
$$+ \kappa_G \Delta t \textstyle\sum_{\nu=k}^{n} e^{-\kappa_S^*\gamma(n-\nu)\Delta t} \max_{(\nu-k)\Delta t\le\tau\le\nu\Delta t} \max_{1\le i\le\ell} \|u^{(i)}(\cdot,\tau)\|_{q_*}\Big].$$

The strong assumptions on the smoothness of the data can be released substantially. They are only introduced here in this form in order to make the statements not too complicated.

With respect to the system (6.2.1) Runge-Kutta methods differ from single-step multiderivative methods only in the treatment of the time-dependent vector $\hat{C}(t)$. However, by Theorem (5.3.12), the error bound of Theorem (6.2.9) contains here the additional term

$$\Delta t^p n\Delta t \textstyle\sum_{k=p^*}^{p} |||A^{p+1-k} M^{1/2} \tilde{U}^{(k)}|||_n.$$

6.3. Error Estimates for Galerkin-Multistep Procedures and Hyperbolic Problems

The Ritz projection $\tilde{u}$ of the exact solution u of the hyperbolic model problem (6.1.6) satisfies

(6.3.1)
$$(\tilde{u}_{tt}(\cdot,t),v) + b(t;\tilde{u}_t(\cdot,t),v) + \chi a(\tilde{u}_t(\cdot,t),v) + a(\tilde{u}(\cdot,t),v)$$
$$= (c(\cdot,t),v) - ([u_{tt} - u_{tt}^{\sim}](\cdot,t),v) - b(t;[u_t - u_t^{\sim}](\cdot,t),v) \ \forall\, v \in G,\ t > 0.$$

If $b \not\equiv 0$ then let $w^{\dagger}(\cdot,t) \in G$ be the projection of $w(\cdot,t) \in H$ with respect to the positive definite and symmetric bilinear form $b(t;\cdot,\cdot)$,

$$b(t;w(\cdot,t) - w^{\dagger}(\cdot,t),v) = 0 \ \forall\, v \in G,$$

and let

(6.3.2) $$A^2 = - M^{-1/2}KM^{-1/2},\ B(t) = - M^{-1/2}N(t)M^{-1/2}.$$

Then the time-dependent part $\tilde{U}$ of $\tilde{u} = \tilde{U}^T\Phi$ satisfies

(6.3.3)
$$M^{1/2}\tilde{U}'' = A^2M^{1/2}\tilde{U} + (B(t) + \chi A^2)M^{1/2}\tilde{U}' + M^{1/2}\hat{C}(t)$$
$$- M^{1/2}\hat{H}_1(t) - M^{-1/2}N(t)H_2^{\dagger}(t)$$

where the matrix N(t) is defined in (6.1.7) and

$$\hat{h}_1(x,t) = [u_{tt} - u_{tt}^{\sim}]^{\wedge}(x,t) = \hat{H}_1(t)^T\Phi(x),$$

$$h_2^{\dagger}(x,t) = [u_t - u_t^{\sim}]^{\dagger}(x,t) = H_2^{\dagger}(t)^T\Phi(x).$$

In the following lemma all outstanding bounds are assembled which are needed in this section.

<u>(6.3.4) Lemma</u>. *Let the solution* u *of* (6.1.6) *be sufficiently smooth and let Assumption* (6.1.2) *be fulfilled. Then*

$$|M^{1/2}\hat{H}_1^{(i)}(t)| \le \kappa_G\Delta x^q\|u^{(i+1)}(\cdot,t)\|_{q_*};$$

$$|M^{-1/2}N(t)H_2^{\dagger(i)}(t)| \le \beta\kappa_G\Delta x^q\|u^{(i+1)}(\cdot,t)\|_{q_*} \text{ if } i = 1 \text{ or } N(t) = N \text{ constant};$$

$$|AM^{1/2}\tilde{U}(t)| \le \Gamma^{1/2}\|u(\cdot,t)\|_s.$$

Proof. The first assertion follows from (6.2.8). For the second assertion it suffices to consider the case i = 1. Because

$$|M^{-1/2}N(t)H_2^\dagger(t)| \le |M^{-1/2}N(t)^{1/2}||N(t)^{1/2}H_2^\dagger(t)|$$

and

$$|M^{-1/2}N(t)^{1/2}| = |N(t)^{1/2}M^{-1/2}| = \max_{|V|=1} V^TM^{-1/2}N(t)M^{-1/2}V$$

$$\le \beta\max_{|V|=1} V^TM^{-1/2}MM^{-1/2}V = \beta$$

this result follows from

$$|N(t)^{1/2}H_2^\dagger(t)|^2 = H_2^\dagger(t)^TN(t)H_2^\dagger(t) = b(t;h_2^\dagger(\cdot,t),h_2^\dagger(\cdot,t))$$

$$\le b(t;h_2(\cdot,t),h_2(\cdot,t)) \le \beta|h_2(\cdot,t)|_0^2$$

$$= \beta|[u_t - u_t^\sim](\cdot,t)|_0^2 \le \beta\kappa_G^2\Delta x^{2q}\|u_t(\cdot,t)\|_{q*}^2 .$$

The third assertion finally follows from

$$|AM^{1/2}\tilde{U}(t)|^2 = \tilde{U}(t)^TK\tilde{U}(t) = a(\tilde{u}(\cdot,t),\tilde{u}(\cdot,t)) \le a(u(\cdot,t),u(\cdot,t)) \le \Gamma\|u(\cdot,t)\|_s^2 .$$

For the solution of the semi-discrete problem (6.1.8) by an indirect multistep method, we have to write the differential system of second order as an explicit differential system of first order,

$$Z' = - M^{-1}K^*(t)Z + C(t),$$

where $Z = (U_G, U_G')^T$, $C(t) = (0, \hat{C}(t))^T$, and

$$K^*(t) = \begin{bmatrix} 0 & -M \\ K & N(t) + \chi K \end{bmatrix},$$

cf. Section 1.4. If we suppose here that the damping does not depend on time then $N(t) = N$, $K^*(t) = K^*$, and the computational device is again (6.1.10),

$$(6.3.5)\quad \sum_{i=0}^k\sigma_i(-\Delta tM^{-1}K^*)T^iV_n^* = - \sum_{i=0}^k\sum_{j=1}^\ell\sigma_{ij}(-\Delta tM^{-1}K^*)\Delta t^jT^iC_n^{(j-1)},\quad n = 0,1,\dots .$$

But now we have $V_n^* = (V_n, W_n)^T$ and the functions $u_{\Delta,1}$ and $u_{\Delta,2}$ with $u_{\Delta,1}(x,n\Delta t) = V_n^T\Phi(x)$, $u_{\Delta,2}(x,t) = W_n^T\Phi(x)$ represent approximations of $u(x,n\Delta t)$ and $u'(x,n\Delta t)$ respectively.

For the error estimation we consider the modified system

(6.3.6) $\quad Z^{*\prime} = A^*Z^* + C^*(t) - H^*(t)$

with the notations

$$Z^* = (M^{1/2}\tilde{U},\ M^{1/2}\tilde{U}')^T,\ C^*(t) = (0,\ M^{1/2}\hat{C}(t))^T,$$

$$H^*(t) = (0,\ M^{1/2}\hat{H}_1(t) + M^{-1/2}NH_2^+(t))^T,$$

and

$$A^* = \begin{bmatrix} 0 & I \\ A^2 & B + \chi A^2 \end{bmatrix},$$

cf. (6.3.2). Then, if (6.3.5) is also written as

(6.3.7) $\quad \sum_{i=0}^{k}\sigma_i(\Delta tA^*)T^iM^{1/2}V_n^* = -\sum_{i=0}^{k}\sum_{j=1}^{\ell}\sigma_{ij}(\Delta tA^*)\Delta t^jT^iM^{1/2}C_n^{*(j-1)},\quad n = 0,1,\ldots,$

the pair (6.3.6), (6.3.7) of analytic equation and numerical method is the same as that considered in Section 1.4 and 1.5:

If the damping in the hyperbolic equation (6.1.6) is independent of a, i.e., $\chi = 0$, and b is independent of space and time,

$$b(t;u,v) = \beta\cdot(u,v),\ \beta \geq 0,$$

then $B(t) = B = -\beta I$ commutes with A^2. If then moreover damping is less than critical damping, $\beta \leq 2\sqrt{\gamma}$, then Theorem (1.4.7) yields immediately the following error bound:

$$|M^{1/2}(\tilde{U}_n - V_n)| + |A|^{-1}|M^{1/2}(\tilde{U}_n' - W_n)|$$

$$\leq \kappa_R[(\gamma+1)/(2\gamma-\sqrt{\gamma}\beta)]^{1/2}\Big[\sum_{i=0}^{k-1}(|M^{1/2}(\tilde{U}_i - V_i)| + |M^{1/2}(\tilde{U}_i' - W_i)|)$$

$$+ \Delta t^p\int_0^{n\Delta t}(|M^{1/2}\tilde{U}^{(p+1)}(\tau)| + |M^{1/2}\tilde{U}^{(p+2)}(\tau)|)d\tau$$

$$+ \kappa_G n\Delta t\max_{0\leq i\leq \ell-1}|||M^{1/2}H^{*(i)}|||_n\Big].$$

If $\chi > 0$ and b does not depend on time then Theorem (5.4.16) yields a similar bound for diagonal and subdiagonal Padé approximants. Both error estimations can be written with the arguments $\tilde{u}$, $u_{\Delta,1}$, $u_{\Delta,2}$, and u instead of $M^{1/2}\tilde{U}$, $M^{1/2}V$, $M^{1/2}W$, and H^* by the fundamental relation (6.2.3) and Lemma (6.3.4). The results are summarized in the following theorems.

<u>(6.3.8) Theorem</u>. *(i) Let the hyperbolic problem* (6.1.6) *and the Galerkin subspace* $G \subset H$ *satisfy Assumption* (6.1.2); *let the exact solution* u *satisfy* $u^{(i)}(\cdot,t) \in H$, $i = 0,\ldots$ $\ldots,\ell+1$, $u^{(\ell+2)}(\cdot,t) \in L^2(\Omega)$ *for* $t > 0$, *and let the Ritz projection* $\tilde{u}$ *be (p+2)-times continuously differentiable with respect to* t.
(ii) Let $\chi = 0$ *in* (6.1.6), $b(t;u,v) = \beta\cdot(u,v)$, *and* $0 \le \beta < 2\sqrt{\gamma}$ *where* γ *is the ellipticity constant.*
(iii) Let the method (6.3.5) *be consistent of order* $p \ge \ell$ *with stability region* S.
(iv) Let $Sp(-\Delta t M^{-1/2}KM^{-1/2}) \subseteq R \subseteq S$ *where* R *is closed in* $\overline{\mathbb{C}}$.
Then for $n = k,k+1,\ldots$,

$$|(\tilde{u} - u_{\Delta,1})(\cdot)_n| + |M^{-1/2}KM^{-1/2}|^{-1/2}|(\tilde{u}_t - u_{\Delta,2})(\cdot)_n|$$

$$\le \kappa_R[(\gamma+1)/(2\gamma-\sqrt{\gamma}\beta)]^{1/2}\Big[\sum_{i=0}^{k-1}(|(\tilde{u} - u_{\Delta,1})(\cdot)_i| + |(\tilde{u}' - u_{\Delta,2})(\cdot)_i|)$$

$$+ \Delta t^p \int_0^{n\Delta t} (|\tilde{u}^{(p+1)}(\cdot,\tau)| + |\tilde{u}^{(p+2)}(\cdot,\tau)|)d\tau$$

$$+ \kappa_G(1+\beta)n\Delta t\Delta x^q \max_{1\le i\le \ell+1} |||u^{(i)}|||_{q_*,n}\Big].$$

<u>(6.3.9) Theorem</u>. *(i) Let assumption (i) of Theorem* (6.3.8) *be fulfilled.*
(ii) Let $\chi \ge 0$ *in* (6.1.6) *and let the bilinear form* b *be independent of time,* $b(t;u,v) = b(u,v)$.
(iii) Let the method (6.3.5) *be a diagonal or subdiagonal Padé approximant of order* p.
Then for $n = 1,2,\ldots$,

$$|(\tilde{u} - u_{\Delta,1})(\cdot)_n| + |(\tilde{u}_t - u_{\Delta,2})(\cdot)_n|$$

$$\le \kappa\gamma^{-1/2}\Big[\Gamma^{1/2}\|(\tilde{u} - u_{\Delta,1})(\cdot,0)\|_S + |(\tilde{u}_t - u_{\Delta,2})(\cdot,0)|$$

$$+ \Delta t^p \int_0^{n\Delta t} (\Gamma^{1/2}\|u^{(p+1)}(\cdot,\tau)\|_S + |\tilde{u}^{(p+2)}(\cdot,\tau)|)d\tau$$

$$+ \kappa_G(1+\beta)n\Delta t\Delta x^q \max_{1\le i\le \ell+1} |||u^{(i)}|||_{q_*,n}\Big].$$

Now we turn to the direct methods studied in Chapter II and consider at first hyperbolic problems (6.1.6) without damping, i.e., $b \equiv 0$ and $\chi = 0$. Obviously, the multistep multiderivative method (2.1.33) and the Nyström type procedure (2.4.9), (2.4.11) yield computational devices for the semi-discrete problem (6.1.8) if we substitute $A^2 = -M^{-1}K$ and $c_n = \hat{C}_n$.

Theorem (2.1.34) and (2.1.39) apply in the same way as above. The result is stated in the following theorem:

(6.3.10) Theorem. *(i) Let the hyperbolic problem* (6.1.6) *and the Galerkin subspace* $G \subset H$ *satisfy Assumption* (6.1.2); *let the exact solution* u *satisfy* $u^{(i)}(\cdot,t) \in H$, $i = 0,\dots,2\ell$, $u^{(2\ell+2)}(\cdot,t) \in L^2(\Omega)$ *for* $t > 0$, *and let the Ritz projection* $\tilde{u}$ *be (p+2)-times continuously differentiable with respect to* t.

(ii) Let $\chi = 0$ *and* $b \equiv 0$.

(iii) Let the method (2.1.33) *be consistent of order* $p \geq 2\ell-1$ *with stability region* S.

(iv) Let $Sp(\Delta t^2 A^2) \subseteq R \subseteq S$ *where* R *is closed in* $\bar{\mathbb{C}}$ *and* $A^2 = -M^{-1/2}KM^{-1/2}$.

Then for $n = k,k+1,\dots,$

$$|(\tilde{u} - u_\Delta)(\cdot)_n| \leq \kappa_R n\Delta t\cdot\Delta t^{-1}\sum_{i=0}^{k-1}|(\tilde{u} - u_\Delta)(\cdot)_i| + \Psi$$

where

$$\Psi = \kappa_R n\Delta t\Big[\Delta t^p \int_0^{n\Delta t} |\tilde{u}^{(p+2)}(\cdot,\tau)|d\tau + \kappa_G n\Delta t\Delta x^q \max_{2\leq i\leq 2\ell} |||u^{(i)}|||_{q_*,n}\Big].$$

If the method is strongly D-stable in $R = [-s, 0]$ *(and* $\kappa_R \geq 1$*) then*

$$|(\tilde{u} - u_\Delta)(\cdot)_n| \leq \kappa_R\Big[(1 + n\Delta t|A|)\sum_{i=0}^{k-2}|(\tilde{u} - u_\Delta)(\cdot)_i| + n\Delta t\Delta t^{-1}\sum_{i=1}^{k-1}|(\tilde{u}-u_\Delta)(\cdot)_i - (\tilde{u}-u_\Delta)(\cdot)_{i-1}|\Big] + \Psi.$$

Theorem (2.4.14) and (2.4.17) lead to the following result where A^2 and Ψ have the same meaning as in Theorem (6.3.10).

(6.3.11) Theorem. *Let the assumption of Theorem* (6.3.10) *be fulfilled for the Nyström type procedure* (2.4.9). *Then for* $n = 1,2,\dots,$

$$\max\{|(\tilde{u} - u_{\Delta,1})(\cdot)_n|,\ \Delta t|(\tilde{u}_t - u_{\Delta,2})(\cdot)_n|\}$$
$$\leq \kappa_R n\Delta t\cdot\Delta t^{-1}[|(\tilde{u} - u_{\Delta,1})(\cdot)_0| + |(\tilde{u}_t - u_{\Delta,2})(\cdot)_0|] + \Psi.$$

If the method is strongly D-stable in $R = [-s, 0]$ *then*

$$|(\tilde{u} - u_{\Delta,1})(\cdot)_n| \leq \kappa_R[(1 + n\Delta t|A|)(\tilde{u} - u_{\Delta,1})(\cdot)_0| + n\Delta t|(\tilde{u}_t - u_{\Delta,2})(\cdot)_0|] + \Psi,$$

$$\Delta t|(\tilde{u}_t - u_{\Delta,2})(\cdot)_n| \leq \Delta t\kappa_R[n\Delta t|A^2||(\tilde{u} - u_{\Delta,1})(\cdot)_0| + (1+n\Delta t|A|)|(\tilde{u}_t - u_{\Delta,2})(\cdot)_0|] + \Psi.$$

In Section 2.3 and 2.4 we have considered linear multistep methods and differential systems of second order with damping. The scheme (2.2.3) yields for the semi-discrete problem (6.1.8) the computational device

(6.3.12) $M\rho_0(T)V_n - \Delta t^2 K\rho_1(T)V_n - \Delta t\sum_{i=0}^k \beta_i(N_{n+i} + \chi K)\tau_i(T)V_n = -\Delta t^2\rho_1(T)C_n^*$, $n = 0,1,\ldots,$

cf. the corresponding scheme (6.1.11) for the differential system of first order (6.1.9). If the bilinear form b does not depend on time then we can write

$$\tau(\zeta) = \sum_{i=0}^k \beta_i\tau_i(\zeta)$$

and obtain

(6.3.13) $M\rho_0(T)V_n - \Delta t^2 K\rho_1(T)V_n - \Delta t(N + \chi K)\tau(T)V_n = -\Delta t^2\rho_1(T)C_n^*$, $\quad n = 0,1,\ldots,$

cf. (2.3.3). For the error estimation again the pair of analytic equation and numerical procedure, (6.1.8) and (6.3.12), is slightly modified into (6.3.3) and

$$\rho_0(T)M^{1/2}V_n + \Delta t^2A^2\rho_1(T)M^{1/2}V_n + \Delta t\sum_{i=0}^k\beta_i(B_{n+i} + \chi A^2)\tau_i(T)M^{1/2}V_n$$

$$= -\Delta t^2\rho_1(T)M^{1/2}\hat{C}_n, \qquad n = 0,1,\ldots,$$

with the notations (6.3.2). For $b(t;u,v) = \beta\cdot(u,v)$ we have the case of orthogonal damping and Theorem (2.3.13) and (2.3.15) yield

<u>(6.3.14) Theorem.</u> *(i) Let assumption (i) of Theorem* (6.3.10) *be fulfilled for* $\ell = 1$.
(ii) Let $\chi \geq 0$ *and* $b(t;u,v) = \beta\cdot(u,v)$.
(iii) Let the method (6.3.13) *be consistent of order* p *with the two-dimensional stability region* S^2.
Let $Sp(\Delta t^2A^2)\times Sp(\Delta t(-\beta I+\chi A^2)) \subseteq R \subseteq S^2$ *where* R *is closed in* $\overline{\mathbb{C}^2}$ *and* $A^2 = -\Delta t^2M^{-1/2}K\cdot M^{-1/2}$.
Then for $n = k,k+1,\ldots,$

$$|(\tilde{u} - u_\Delta)(\cdot)_n| \leq \Xi + \kappa_R n\Delta t\left[\Delta t^p\int_0^{n\Delta t}(|\tilde{u}^{(p+2)}(\cdot,\tau)| + \psi(\tau))d\tau + \kappa_G n\Delta t\Delta x^q |||u_{tt}|||_{q_*,n}\right]$$

where

$$\Xi = \kappa_R n\Delta t\cdot\Delta t^{-1}\sum_{i=0}^{k-1}|(\tilde{u} - u_\Delta)(\cdot)_i|$$

and

$$\psi(\tau) = (\beta + \chi|A^2|)|\tilde{u}^{(p+1)}(\cdot,\tau)|.$$

If the method is strongly D-stable in $R = [-s,\ 0]\times[-r,\ 0]$ *(and* $\kappa_R \geq 1$*) then*

$$\Xi = \kappa_R[(1 + n\Delta t(\beta + \chi^{1/2}|A|))\sum_{i=0}^{k-2}|(\tilde{u} - u_\Delta)(\cdot)_i| +$$

$$+ n\Delta t\cdot\Delta t^{-1}\textstyle\sum_{i=1}^{k-1}|(\tilde{u} - u_\Delta)(\cdot)_i - (\tilde{u} - u_\Delta)(\cdot)_{i-1}|]$$

If $\tau(\zeta)$ *has exact degree* k *and*

$$|\tau(T)u(t) - \Delta t\rho_1(T)u'(t)| \le \kappa\Delta t^{p+2} \qquad \forall\, u \in C^{p+2}(\mathbb{R};\mathbb{R}^m)$$

then $\psi(\tau) \equiv 0$.

If the bilinear form b in the hyperbolic model problem (6.1.6) does not depend on time but depends on the space variable $x \in \mathbb{R}^r$ such that the matrix $N = [b(\phi_\mu,\phi_\nu)]_{\mu,\nu=1}^m$ does no longer commute with the stiffness matrix K then Theorem (2.2.10) yields the following result where suppose that $\chi = 0$ because K is an ill-conditioned matrix.

<u>(6.3.15) Theorem.</u> *(i) Let assumption (i) of Theorem* (6.3.10) *be fulfilled for* $\ell = 1$.
(ii) Let $\chi = 0$ *and let* b *be independent of time,* $b(t;u,v) = b(u,v)$.
(iii) Let the method (6.3.13) *be consistent of order* p *with stability region* S.
(iv) Let $Sp(\Delta t^2A^2) \subseteq R \subseteq S$ *where* R *is closed in* $\overline{\mathbb{C}}$ *and* $A^2 = - M^{-1/2}KM^{-1/2}$.
(v) Let $\tau(\zeta)$ *be a polynomial of degree not greater than* k-1 *and let* $\rho_0(0) + \eta^2\rho_1(0) \neq 0\ \forall\, \eta^2 \in R$.
Then for $n = k,k+1,\ldots,$

$$|(\tilde{u} - u_\Delta)(\cdot)_n| \le \kappa_R n\Delta t\exp\{\kappa_R^*\beta n\Delta t\}\Big[\Delta t^{-1}\textstyle\sum_{i=0}^{k-1}|(\tilde{u} - u_\Delta)(\cdot)_i|$$
$$+ \Delta t^p\int_0^{n\Delta t}(|\tilde{u}^{(p+2)}(\cdot,\tau)| + \beta|\tilde{u}^{(p+1)}(\cdot,\tau)|)d\tau + \kappa_G n\Delta t\Delta x^q|||u_{tt}|||_{q_*,n}\Big].$$

Finally, if b has the general time-dependent form stipulated in Section 6.1 then Theorem (2.2.18) applies. However, we assume again that $\chi = 0$ because in the other case the exponential multiplication factor grows up if the parameter Δx of the space discretization becomes small.

<u>(6.3.16) Theorem.</u> *Let assumption (i), (iii), and (iv) of Theorem* (6.3.15) *be fulfilled, let* $\chi = 0$ *in* (6.1.6), *and let the method* (6.3.12) *fulfil assumption (ii) and (iv) of Theorem* (2.2.18) *for* $R = [-s, 0]$ *and* $B(t) = -M^{-1/2}N(t)M^{-1/2}$. *Then for* $n = k,k+1,\ldots,$

$$|(\tilde{u} - u_\Delta)(\cdot)_n| \le \textstyle\sum_{i=0}^{k-2}|(\tilde{u} - u_\Delta)(\cdot)_i|$$
$$+ \kappa_R n\Delta t\exp\{\kappa_R^*\beta n\Delta t\}\Big[|A|\textstyle\sum_{i=0}^{k-2}|(\tilde{u} - u_\Delta)(\cdot)_i| + \Delta t^{-1}\sum_{i=1}^{k-1}|(\tilde{u} - u_\Delta)(\cdot)_i - (\tilde{u} - u_\Delta)(\cdot)_{i-1}|$$
$$+ \Delta t^p\int_0^{n\Delta t}(|\tilde{u}^{(p+2)}(\cdot,\tau)| + \beta|\tilde{u}^{(p+1)}(\cdot,\tau)|)d\tau + \kappa_G n\Delta t\Delta x^q|||u_{tt}|||_{q_*,n}\Big].$$

Appendix

A.1. Auxiliary Results on Algebraic Functions

Let $\pi(\zeta,\eta)$ be a polynomial in ζ and η with real or complex coefficients,

(A.1.1) $\pi(\zeta,\eta) = \sum_{i=0}^{k}\sigma_i(\eta)\zeta^i = \sum_{j=0}^{\ell}\rho_j(\zeta)\eta^j, \ \sigma_k(\eta) \not\equiv 0.$

Then the roots $\zeta_i: \eta \to \zeta_i(\eta)$, $i = 1,\dots, k$, are regular functions in $\mathbb{C}$ with exception of the critical points. These are the singularities η_0 where $\sigma_k(\eta_0) = 0$ and the branching points of the algebraic function ζ defined by

(A.1.2) $\pi(\zeta(\eta),\eta) \equiv 0,$

i.e., the points η_1 in which some roots $\zeta_i(\eta_1)$ of $\pi(\zeta,\eta_1)$ coincide. Obviously, there exist at most ℓ poles and if $\pi(\zeta,\eta)$ is irreducible, i.e., if there exist no polynomials $\pi_1(\zeta,\eta)$ and $\pi_2(\zeta,\eta)$ such that $\pi(\zeta,\eta) = \pi_1(\zeta,\eta)\pi_2(\zeta,\eta)$, then there is only a finite number of branching points. See e.g. Ahlfors [53] and Behnke and Sommer [65]. $\eta_0 \in \overline{\mathbb{C}}$ is a removable singularity or a branching point iff all branches ζ_i of ζ remain bounded in a neighborhood of η_0. Accordingly, the behavior of ζ at the point $\eta = \infty$ is ruled by the following simple result.

(A.1.3) Lemma. *Let the coefficients* σ_i *of* (A.1.1) *be polynomials of degree* $\deg(\sigma_i)$. *Then all roots of* $\pi(\zeta,\eta)$ *are bounded in a neighborhood of* $\eta = \infty$ *iff*

$$\deg(\sigma_k) \geq \max_{0\leq i\leq k-1}\deg(\sigma_i).$$

Proof. The rational functions $\sigma_i(\eta)/\sigma_k(\eta)$ are the elementary symmetric functions of $\zeta_1(\eta),\dots, \zeta_k(\eta)$ by Vieta's Root Criterium hence they are bounded near $\eta = \infty$ if $\zeta_1(\eta)$, $\dots, \zeta_k(\eta)$ are bounded. But $\lim_{\eta\to\infty}|\sigma_i(\eta)/\sigma_k(\eta)| < \infty$ holds only if $\deg(\sigma_i) \leq \deg(\sigma_k)$ therefore the condition is necessary. On the other side, if the condition is fulfilled and if $\ell = \deg(\sigma_k)$ then the polynomial $\eta^\ell\pi(\zeta,\eta^{-1})$ in ζ has polynomial coefficients in η and the leading coefficient $\eta^\ell\sigma_k(\eta^{-1})$ is unequal zero for $\eta = 0$. Thus the roots $\zeta_i(\eta^{-1})$ of $\pi(\zeta,\eta^{-1})$ cannot possibly have a genuine singularity in $\eta = 0$. This proves the sufficiency.

If some roots $\zeta_i(\eta)$ of (A.1.1), say r, coalesce in the point η_1 which is not a pole then these branches of the algebraic function ζ are regular in a neighborhood of η_1

or they behave locally like $(\eta - \eta_1)^{p/q}$ where $p,q \in \mathbb{N}$ and $1 \le q \le r \le k$. This is a classical result of analytic function theory which reads more exactly as follows (cf. e.g. Ahlfors [53], Hille [62], and Behnke and Sommer [65]).

<u>(A.1.4) Theorem.</u> *If* $\sigma_k(\eta_1) \neq 0$ *and if* r *roots of* $\pi(\zeta,\eta)$ *coincide in* η_1, *say* $\zeta_1(\eta_1) = \zeta_2(\eta_1) = \dots = \zeta_r(\eta_1)$, *then there exists a neighborhood* $\mathcal{N}$ *of zero such that – possibly after some permutation –*

$$\text{(A.1.5)}\quad \zeta_\nu(\eta) = \zeta_1(\eta_1) + \sum_{\mu=p}^{\infty}\phi_\mu[e^{2\pi i\nu/q}(\eta - \eta_1)^{1/q}]^\mu, \qquad \eta - \eta_1 \in \mathcal{N},\ \nu = 0,\dots,q-1,$$

where $p,q \in \mathbb{N}$ *have no common factor,* $1 \le q \le r \le k$, *and* $(\eta - \eta_1)^{1/q}$ *is a fixed branch of* $\xi^q - (\eta - \eta_1) = 0$.

If a simply connected domain $\mathfrak{D}$ is given which contains no branching points or singularities in its interior then this representation shows that all roots of $\pi(\zeta,\eta)$ can be numbered in such a way that they are continuous in $\mathfrak{D}$.

Let

$$S^* = \{\eta \in \overline{\mathbb{C}},\ |\zeta_i(\eta)| \le 1,\ i = 1,\dots,k\}$$

be the general 'stability region' of $\pi(\zeta,\eta)$. We shall show in this section that for $\eta_1 \in \partial S^*$ the values of p, q, r, and ϕ_p associated with unimodular roots in η_1 by (A.1.5) determine the shape of S^* near η_1. But at first we derive some auxiliary results for the following case:

Let $\pi(\zeta,\eta^2)$ be the *characteristic* polynomial of a convergent multistep multiderivative method for differential systems of *second* order then there exist at least two roots, $\zeta_1(\eta)$ and $\zeta_2(\eta)$, which coalesce to a double roots of modulus one for $\eta = 0$. Without loss of generality let all pairs with this property be ζ_j, ζ_{j+1}, $j = 1,3,\dots,k_*-1$, $k_* < k$. (As the method is supposed to be convergent $\pi(\zeta,0) = \rho_0(\zeta)$ can have only simple or double roots of modulus one.) Then (A.1.5) yields

$$\text{(A.1.6)}\quad \zeta_{j,j+1}(\eta) = \zeta_j(0) \pm \hat{\chi}_j\eta + \mathcal{O}(|\eta|^2), \qquad \eta \to 0,$$

where

$$\text{(A.1.7)}\quad \hat{\chi}_j^2 = -2\rho_1(\zeta_j(0))/\rho_0''(\zeta_j(0))$$

and hence $\hat{\chi}_j^2 \neq 0$ if the method is linear. Recall that the stability region S is defined here with respect to η^2 (Definition (2.1.14)).

<u>(A.1.8) Lemma.</u> *Let* $[-s, 0] \subset S$, $0 < s \le \infty$, *and let* $\hat{\chi}_j^2 \neq 0$ *for* $j = 1,3,\dots,k_*-1$. *Then*

$$\max_{1\le i\le k_*}|\zeta_i(0) - \zeta_i(\eta)| \le \kappa_s|\eta|, \qquad -s \le \eta^2 \le 0.$$

Proof. Because of (A.1.5) and (A.1.6) we can write

$$\text{(A.1.9)} \qquad \zeta_{j,j+1}(\eta) = \zeta_j(0) \pm \hat{\chi}_j\eta(1 + \psi_j(\pm\eta))$$

where $\psi_j(\eta) = (\zeta_j(\eta) - \zeta_j(0))/\hat{\chi}_j\eta - 1$ is bounded in $[-s, 0]$ because $|\zeta_j(\eta)| \le 1$ in this interval, and $\psi_j(0) = 0$. Accordingly, the assertion follows with

$$\kappa_s = \max_{j=1,3,\ldots,k_*-1}\sup_{-s\le\eta^2\le 0}|\hat{\chi}_j||1 + \psi_j(\pm\eta)| < \infty.$$

<u>(A.1.10) Lemma</u>. *Let the assumption of Lemma* (A.1.8) *be fulfilled but let all roots* $\zeta_i(\eta)$ *of* $\pi(\zeta,\eta^2)$ *be simple for* $-s \le \eta^2 \le 0$ *with exception of* $\eta^2 = 0$. *Then*

(i) $\qquad \max_{1\le i\le k}|\zeta_i(\eta_1) - \zeta_i(\eta_2)| \le \kappa_s|\eta_1 - \eta_2|, \qquad -s \le \eta_\nu^2 \le 0,\ \nu = 1,2,\ s < \infty.$

(ii) $\qquad \max_{j=1,3,\ldots,k_*-1}\{|\zeta_j(\eta) - \zeta_{j+1}(\eta)|^{-1}\} \le \kappa_s\max\{1,|\eta|^{-1}\}, \ -s \le \eta^2 \le 0,\ s \le \infty.$

Proof. (i) If $\zeta_i(0)$ is a simple root of $\pi(\zeta,0) = \rho_0(\zeta)$ then $|\zeta_i'(\eta)|$ is bounded in the finite interval $[-s, 0]$ and the assertion is an immediate consequence of the Mean Value Theorem. If $\zeta_j(0)$ is a double root of $\pi(\zeta,0)$ then it can be written in the form (A.1.9) where now the function ψ_j is regular on the entire interval $[-s, 0]$, i.e., in an open domain containing $[-s, 0]$, because $\zeta_j(\eta)$ is regular in $[-s, 0)$ by assumption. Hence the assertion follows again by the Mean Value Theorem applied to (A.1.9)
(ii) By (A.1.6) there is a $\delta > 0$ and a $\kappa_1 > 0$ such that

$$|\zeta_j(\eta) - \zeta_{j+1}(\eta)| \ge \kappa_1|\eta|, \qquad -\delta \le \eta^2 \le 0,\ j = 1,3,\ldots,k_*-1,$$

and in $[-s, -\delta]$ all roots are bounded away from each other by assumption hence

$$\kappa_s = \min_{j=1,3,\ldots,k_*-1}\min_{-s\le\eta^2\le-\delta}|\zeta_j(\eta) - \zeta_{j+1}(\eta)| > 0.$$

Both bounds together yield

$$|\zeta_j(\eta) - \zeta_{j+1}(\eta)| \ge \min\{\kappa_s,\kappa_1\}\min\{1,|\eta|\}, \quad -s \le \eta^2 \le 0,\ j = 1,3,\ldots,k_*-1.$$

Next, we have to prove a modification of Lemma (A.1.8) for linear multistep methods and differential systems with orthogonal damping. In this case the characteristic polynomial has the form

(A.1.11) $\pi(\zeta,\eta^2,\mu) = \rho_0(\zeta) + \eta^2\rho_1(\zeta) + \mu\tau(\zeta).$

The method defined by this polynomial is consistent iff (2.3.8) is fulfilled, i.e.,

(A.1.12) $\rho_0(1) = \rho_0'(1) = \rho_0''(1) + 2\rho_1(1) = \tau(1) = \tau'(1) - \rho_1(1) = 0,$

and if $0 \in S$ for $\mu = 0$ then $\rho_0''(1) = -2\rho_1(1) \neq 0$. Recall that the stability region S^2 is defined here with respect to (η^2,μ) (Definition (2.3.12)). By a slight modification of Theorem (A.1.4) the principal roots $\zeta_1(\eta,\mu)$ and $\zeta_2(\eta,\mu)$ are now regular functions of η and μ in a neighborhood of $(\eta,\mu) = (0,0)$ and we obtain instead of (A.1.9)

$$\zeta_{1,2}(\eta,\mu) = 1 + (\pm\hat{\chi}\eta + \omega\mu)(1 + \psi(\pm\eta,\mu))$$

where $\hat{\chi}^2 = -\rho_1(1)/\rho_0''(1)$ as in (A.1.7) and $\omega = 1/\tau'(1)$. Again ψ denotes a function which is regular in a neighborhood of $(\eta,\mu) = (0,0)$ with $\psi(0,0) = 0$ and which is bounded in the stability region S^2. Hence we can state:

<u>(A.1.13) Lemma.</u> *If* (A.1.12) *holds with* $\rho_1(1) \neq 0$ *then the principal roots of a method defined by* (A.1.11) *satisfy*

$$|1 - \zeta_{1,2}(\eta,\mu)| \leq \kappa_{r,s}(|\eta| + |\mu|) \qquad \forall\ (\eta^2,\mu) \in [-s,\ 0]\times[-r,\ 0] \subset S^2.$$

After these estimations we now assume again that $\pi(\zeta,\eta)$ is an *irreducible* but otherwise *arbitrary* polynomial (A.1.1) of degree k in ζ and degree ℓ in η and consider the behavior of a root $\zeta_i(\eta)$ in a neighborhood of a point $\eta^* \in \mathbb{C}$ where $|\zeta_i(\eta^*)| = 1$. For simplicity we set $\eta^* = 0$ and write $\xi(\eta) = \zeta_i(\eta)$, $\xi(0) = \xi^*$, i.e. $|\xi^*| = 1$. Then Theorem (A.1.4) yields in a neighborhood of $\eta = 0$

(A.1.14) $\xi(\eta) = \xi^*(1 + \chi\eta^{p/q} + \mathcal{O}(\eta^s)),\qquad \eta \to 0,\ s > p/q,$

and a simple calculation shows that

(A.1.15) $|\xi(\eta)| = 1 + \mathrm{Re}(\chi\eta^{p/q}) + \mathcal{O}(|\eta|^{\min\{2p/q,s\}}),\qquad \eta \to 0.$

In this equation we insert

(A.1.16) $\chi\eta^{p/q} = \rho e^{i\theta},\qquad 0 \leq \theta < 2\pi,\ \rho > 0,$

and obtain

$$|\xi(\eta)| = 1 + \rho\cos\theta + \mathcal{O}(\rho^{\min\{2,s^*\}}), \qquad \eta \to 0,\ s^* > 1.$$

Hence, if $\varepsilon > 0$ is sufficiently small then there exists a $\rho_\varepsilon > 0$ such that

$$|\xi(\eta)| > 1 \text{ if } |\theta| \le \frac{\pi}{2} - \varepsilon,\ |\xi(\eta)| < 1 \text{ if } |\theta - \pi| \le \frac{\pi}{2} - \varepsilon,\ 0 < \rho < \rho_\varepsilon.$$

But from (A.1.16) we find that

$$\eta = (\rho/\chi)^{q/p} e^{i\theta q/p} e^{2\pi ijq/p}, \qquad j = 0,1,\ldots,p-1,$$

and, accordingly,

$$\arg\eta = \frac{q}{p}(-\arg\chi + \theta + 2\pi j).$$

We write ε for $\varepsilon q/p$ and assemble the result in the following lemma which is stated at once for general $\eta^* \in \mathbb{C}$; cf. also Jeltsch [77] and Wanner, Hairer, and Norsett [78a].

(A.1.17) Lemma. *Let*

(A.1.18) $$\xi(\eta) = \xi^*(1 + \chi(\eta - \eta^*)^{p/q} + \mathcal{O}((\eta - \eta^*)^s)), \qquad \eta \to \eta^*,\ s > p/q,$$

be a root of $\pi(\zeta,\eta)$ *with* $|\xi^*| = 1$, $\chi \ne 0$, *and* $p,q \in \mathbb{N}$ *having no common factor. Then there exists for each small* $\varepsilon > 0$ *a* $\rho_\varepsilon > 0$ *and branches* $\zeta_\mu(\eta)$, $\zeta_\nu(\eta)$ *of* (A.1.18) *such that* $|\zeta_\mu(\eta)| > 1$ *for*

(A.1.19) $$\eta = \eta^* + \rho e^{i\theta},\ 0 < \rho < \rho_\varepsilon,\ \left|\theta - \frac{q(2j\pi - \arg\chi)}{p}\right| \le \frac{q\pi}{2p} - \varepsilon,\ j = 0,1,\ldots,p-1,$$

and $|\zeta_\nu(\eta)| < 1$ *for*

(A.1.20) $$\eta = \eta^* + \rho e^{i\theta},\ 0 < \rho < \rho_\varepsilon,\ \left|\theta - \frac{q((2j+1)\pi - \arg\chi)}{p}\right| \le \frac{q\pi}{2p} - \varepsilon,\ j = 0,1,\ldots,p-1.$$

The number q-1 is called the *ramification index* of the roots (A.1.18). If $q > 2$ then the angular domains (A.1.19) overlap each other. For $q = 2$ only the half-rays

$$\eta = \eta^* + \rho e^{i\theta},\ \theta = \frac{(4j+1)\pi - 2\arg\chi}{p}, \qquad j = 0,1,\ldots,p-1,$$

are not contained asymptotically in the set defined by (A.1.19), and for $q = 1$ the angular domains (A.1.19) and (A.1.20) alternate for increasing θ. Thus we can state the following corollary where $[x]$ denotes the largest integer not greater than x.

(A.1.21) Corollary. *Let* $\eta^* \in \partial S^*$, *let* $\zeta_i(\eta)$, $i = 1,\dots,k_+$, *be the roots of* $\pi(\zeta,\eta)$ *with* $|\zeta_i(\eta^*)| = 1$, *and let* $q_i - 1$ *be the ramification index of* ζ_i *in* η^* *and* χ_i *the growth parameter defined by* (A.1.18).

(i) If $q_i > 2$ *for some* i *then there exists a disk* $\mathfrak{D}$ *with center* η^* *such that*

$$\mathfrak{D} \setminus \{\eta^*\} \subset \overline{\mathbb{C}} \setminus S^*.$$

(ii) If $q_i = 2$ *for some* i *then there exists no angular domain*

$$\mathcal{A}(\alpha,\beta,\rho) = \{\eta \in \mathbb{C},\ 0 < |\eta - \eta^*| < \rho,\ |\beta - \arg(\eta - \eta^*)| \leq \alpha\}$$

with $\alpha > 0$ *and* $\rho > 0$ *such that* $\mathcal{A}(\alpha,\beta,\rho) \subset S^*$.

(iii) For each ε, $0 < \varepsilon \leq \alpha$, *there exists an angular domain* $\mathcal{A}(\alpha - \varepsilon,\beta,\rho_\varepsilon) \subset S^*$ *iff* $q_i = 1$, $1 \leq p_i \leq \min\{\ell,[\pi/2\alpha]\}$, *and*

$$\text{(A.1.22)}\qquad |\pi - p_i\beta - \arg\chi_i| \leq \frac{\pi}{2} - p_i\alpha, \qquad i = 1,\dots,k_*.$$

Proof. We have only to verify the third assertion. $1 \leq p \leq [\pi/2\alpha]$ and (A.1.22) follow immediately from (A.1.20). In order to show $p \leq \ell$ we observe that

$$\zeta(\eta) = \zeta^*(1 + \chi\eta^p + \mathcal{O}(\eta^{p+1})) \qquad \eta \to 0,$$

holds iff the algebraic function $\eta(\zeta)$ defined by $\pi(\zeta,\eta(\zeta)) = 0$ has p branches $\eta_i(\zeta)$ of the form

$$\eta_i(\zeta) = \left(\frac{\zeta - \zeta^*}{\chi\zeta^*}\right)^{1/p} + \mathcal{O}((\zeta - \zeta^*)^{\tilde{s}}), \qquad \zeta - \zeta^* \to 0,\ \tilde{s} > 1/p,$$

according to the chosen branch of $\eta^p - ((\zeta - \zeta^*)/\chi\zeta^*) = 0$. But $\eta(\zeta)$ cannot have more than ℓ branches hence $1 \leq p \leq \ell$.

(A.1.21)(iii) is illustrated by Example (A.4.7). In the situation of (A.1.21)(ii) always two segments of the boundary curve ∂S^* of S^* emanating from η^* are tangent to each other in η^* (or coincide near η^*). This is illustrated in a particular way by the examples in Appendix A.6.

The general relation between p, q, and the multiplicity r of ξ^* in (A.1.18) is somewhat complicated because the irreducibility of $\pi(\zeta,\eta)$ does not imply the *pairwise* irreducibility of the polynomials $\rho_j(\zeta)$, $j = 1,\dots,\ell$, with exception of $\ell = 1$, i.e., of linear multistep methods. However, if $r = 1$, as always in single step methods, then obviously $q = 1$. The following two cases concern single step methods and *linear* multistep methods and do not involve the Puiseux diagram. But, naturally, they are contained in the general result, Lemma (A.1.40).

(A.1.23) Corollary. *Let ξ^* be a simple root of $\pi(\zeta,\eta^*)$. Then, in* (A.1.18), $q = 1$ *and*

$$p = \min\{\nu \in \mathbb{N}, \frac{\partial^\nu \pi}{\partial \eta^\nu}(\xi^*,\eta^*) \neq 0\}.$$

Proof. Here we obtain by a Taylor expansion

$$\chi = \xi^{*-1}\zeta^{(p)}(\eta^*)/p!, \qquad \xi^* = \zeta(\eta^*),$$

and a differentiation of $\pi(\zeta(\eta),\eta)$ with respect to η yields omitting the argument

$$\frac{\partial \pi}{\partial \zeta}\zeta^{(p)} + \frac{\partial^p \pi}{\partial \eta^p} = 0$$

if $[\partial\pi^\nu/\partial\eta^\nu](\zeta^*,\eta^*) = 0$, $\nu = 0,1,\ldots,p-1$.

(A.1.24) Corollary. *Let ξ^* be an r-fold root of $\pi(\zeta,\eta^*)$. Then, in* (A.1.18), $q = r$ *and* $p = 1$ *if*

(A.1.25) $\frac{\partial \pi}{\partial \eta}(\xi^*,\eta^*) \neq 0.$

Proof. Let without loss of generality

$$\zeta_1(\eta^*) = \ldots = \zeta_r(\eta^*) = \xi^*$$

and let $\tilde\eta$ be a fixed branch of $\tilde\eta^q = \eta - \eta^*$ in a neighborhood of $\tilde\eta = 0$ with exception of an arbitrary but fixed half-line with endpoint in $\tilde\eta = 0$. Then there are by Theorem (A.1.4) $1 \le q \le r$ branches $\zeta_j(\eta)$, $j = 1,\ldots,q$, which can be written in this set, $\mathcal{N}$, as regular functions $\tilde\zeta_j$ in $\tilde\eta$,

$$\tilde\zeta_j(\tilde\eta) = \sum_{\mu=0}^\infty \phi_\mu (e^{2\pi i j/q}\tilde\eta)^\mu, \qquad j = 1,\ldots,q,$$

where $\phi_0 = \zeta_1(\eta^*) = \xi^*$. For the computation of $\phi_1 = \tilde\zeta_1'(0)$ we write

$$\tilde\pi(\zeta,\tilde\eta) = \pi(\zeta,\tilde\eta^q + \eta^*) = \sum_{j=0}^{\ell}\rho_j(\zeta)(\tilde\eta^q + \eta^*)^j = \sum_{j=0}^{\ell}\tilde\rho_j(\zeta)\tilde\eta^{qj}$$

then the values $\tilde\zeta_j(\tilde\eta)$, $j = 1,\ldots,q$, are roots of $\tilde\pi(\zeta,\tilde\eta)$ for $\tilde\eta \in \mathcal{N}$. Because $\zeta_1(\eta^*)$ is an r-fold root of $\pi(\zeta,\eta^*)$ on the one side and because $\tilde\pi(\zeta,\tilde\eta)$ is a function of $\tilde\eta^q$ on the other side we obtain

(A.1.26) $[(\frac{\partial}{\partial\zeta}\tilde\zeta' + \frac{\partial}{\partial\tilde\eta})^q\tilde\pi](\tilde\zeta_1(0),0) = \tilde\zeta'(0)^q[\frac{\partial^q}{\partial\zeta^q}\tilde\pi](\tilde\zeta_1(0),0) + [\frac{\partial^q}{\partial\tilde\eta^q}\tilde\pi](\tilde\zeta_1(0),0) = 0.$

But

$$[\frac{\partial^q}{\partial\zeta^q}\tilde{\pi}](\tilde{\zeta}_1(0),0) = [\frac{\partial^q}{\partial\zeta^q}\pi](\zeta_1(\eta^*),\eta^*) \begin{cases} = 0, \ 1 \le q < r, \\ \ne 0, \ q = r \end{cases}$$

and

$$[\frac{\partial^q}{\partial\tilde{\eta}^q}\tilde{\pi}](\tilde{\zeta}_1(0),0) = q!\tilde{\rho}_1(\zeta_1(\eta^*)) = q!\sum_{j=0}^{\ell}\rho_j(\zeta_1(\eta^*))j\eta^{*j-1} = [\frac{\partial}{\partial\eta}\pi](\zeta_1(\eta^*),\eta^*)q!$$

hence $q < r$ leads to a contradiction by (A.1.26). Therefore we have $q = r$ and then (A.1.26) yields $\tilde{\zeta}_1'(0) \ne 0$ which implies $p = 1$ in the expansion (A.1.18).

Now we observe that in linear multistep methods the irreducibility of $\pi(\zeta,\eta)$ implies $[\partial\pi/\partial\eta](\zeta_1(\eta^*),\eta^*) = \rho_1(\zeta_1(\eta^*)) \ne 0$ and obviously the same conclusions can be drawn for characteristic polynomials $\pi(\zeta,\eta^2)$ of linear multistep methods for differential systems of second order replacing (A.1.25) with

(A.1.25)* $\dfrac{\partial\pi}{\partial(\eta^2)}(\zeta_1(\eta^*),\eta^{*2}) = \rho_1(\zeta_1(\zeta^*)) \ne 0$

because it doesn't matter whether we write η^2 instead of η. Thus a combination of Corollary (A.1.21) and Corollary (A.1.24) yields for *linear* multistep methods with irreducible characteristic polynomial:

(i) If S is the stability region of a method for differential systems of first order (cf. Definition (1.2.7)) then $\overline{S} \setminus S$ consists of the points $\eta^* \in \overline{\mathbb{C}}$ where $\pi(\zeta,\eta)$ has double unimodular roots. In a point $\eta^* \in \overline{S} \setminus S$ both segments of ∂S emanating from η^* are tangent to each other by Corollary (A.1.21)(ii).

(ii) If S is the stability region of a method for differential systems of second order (cf. Definition (2.1.14)) then $\overline{S} = S$. Therefore, the constants κ_R in Theorem (2.1.34) and (2.2.10) and κ_s in Theorem (2.1.39) depend only on the data of the method. If $[-s, 0]$ is a periodicity interval then it is a subset of ∂S and double unimodular roots can only lie in the endpoints of the interval. Hence the Frobenius matrix of the method is diagonable in the interior of the periodicity interval, cf. assumption (iii) of Theorem (3.3.3).

Following Hensel and Landsberg [02] or Hille [62] let now

(A.1.27) $\sigma_i(\eta) = \sum_{j=n_i}^{\ell}\alpha_{ji}\eta^j, \ \alpha_{n_i,i} \ne 0, \ \alpha_{0k} \ne 0$ (cf. (1.1.5)) $\qquad i = 0,\dots,k,$

and write instead of (A.1.18) near $\eta = 0$

(A.1.28) $\xi(\eta) = \sum_{\mu=0}^{\infty}x_\mu\eta^{\varepsilon_\mu}, \ x_0 \ne 0, \qquad \varepsilon_0 < \varepsilon_1 < \dots .$

For the computation of ε_0 we substitute (A.1.27) and (A.1.28) into $\pi(\xi(\eta),\eta)$ and assemble equal powers of η then we obtain

$$\text{(A.1.29)}\quad \pi(\xi(\eta),\eta) = c_0\eta^{\gamma_0} + c_1\eta^{\gamma_1} + \dots, \qquad \gamma_0 < \gamma_1 < \dots .$$

If (A.1.28) represents some root of $\pi(\zeta,\eta)$ near $\eta = 0$ then the coefficients c_ν must disappear identically. Assembling on the other side all components whose exponent contains ε_0 we find

$$\text{(A.1.30)}\quad \begin{aligned}&\pi(\xi(\eta),\eta)\\ &= \alpha_{n_0,0}\eta^{n_0} + \alpha_{n_1,1}x_0\eta^{n_1+\varepsilon_0} + \alpha_{n_2,2}x_0^2\eta^{n_2+2\varepsilon_0} + \dots + \alpha_{n_k,k}x_0^k\eta^{n_k+k\varepsilon_0} + \dots .\end{aligned}$$

$c_0\eta^{\gamma_0}$ is obtained from this equation by collecting all terms in the sum with the same minimal exponent, say $n_g + g\varepsilon_0$,

$$n_g + g\varepsilon_0 = \min_{0\le i\le k}\{n_i + i\varepsilon_0\}, \qquad g \in \mathbb{N} \cup \{0\}.$$

If g is the only number with this property then

$$c_0\eta^{\gamma_0} = \alpha_{n_g,g}x_0^g\eta^{n_g+g\varepsilon_0}$$

and thus c_0 does not disappear hence ε_0 must be chosen in a way that at least two numbers, say g and h, satisfy

$$\gamma_0 = n_g + g\varepsilon_0 = n_h + h\varepsilon_0 = \min_{0\le i\le k}\{n_i + i\varepsilon_0\}, \qquad g \ne h.$$

Then we have

$$\varepsilon_0 = (n_g - n_h)/(h - g)$$

and, by this way, a finite number of admissible values for ε_0 can be derived. A more geometrical version of this idea is the *Puiseux diagram:*

Let in a (x,y)-plane

$$z_i = (i,n_i),\ z_i^* = (0,n_i + i\cdot\mathrm{tg}(\pi - \phi)), \qquad i = 0,\dots,k.$$

z_i^* is the intersection of the ordinate axis and a straight line through z_i with the angle ϕ with respect to the positive real line. The ordinates of z_i^* are the exponents in (A.1.30) therefore we have to compute all chords C_ν with angle ϕ_ν, $\nu = 1,\dots,s$, through at least two points z_i such that all other points lie on or above C_ν. Then $\varepsilon_0^{(\nu)} = \mathrm{tg}(\pi - \phi_\nu)$, $\nu = 1,\dots,s$, are the admissible exponents.

Let us now assume that C_ν is such a lower boundary chord of the point set $\{z_i\}$

and that

$$\{z_{i_0}, z_{i_1}, \ldots, z_{i_\nu}\} \subset C_\nu, \qquad i_0 < i_1 < \ldots < i_\nu,$$

then

$$\gamma_0 = n_{i_0} + i_0\varepsilon_0 = n_{i_1} + i_1\varepsilon_0 = \ldots = n_{i_\nu} + i_\nu\varepsilon_0$$

whereas all other points z_i lie above but not on C_ν. Accordingly, the coefficient c_0 belonging to γ_0 in (A.1.29) is by (A.1.30)

$$\text{(A.1.31)} \quad c_0 = \alpha_{n_{i_0},i_0} x_0^{i_0} + \alpha_{n_{i_1},i_1} x_0^{i_1} + \ldots + \alpha_{n_{i_\nu},i_\nu} x_0^{i_\nu} = \tilde{P}_\nu(x_0),$$

i.e., $x_0 \neq 0$ must be a root of the polynomial of degree $i_\nu - i_0$,

$$\text{(A.1.32)} \quad P_\nu(x) = \alpha_{n_{i_0},i_0} + \alpha_{n_{i_1},i_1} x^{i_1-i_0} + \ldots + \alpha_{n_{i_\nu},i_\nu} x^{i_\nu-i_0}.$$

By this way the initial terms in (A.1.28) of all roots of $\pi(\zeta,\eta)$ can be found.

After these preliminaries we turn to our actual problem namely to compute p and x_p^* in

$$\text{(A.1.33)} \quad \xi(\eta) = \xi^* + \textstyle\sum_{\mu=p}^{\infty} x_\mu^* \eta^\mu \equiv \xi^* + \Phi(\eta)$$

where $\pi(\xi^*,0) = 0$ and $x_\mu^* \neq 0$. For this we have to modify slightly the above device. A substitution of (A.1.33) into $\pi(\xi(\eta),\eta)$ yields

$$\text{(A.1.34)} \quad \pi(\xi^* + \Phi(\eta),\eta) = \textstyle\sum_{i=0}^{k}\sigma_i(\eta)\sum_{j=0}^{i}\binom{i}{j}\xi^{*i-j}\Phi(\eta)^j$$

$$= \textstyle\sum_{i=0}^{k}[\sum_{m=i}^{k}\sigma_m(\eta)\binom{m}{i}\xi^{*m}](\Phi(\eta)/\xi^*)^i \equiv \sum_{i=0}^{k}\sigma_i^*(\eta)(\Phi(\eta)/\xi^*)^i$$

where

$$\sigma_i^*(\eta) = \textstyle\sum_{m=i}^{k}(\sum_{j=0}^{\ell}\alpha_{jm}\eta^j)\binom{m}{i}\xi^{*m} = \sum_{j=0}^{\ell}[\sum_{m=i}^{k}\alpha_{jm}\binom{m}{i}\xi^{*m}]\eta^j.$$

But

$$\frac{\partial^i}{\partial\zeta^i}\rho_j(\zeta) = \textstyle\sum_{m=i}^{k}\binom{m}{i}i!\,\alpha_{jm}\zeta^{m-i}$$

hence

$$\text{(A.1.35)} \quad \sigma_i^*(\eta) = \textstyle\sum_{j=0}^{\ell}(\rho_j^{(i)}(\xi^*)\frac{\xi^{*i}}{i!})\eta^j \equiv \sum_{j=n_i}^{\ell}\alpha_{ji}^*\eta^j, \qquad i = 0,\ldots,k.$$

Now we observe that

$$\rho_j^{(i)}(\xi^*) = \frac{\partial^{i+j}\pi}{\partial\zeta^i\partial\eta^j}(\xi^*,0)/j!$$

therefore we obtain for the crucial values n_i

(A.1.36) $n_i = \min\{j \in \mathbb{N},\ [\partial^{i+j}\pi/\partial\zeta^i\partial\eta^j](\xi^*,0) \neq 0\}, \qquad i = 0,\ldots,k.$

With respect to (A.1.33), the Puiseux diagram suggests the following notations:

(A.1.37) $i_\nu = \max\{\arg\min\{(n_i - n_{i_{\nu-1}})/(i - i_{\nu-1}),\ i > i_{\nu-1}\}\},\ i_0 = 0,$

(A.1.38) $p_\nu = -(n_{i_\nu} - n_{i_{\nu-1}})/(i_\nu - i_{\nu-1}),$

and

(A.1.39) $I_\nu = \{j \in \mathbb{N},\ (n_j - n_{i_{\nu-1}})/(j - i_{\nu-1}) = -p_\nu\},$

say for $\nu = 1,\ldots,\nu^*$. Then Corollary (A.1.21)(iii) and the Puiseux diagram applied to (A.1.34) and (A.1.35) yield (cf. also Jeltsch [76a, 77]):

<u>(A.1.40) Lemma</u>. *For each* ε, $0 < \varepsilon \le \alpha$, *there exists a* $\rho_\varepsilon > 0$ *such that*

$$\{\eta \in \mathbb{C},\ 0 < |\eta| \le \rho_\varepsilon,\ |\pi - \arg\eta| \le \alpha - \varepsilon\} \subset S^*, \qquad \alpha > 0,$$

iff for each unimodular root ξ^* *of* $\pi(\zeta,0) = \rho_0(\zeta)$ *the associated* i_ν, p_ν, *and* I_ν *satisfy for* $\nu = 1,\ldots,\nu^*$

(i) $1 \le p_\nu \le \min\{\ell, [\pi/2\alpha]\}$ *and* $p_\nu \in \mathbb{N}$,

(ii) $|(1 - p_\nu)\pi - \arg\chi| \le \frac{\pi}{2} - p_\nu\alpha$ *for all roots* $\chi \neq 0$ *of the polynomial*

$$\textstyle\sum_{i \in I_\nu}\left[[\partial^{i+n_i}\pi/\partial\zeta^i\partial\eta^{n_i}](\xi^*,0)\right]\frac{\xi^{*i}}{i!\,n_i!}\,\chi^i.$$

Naturally, these necessary and sufficient *algebraic* conditions for S^* having an edge of angle 2α in $\eta = 0$ can be carried over to general $0 \neq \eta^* \in \partial S^*$ without difficulty. The next result concerns the case of a disk instead of an angular domain; cf. also Jeltsch [77].

<u>(A.1.41) Lemma</u>. *There exists a* $\rho > 0$ *such that*

$$\mathfrak{D}_\rho = \{\eta \in \mathbb{C},\ |\eta + \rho| \le \rho\} \subset S^*$$

iff every root $\zeta_j(\eta)$ *of* $\pi(\zeta,\eta)$ *with* $|\zeta_j(0)| = 1$ *has near* $\eta = 0$ *the form*

(A.1.42) $\zeta_j(\eta) = \xi^*(1 + \chi\eta + \mathcal{O}(\eta^s)),\ |\xi^*| = 1, \qquad \chi > 0,\ s \ge 2.$

Proof. If (A.1.42) holds then a substitution of $\eta = \rho(e^{i\theta} - 1)$, $\rho > 0$, $0 \leq \theta < 2\pi$, yields by (A.1.15)

$$|\zeta_j(\eta)| = 1 - \chi\rho(1 - \cos\theta) + \mathcal{O}([\rho(1 - \cos\theta)^{1/2}]^s), \qquad \chi > 0,$$

because $|\eta|^2 = 2\rho^2(1 - \cos\theta)$. This proves that $\mathfrak{D}_\rho \subset S^*$ for $s \geq 2$ if $\rho > 0$ is sufficiently small. On the other side, let $\mathfrak{D}_\rho \subset S^*$ then ∂S^* has in $\eta = 0$ an 'edge' of angle π and, by Corollary (A.1.21)(iii), $\zeta_j(\eta)$ must be a branch of $\xi(\eta)$ having near $\eta = 0$ the form

$$\xi(\eta) = \xi^*(1 + \chi\eta + \mathcal{O}(\eta^s)), \qquad \chi > 0,\ s > 1.$$

More exactly, let $s = p/q > 1$ where $p \in \mathbb{N}$ and $1 < q \in \mathbb{N}$ have no common factor, let the η-plane be cut along the positive real axis, and let $\eta^{1/q}$ be henceforth in this proof that branch of $\zeta^q - \eta = 0$ with $(-1)^{1/q} = e^{i\pi/q}$. Then we have after a suitable renumeration by Theorem (A.1.4)

$$\zeta_j(\eta) = \xi^*(1 + \chi\eta + \textstyle\sum_{\mu=q+1}^{\infty}\phi_\mu[e^{2\pi ij/q}\eta^{1/q}]^\mu), \qquad j = 1,\dots,q,$$

or, writing $\phi_\mu = \chi_\mu e^{i\psi_\mu}$, $\chi_\mu = |\phi_\mu| \geq 0$, and

$$\Phi(j,\mu,\omega) = \psi_\mu + (2\pi j + \omega)\mu/q,$$

$$\zeta_j(\eta) = \xi^*(1 + \chi\eta + \textstyle\sum_{\mu=q+1}^{\infty}\chi_\mu|\eta|^{\mu/q}e^{i\Phi(j,\mu,\arg\eta)}), \qquad j = 1,\dots,q,$$

and hence

$$|\zeta_j(\eta)|^2 = 1 + 2\chi\mathrm{Re}\,\eta + \chi^2|\eta|^2 + 2\textstyle\sum_{\mu=q+1}^{2q}\chi_\mu|\eta|^{\mu/q}\cos(\Phi(j,\mu,\arg\eta)) + \mathcal{O}(|\eta|^{s^*}),\ s^* > 2.$$

Into this equation we substitute again $\eta = \rho(e^{i\theta} - 1)$ and observe that for these circles

$$\mathrm{Re}\,\eta = -|\eta|^2/2\rho$$

then we obtain

$$|\zeta_j(\eta)|^2 = 1 + 2\textstyle\sum_{\mu=q+1}^{2q-1}\chi_\mu|\eta|^{\mu/q}\cos(\Phi(j,\mu,\arg\eta))$$
$$+ |\eta|^2(\chi^2 - \chi\rho^{-1} + 2\chi_{2q}\cos(\psi_{2q} + 2\arg\eta)) + \mathcal{O}(|\eta|^{s^*}), \qquad s^* > 2.$$

Obviously, for every fixed $\rho > 0$ the terms in the sum dominate here the other terms in $|\eta|$ for $\theta \to 0$. Thus, if $\mathfrak{D}_\rho \subset S^*$ for some $\rho > 0$ then we have either $\chi_\mu = 0$, $\mu =$

q+1,...,2q-1, or if $\chi_\mu = 0$, $\mu = q+1,\ldots,q+\nu-1$, and $\chi_{q+\nu} > 0$ then

(A.1.43) $\cos(\Phi(j,q+\nu,\arg\eta)) < 0, \qquad \theta \to 0,\ j = 1,\ldots,q.$

But

$$\arg\eta \to \pi/2 \text{ if } \theta \to 0_+ \text{ and } \arg\eta \to 3\pi/2 \text{ if } \theta \to 0_-$$

hence, writing

$$\Psi_1(\psi,j,\nu) = \psi + 2\pi j\frac{\nu}{q} + \frac{\pi}{2}\frac{q+\nu}{q},\quad \Psi_2(\psi,j,\nu) = \psi + 2\pi j\frac{\nu}{q} + \frac{3\pi}{2}\frac{q+\nu}{q},$$

we obtain from (A.1.43) that $\cos(\Psi_1(\psi,j,\nu)) \le 0$ and $\cos(\Psi_2(\psi,j,\nu)) \le 0$. Accordingly, in order to prove that $\mathfrak{D}_\rho \subset S^*$ for some $\rho > 0$ implies $\chi_\mu = 0$, $\mu = q+1,\ldots,2q-1$, we have to prove that for all $\nu \in \{1,\ldots,q-1\}$ and all $\psi \in [0, 2\pi)$ there exists a $j \in \{1,\ldots,q\}$ such that

(A.1.44) $\cos(\Psi_1(\psi,j,\nu)) > 0$

or

(A.1.45) $\cos(\Psi_2(\psi,j,\nu)) > 0.$

For this let ν be given and let $\tilde\nu$, $\tilde q$ satisfy $\tilde\nu/\tilde q = \nu/q$ but having no common divisor. Then apparently $1 \le \tilde\nu < \tilde q \le q$ and $1 < \tilde q$. We consider two cases:
(i) If $\tilde q > 2$ then $\Psi_1(\psi,j,\nu)$ differ by multiples of $2\pi/\tilde q$ and all different multiples modulo 2π occur for $j = 1,\ldots,q$. Hence there is at least one $j \in \{1,\ldots,q\}$ such that (A.1.44) is fulfilled.
(ii) If $\tilde q = 2$ then $\tilde\nu = 1$ and $(q+\nu)/q = 1 + (\tilde\nu/\tilde q) = 3/2$. In this case $\Psi_1(\psi,j,\nu)$ and $\Psi_2(\psi,j,\nu)$ have only different values for $j = 1,2$ and we obtain modulo 2π

$$\Psi_1(\psi,1,\nu) = \psi - \frac{\pi}{4},\quad \Psi_1(\psi,2,\nu) = \psi + \frac{3\pi}{4}$$

$$\Psi_2(\psi,1,\nu) = \psi - \frac{3\pi}{4},\quad \Psi_2(\psi,2,\nu) = \psi + \frac{\pi}{4}.$$

Hence there exists also here a j such that (A.1.44) or (A.1.45) is fulfilled.

We conclude this section with an *algebraic* characterization of the 'disk stability' near $\eta = 0$ described in the last lemma. For this we need some further aids and define

$$\mathcal{H}_\lambda = \{0 \ne z \in \mathbb{C},\ |\arg z - \arg\lambda| < \pi/2\},$$

$\overline{\mathcal{H}}_\lambda$ denoting the closed hull of $\mathcal{H}_\lambda$.

(A.1.46) Theorem. (Lucas.) *Let* $p(z) = \sum_{\nu=0}^{n} a_\nu z^\nu$ *be a non-constant polynomial then all roots of* $p'(z)$ *lie in the convex hull* $\mathcal{K}$ *of the set of the roots of* $p(z)$. *If the roots of* $p(z)$ *are not collinear, no root of* $p'(z)$ *lies on the boundary of* $\mathcal{K}$ *unless it is a multiple root of* $p(z)$.

Proof. See Marden [66 , p. 22].

(A.1.47) Lemma. (Jeltsch[77].) *Let* $p(z) = \sum_{\nu=0}^{n} a_\nu z^\nu$ *satisfy for* $0 \le k < m - 1 < n$

$$a_k \ne 0,\ a_\nu = 0,\ \nu = k + 1,\dots,m - 1,\ a_m \ne 0.$$

Then there exists for every $0 \ne \lambda \in \mathbb{C}$ *a root* $z^* \in \overline{\mathcal{K}}_\lambda \setminus \{0\}$ *of* $p(z)$ *and* $z^* \in \mathcal{K}_\lambda$ *for* $k < m - 2$.

Proof. Assume that all roots of $p(z)$ lie in $\mathbb{C} \setminus (\overline{\mathcal{K}}_\lambda \setminus \{0\})$ then by Theorem (A.1.46) all roots of $p^{(k)}(z)$ lie in this set. But then all roots of

$$q(z) = z^{n-k} p(1/z) = b_{n-k} z^{n-k} + b_{n-m} z^{n-m} + \dots + b_0$$

lie in $\mathbb{C} \setminus (\overline{\mathcal{K}}_{1/\lambda} \setminus \{0\})$ and hence all roots of $q^{(n-m)}(z)$ lie in this latter set. We obtain by assumption

$$q^{(n-m)}(z) = c_1 z^{m-k} + c_0, \qquad c_0 \ne 0,\ c_1 \ne 0,\ m - k \ge 2,$$

thus at least one root of $q^{(n-m)}(z)$ lies outside

$$\mathbb{C} \setminus (\overline{\mathcal{K}}_{1/\lambda} \setminus \{0\}) = \{z \in \mathbb{C},\ |\arg z - \arg(1/\lambda)| > \pi/2\} \cup \{0\}$$

which is a contradiction. In the same way it is shown that a root of $p(z)$ lies in $\mathcal{K}_\lambda$ for $k < m - 2$.

Now we introduce the polynomials

$$\text{(A.1.48)} \quad Q_{ij}(\xi,\chi) = \sum_{s=t}^{u} \rho_{i-s}^{(s)}(\xi)\xi^s \binom{s}{j} \chi^{s-j}/s!, \qquad t = \max\{i - \ell, j\},\ u = \min\{k, i\},$$

and deduce the algebraic version of the disk Lemma (A.1.41) in two steps:

(A.1.49) Lemma. (Jeltsch [76a].) *Let* ξ^* *be a unimodular root of* $\pi(\zeta,0) = \rho_0(\zeta)$ *with multiplicity* r. *Then every root* $\zeta_i(\eta)$ *with* $\zeta_i(0) = \xi^*$ *has near* $\eta = 0$ *the form*

$$\text{(A.1.50)} \quad \xi(\eta) = \xi^*(1 + \chi\eta + \mathcal{O}(\eta^s)), \qquad \chi > 0,\ s > 1,$$

iff
(i) ξ^* *is a* $(r-j)$*-fold root of* $\rho_j(\zeta)$, $j = 1,\dots,r$, *and* $\rho_j(\xi^*) \neq 0$, $j = r+1,\dots,\ell$,
(ii) all roots of $Q_{r0}(\xi^*,\chi)$ *are real and positive.*

Proof. We have to reconsider the Puiseux diagram for the polynomial (A.1.34),

$$\tilde{\pi}(\Phi(\eta),\eta) = \pi(\xi^* + \Phi(\eta),\eta) = \sum_{i=0}^{k}\sum_{j=0}^{\ell}\rho_j^{(i)}(\xi^*)\frac{\xi^{*i}}{i!}\eta^j(\Phi(\eta)/\xi^*)^i$$

with the r-fold root $\Phi(0) = 0$ for $\eta = 0$. If (i) holds then (A.1.36) yields $n_i = r - i$ and we obtain two lower boundary chords in the Puiseux diagram as the following figure describes for $r = 4$ and $k = 7$.

(A.1.51) Figure:

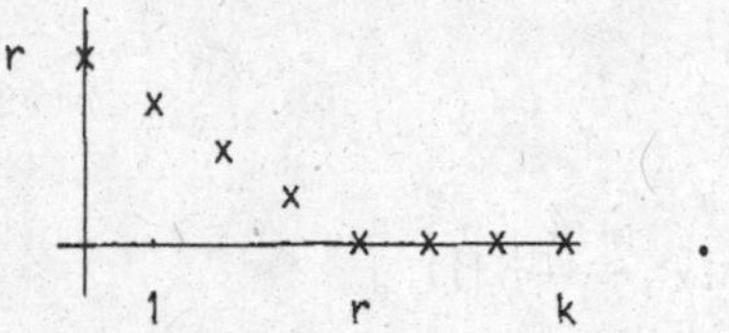

The chord with the ascent rate $-p = -1$ belongs to the root $\Phi(0) = 0$ and the chord with the ascend rate zero belongs to the $k - r$ nonzero roots of $\pi(\Phi(\eta),\eta)$. In the first case the growth parameters χ are the roots of the polynomial (A.1.32),

$$\text{(A.1.52)} \quad \sum_{i=0}^{r}\rho_{n_i}^{(i)}(\xi^*)\frac{\xi^{*i}}{i!}\chi^i = \sum_{i=0}^{r}\rho_{r-i}^{(i)}(\xi^*)\frac{\xi^{*i}}{i!}\chi^i = Q_{r0},$$

which are by assumption (ii) real and positive. Therefore (A.1.50) holds. On the other side, if (A.1.50) holds for all roots $\zeta_i(\eta)$ with $\zeta_i(0) = \xi^*$ then the Puiseux diagram must necessarily have exactly the two lower boundary chords described in Figure (A.1.51) and $n_i \geq r - i$, $i = 1,\dots,r-1$. But as all roots χ of the polynomial (A.1.52) are real and positive by assumption, this polynomial cannot have a zero coefficent by Lemma (A.1.47). This implies that $\rho_{r-i}^{(i)}(\xi^*) \neq 0$, i.e., $\rho_i^{(r-i)}(\xi^*) \neq 0$, $i = 0,1,\dots,r$. Hence (i) and (ii) are necessary conditions for the expansion (A.1.50).

<u>(A.1.53) Lemma</u>. (Jeltsch [79].) *There exists a* $\rho > 0$ *such that*

$$\mathfrak{D}_\rho = \{\eta \in \mathbb{C},\ |\eta + \rho| \leq \rho\} \subset S^*$$

iff the following three conditions are fulfilled for unimodular roots ξ^* *of* $\pi(\zeta,0)$:
(i) $r_j = r_0 - j$, $j = 0,\dots,r_0$, *where* r_j *denotes the multiplicity of* ξ^* *as a root of* $\rho_j(\zeta)$ *and* $\rho_j(\xi^*) \neq 0$, $j = r_0+1,\dots,\ell$.
(ii) all roots of $Q_{r_00}(\xi^*,\chi)$ *are real and positive.*
(iii) If χ^* *is a root of* $Q_{r_00}(\xi^*,\chi)$ *of multiplicity* $\kappa \geq 2$ *then*

$$Q_{ij}(\xi^*,\chi^*) = 0$$

for all integers i,j *with* $r_0 < i < -j + \kappa + r_0$, $j = 0,\dots,\kappa - 2$.

Proof. Notice that by the last result $1 \leq \kappa \leq r_0$ roots $\zeta_i(\eta)$ have the form

$$\xi(\eta) = \xi^*(1 + \chi^*\eta + \Psi(\eta^{1/\kappa})),$$

iff the polynomial (A.1.52) has a root χ^* of multiplicity κ. Hence, because of the disk lemma (A.1.41) and the above lemma we have only to show that the third condition is necessary and sufficient for $s \geq 2$ in (A.1.50). For this we substitute

$$\xi(\eta) = \xi^*(1 + \eta[\chi^* + \Psi(\eta)])$$

into $\pi(\zeta,\eta) = 0$ and obtain

$$\begin{aligned} \pi(\xi(\eta),\eta) &= \textstyle\sum_{j=0}^{\ell}\eta^j p_j(\xi^*(1 + \eta[\chi^* + \Psi(\eta)])) \\ &= \textstyle\sum_{j=0}^{k}\sum_{i=0}^{\ell+k}Q_{ij}(\xi^*,\chi^*)\eta^i\Psi(\eta)^j. \end{aligned} \tag{A.1.54}$$

As $r_j = r_0 - j$ we have for all j

$$Q_{ij}(\xi^*,\chi^*) = 0, \qquad i < r_0,$$

and we find easily that for all i and j

$$\frac{\partial}{\partial\chi}Q_{ij}(\xi,\chi) = (j + 1)Q_{i,j+1}(\xi,\chi).$$

Thus, if χ^* is a root of $Q_{r_0 0}(\xi^*,\chi)$ of multiplicity $\kappa \geq 2$ then

$$Q_{r_0 0}(\xi^*,\chi^*) = 0,\ j = 0,\dots,\kappa - 1,\ Q_{r_0\kappa}(\xi^*,\chi^*) \neq 0,$$

and (A.1.54) yields after division by η^{r_0}

$$\begin{aligned} &Q_{r_0\kappa}(\xi^*,\chi^*)\Psi(\eta)^\kappa + \textstyle\sum_{j=\kappa+1}^{\infty}Q_{r_0 j}(\xi^*,\chi^*)\Psi(\eta)^j \\ &+ \textstyle\sum_{i=1}^{\infty}\sum_{j=1}^{\infty}Q_{r_0+i,j}(\xi^*,\chi^*)\eta^i\Psi(\eta)^j = 0. \end{aligned}$$

Applying here once more the Puiseux diagram we find that condition (iii) is necessary and sufficient for $\Psi(\eta)$ having the form

$$\Psi(\eta) = \omega\eta + \mathcal{O}(\eta^{s^*}), \qquad s^* > 1,$$

with some $\omega \in \mathbb{C}$ being nonzero or not. A substitution of this result into (A.1.54) yields

$$\xi(\eta) = \xi^*(1 + \chi^*\eta + \omega\eta^2 + \mathcal{O}(\eta^s)), \qquad s > 2,$$

which is the necessary and sufficient condition of the disk lemma (A.1.41).

All conditions of this lemma are empty if $\pi(\zeta,0)$ has no unimodular roots at all. The third condition is empty if $Q_{r_0 0}(\xi^*,\chi)$ has only simple roots. As $Q_{r0}(\xi^*,\chi)$ is a linear polynomial for $\ell = 1$ or $k = 1$, the third condition can be omitted for linear multistep methods or single step multiderivative methods. From the present stability point of view the latter class encloses here also the Runge-Kutta methods.

A.2. Auxiliary Results on Frobenius and Vandermonde Matrices

The Frobenius matrix $F_\pi(\eta)$ associated with the polynomial $\pi(\zeta,\eta)$ is defined in (1.2.5). This matrix has the characteristic polynomial $\pi(\zeta,\eta)$,

$$\sigma_k(\eta)\det(\zeta I - F_\pi(\eta)) = \pi(\zeta,\eta),$$

and so the roots of $\pi(\zeta,\eta)$ are the eigenvalues of $F_\pi(\eta)$. If $\pi(\zeta,\eta)$ has k distinct roots then $F_\pi(\eta)$ is therefore diagonable but the converse is also true; see e.g. Stoer and Bulirsch [80 , Theorem (6.3.4)]. Omitting the argument η, a diagonable Frobenius matrix F_π has the Jordan canonical decomposition

$$\text{(A.2.1)} \qquad F_\pi = W Z W^{-1}$$

where $Z = (\zeta_1, \dots, \zeta_k)$ is the diagonal matrix of the eigenvalues of F_π and W is a Vandermonde matrix,

$$\text{(A.2.2)} \qquad W = \begin{bmatrix} 1 & \cdots & 1 \\ \zeta_1 & \cdots & \zeta_k \\ \vdots & & \vdots \\ \zeta_1^{k-1} & \cdots & \zeta_k^{k-1} \end{bmatrix}, \quad \det(W) = \prod_{i>j}(\zeta_i - \zeta_j).$$

Let W_{ji} result from W by cancelling the j-th row and i-th column then we obtain by Cramer's rule

(A.2.3) $$W^{-1} = [w^*_{ij}]^k_{i,j=1},\ w^*_{ij} = (-1)^{i+j}\det(W_{ji})/\det(W).$$

The elements of W and W^{-1} are thus rational functions of $\zeta_1,\dots,\zeta_k$ without singularities if no roots $\zeta_i(\eta)$ coalesce in some point η of the considered domain. Observing that

$$W(\eta_1)^{-1}W(\eta_2) = I + (W(\eta_1)^{-1} - W(\eta_2)^{-1})W(\eta_2)$$

we can state the following result.

<u>(A.2.4) Lemma.</u> *Let* $R \subset \mathbb{C}$ *be a closed domain and let* $\zeta_1,\dots,\zeta_k$ *be* k *distinct holomorphic functions in* R *with* $|\zeta_i(\eta)| \leq 1$. *Then the associated Vandermonde matrix satisfies*

$$\sup_{\eta\in R}|W(\eta)| \leq \kappa,\ \sup_{\eta\in R}|W(\eta)^{-1}| \leq \kappa_R,$$

$$|W(\eta_1)^{-1}W(\eta_2)| \leq 1 + \kappa_R \max_{1\leq i\leq k}|\zeta_i(\eta_1) - \zeta_i(\eta_2)| \qquad \forall\ \eta_1,\ \eta_2 \in R.$$

Let now s_ω be the ω-th elementary symmetric function in the k variables $\zeta_1,\dots,\zeta_k$,

$$s_\omega(\zeta_1,\dots,\zeta_k) = \textstyle\sum_{1\leq\nu_1<\nu_2<\dots<\nu_\omega\leq k}\zeta_{\nu_1}\zeta_{\nu_2}\cdots\zeta_{\nu_\omega},\ 1 \leq \omega \leq k,\ s_0 \equiv 1,$$

and let

$$s^\mu_\omega = s_\omega(\zeta_1,\dots,\zeta_{\mu-1},\zeta_{\mu+1},\dots,\zeta_k),$$

$$s^{\mu\nu}_\omega = s_\omega(\zeta_1,\dots,\zeta_{\mu-1},\zeta_{\mu+1},\dots,\zeta_{\nu-1},\zeta_{\nu+1},\dots,\zeta_k),\ 1 \leq \mu < \nu \leq k,$$

be the elementary symmetric functions in the indicated k-1 and k-2 variables respectively. We write henceforth briefly

$$s_\omega(\eta) = s_\omega(\zeta_1(\eta),\dots,\zeta_k(\eta)).$$

Then, e.g. by Muir [60] or Gautschi [62], the elements w^*_{ij} of the inverse of the Vandermonde matrix W can be written as

(A.2.5) $$w^*_{ij} = (-1)^{j-1}s^i_{k-j}/\prod_{\substack{\nu=1\\ \nu\neq i}}^{k}(\zeta_\nu - \zeta_i).$$

The following two lemmas of Hackmack [81] concern again the characteristic polynomial $\pi(\zeta,\eta^2)$ of a convergent multistep multiderivative method for a differential system of second order; cf. (2.1.4). The first one is a modification of the third assertion in Lemma (A.2.4) for the case where some roots $\zeta_i(\eta)$ coincide in the point $\eta = 0$.

(A.2.6) Lemma. *Let* $\zeta_i(\eta)$, $i = 1,\dots,k$, *fulfil the assumption of Lemma* (A.1.10). *Then the associated Vandermonde matrix* $W(\eta)$ *satisfies for* $-s \le \eta_j^2 \le 0$, $j = 1,2$,

$$|W(\eta_1)^{-1}W(\eta_2)| \le 1 + \kappa_s \max\{1,|\eta_1|^{-1}\}|\eta_1 - \eta_2|, \qquad s < \infty,$$

$$|W(\eta_1)^{-1}W(\eta_2)| \le 1 + \kappa_s \max\{1,|\eta_1|^{-1}\}\max_{1\le i\le k}|\zeta_i(\eta_1) - \zeta_i(\eta_2)|, \qquad s = \infty.$$

Proof. Let δ_{ij} be the Kronecker symbol and let briefly

$$W(\eta_1)^{-1}W(\eta_2) = [u_{ij}]_{i,j=1}^k.$$

We observe that

$$\textstyle\sum_{\ell=1}^k w_{i\ell}^* \zeta_j^{\ell-1} = \delta_{ij}$$

and thus can write by (A.2.5)

$$u_{ij} = \delta_{ij} + \textstyle\sum_{\ell=1}^k (w_{i\ell}^*(\eta_1)\zeta_j(\eta_2)^{\ell-1} - w_{i\ell}^*(\eta_1)\zeta_j(\eta_1)^{\ell-1})$$

$$= \delta_{ij} + [\prod_{\nu\ne i}(\zeta_\nu(\eta_1) - \zeta_i(\eta_1))^{-1}]\textstyle\sum_{\ell=1}^k (-1)^{\ell-1} s_{k-\ell}^i(\eta_1)(\zeta_j(\eta_2)^{\ell-1} - \delta_{ij}\zeta_j(\eta_1)^{\ell-1}).$$

We now consider two cases:

(i) If $i \ne j$ then we obtain by Vieta's root criterium

$$\textstyle\sum_{\ell=1}^k (-1)^{\ell-1} s_{k-\ell}^i(\eta_1)\zeta_j(\eta_2)^{\ell-1}$$

$$= (\zeta_1(\eta_1) - \zeta_j(\eta_2))\cdots(\zeta_{i-1}(\eta_1) - \zeta_j(\eta_2))(\zeta_{i+1}(\eta_1) - \zeta_j(\eta_2))\cdots(\zeta_k(\eta_1) - \zeta_j(\eta_2)).$$

Therefore we have in this case

$$\text{(A.2.7)} \qquad u_{ij} = \prod_{\nu\ne i}[(\zeta_\nu(\eta_1) - \zeta_j(\eta_2))/(\zeta_\nu(\eta_1) - \zeta_i(\eta_2))].$$

As stipulated in Lemma (A.1.8) and (A.1.10), the roots confluenting pairwise in $\eta = 0$ are numbered by i and $i+1$, $i = 1,3,\dots,k_*-1$, $2 \le k_* \le k$. If $i > k_*$ in (A.2.7) then $\zeta_i(\eta)$ is a simple root in $[-s, 0]$ and we find easily that

$$\text{(A.2.8)} \qquad |u_{ij}| \le \kappa_s|\zeta_j(\eta_1) - \zeta_j(\eta_2)|,\ i \ne j, \qquad i > k_*.$$

If $i \le k_*$ then we deduce from (A.2.7)

$$\text{(A.2.9)}\quad |u_{ij}| \le \begin{cases} \kappa_s|\zeta_j(\eta_1) - \zeta_j(\eta_2)|/|\zeta_i(\eta_1) - \zeta_{i+1}(\eta_2)|, & i = 1,3,\ldots,k_*-1 \\ \kappa_s|\zeta_j(\eta_1) - \zeta_j(\eta_2)|/|\zeta_{i-1}(\eta_1) - \zeta_i(\eta_2)|, & i = 2,4,\ldots,k_*, \end{cases} \qquad i \ne j.$$

(ii) If $i = j$ then we have

$$u_{ii} = 1 + [\prod_{\nu\ne i}(\zeta_\nu(\eta_1) - \zeta_i(\eta_1))^{-1}]\sum_{\ell=0}^{k}(-1)^{\ell-1}s^i_{k-\ell}(\eta_1)(\zeta_i(\eta_2)^{\ell-1} - \zeta_i(\eta_1)^{\ell-1}).$$

But the elementary symmetric functions s^i_ω are bounded in the interval $[-s, 0]$ for finite and infinite s, and

$$|a^\ell - b^\ell| = |a - b||\sum_{\nu=0}^{\ell-1}a^{\ell-\nu-1}b^\nu|.$$

Thus, if $i > k_*$ then we obtain here the bound (A.2.8) for $|u_{ii} - 1|$ and $i = j$, and if $i \le k_*$ then we obtain the bound (A.2.9) for $|u_{ii} - 1|$ and $i = j$ in a similar way as above.

An estimation of (A.2.8) and (A.2.9) by means of Lemma (A.1.10) finally proves the assertion.

(A.2.10) Lemma. *Let* $\zeta_i(\eta)$, $i = 1,\ldots,k$, *fulfil the assumption of Lemma* (A.1.10). *Then the associated Vandermonde matrix* $W(\eta)$ *and arbitrary* $Q \in \mathbb{C}^k$ *satisfy*

$$|W(\eta)^{-1}Q| \le \kappa_s(|Q| + |\eta|^{-1}\max_{|\zeta_i(0)|=1}|\zeta_i(0)Q^0_{k-1} - Q^0_k|),\ -s \le \eta^2 \le 0,\ s \le \infty,$$

where $Q^0_n = (q_{n-k+2},\ldots,q_n)^T$.

Proof. We write $s^{\mu\nu}_{-1} = 0$ and $s^{\mu\nu}_0 = 1$ and observe that

$$\text{(A.2.11)}\quad s^{\ell,\ell+1}_{k-2}\zeta_{\ell+1} = s^\ell_{k-1}, \qquad \ell = 1,\ldots,k-1,$$

$$\text{(A.2.12)}\quad s^{\ell-1,\ell}_{k-2}\zeta_{\ell-1} = s^\ell_{k-1}, \qquad \ell = 2,\ldots,k,\ k = 2,3,\ldots,$$

and

$$\text{(A.2.13)}\quad s^{\ell,\ell+1}_{k-i} + s^{\ell,\ell+1}_{k-i-1}\zeta_{\ell+1} = s^\ell_{k-i}, \qquad \ell = 1,\ldots,k-1,$$

$$\text{(A.2.14)}\quad s^{\ell-1,\ell}_{k-i} + s^{\ell-1,\ell}_{k-i-1}\zeta_{\ell-1} = s^\ell_{k-i}, \qquad \ell = 2,\ldots,k,\ i = 2,\ldots,k\ ,\ k = 2,3,\ldots\ .$$

Let

$$u_i(Q) = \sum_{j=1}^{k}w^*_{ij}(\eta)q_j, \qquad i = 1,\ldots,k,$$

be the i-th element of $W(\eta)^{-1}Q$ then we obtain by (A.2.5)

$$\text{(A.2.15)}\quad u_i(Q) = [\prod_{\nu\neq i}(\zeta_\nu(\eta) - \zeta_i(\eta))^{-1}]\sum_{j=1}^{k}(-1)^{j-1}s_{k-j}^{i}(\eta)q_j.$$

If $i > k_*$, i.e., if $\zeta_i(\eta)$ is a simple roots throughout $[-s, 0]$, then we find easily by Schwarz's inequality that

$$\text{(A.2.16)}\quad |u_i(Q)| \leq \tilde{\kappa}_s|Q|.$$

If $i < k_*$ then a substitution of (A.2.11) and (A.2.13) into (A.2.15) yields

$$u_i(Q) = [\prod_{\nu\neq i}(\zeta_\nu(\eta) - \zeta_i(\eta))^{-1}]\sum_{j=2}^{k}(-1)^{j}s_{k-j}^{i,i+1}(\eta)(\zeta_{i+1}(\eta)q_{j-1} - q_j)$$

and an application of Lemma (A.1.10)(ii) leads to

$$|u_i(Q)| \leq \kappa_s\max\{1,|\eta|^{-1}\}\sum_{j=2}^{k}|\zeta_{i+1}(\eta)q_{j-1} - q_j|.$$

But $|\zeta_{i+1}(\eta)| \leq 1$ hence if $|\eta| \geq 1$ then (A.2.16) holds. If $|\eta| < 1$ then we have because $i+1 \leq k_*$ by Lemma (A.1.8)

$$\begin{aligned}|\zeta_{i+1}(\eta)q_{j-1} - q_j| &= |(\zeta_{i+1}(\eta) - \zeta_{i+1}(0))q_{j-1} + \zeta_{i+1}(0)q_{j-1} - q_j| \\ &\leq \kappa_s|\eta||q_{j-1}| + |\zeta_{i+1}(0)q_{j-1} - q_j|.\end{aligned}$$

Therefore we obtain in this case

$$\text{(A.2.17)}\quad |u_i(Q)| \leq \tilde{\kappa}_s(|Q| + |\eta|^{-1}|\zeta_i(0)Q_{k-1}^0 - Q_k^0|)$$

If $i = k_*$ then (A.2.16) and (A.2.17) follow using the formulas (A.2.12) and (A.2.14) instead of (A.2.11) and (A.2.13).

From (A.2.16) and (A.2.17) the assertion follows with $\kappa_s = \tilde{\kappa}_s\sqrt{k}$.

A.3. A Uniform Boundedness Theorem

The main tool of the first two chapters was a theorem on the uniform boundedness of powers of Frobenius matrices whose argument varies in a closed subset R of the stability region S. With respect to methods for differential systems of first order this result has been proved by Gekeler [79] using the fact that in Frobenius matrices the eigenvectors can be expressed explicitely by the eigenvalues. Similar results have been obtained by Crouzeix [80] and LeRoux [79a] in different ways. In the sequel we represent an estimation of a very general form which is due to Crouzeix and Raviart [80].

Let A be a (m,m)-matrix and let $\lambda_1,\ldots,\lambda_\ell$, $\ell \le m$, be the different eigenvalues of A. Then the characteristic polynomial of A can be written in the form

$$\det(A - \lambda I) = (-1)^m \prod_{i=1}^{\ell} (\lambda - \lambda_i)^{m_i}$$

and the integer m_i is called the (algebraic) multiplicity of the eigenvalue λ_i; see e.g. Stoer and Bulirsch [80 , p. 316]. Let now $D \subset \bar{\mathbb{C}}$ be an open domain and let A: $D \ni \zeta \to A(\zeta) \in \mathbb{C}^{m\times m}$ be a matrix-valued function. If $\tilde{A}(\zeta) \equiv A(\zeta^{-1})$ exists for $\zeta = 0$ then we write $A(\infty) = \tilde{A}(0)$. We denote by spr(A) the spectral radius of the matrix A and define for $R \subset D$

$$m^*(R) = \begin{cases} 1 & \text{if } \sup_{\zeta\in R} \mathrm{spr}(A(\zeta)) < 1 \\ \sup_{\zeta\in R} \max_{1\le i\le \ell} \max_{|\lambda_i(\zeta)|\ge 1} \{m_i(\zeta)\} & \text{if } \sup_{\zeta\in R} \mathrm{spr}(A(\zeta)) \geqq 1. \end{cases}$$

Then the following theorem is a slight modification of Crouzeix and Raviart [80 , Theorem 8.1].

<u>(A.3.1) Uniform Boundedness Theorem</u>. *(i) Let* $D \subset \bar{\mathbb{C}}$ *be open and let* A: $D \to \mathbb{C}^{m\times m}$ *be a continuous matrix-valued function.*
(ii) Let $R \subset D$ *be closed in* $\bar{\mathbb{C}}$ *and let* $\sup_{\zeta\in R} \mathrm{spr}(A(\zeta)) \le 1$.
Then

$$\sup_{\zeta\in R} \sup_{n\in\mathbb{N}} \| n^{1-m^*(R)} A(\zeta)^n \| \le \kappa_R.$$

The proof follows Crouzeix and Raviart [80] and is partitioned into several steps:

<u>(A.3.2) Lemma</u>. *For every (m,m)-matrix* A *there exists a unitary matrix* U *such that* $R = U^H A U$ *is an upper-triangular matrix.*

Proof. See e.g. Stoer and Bulirsch [80 , Theorem 6.4.1].

<u>(A.3.3) Lemma</u>. *Let* A: $D \to \mathbb{C}^{m\times m}$ *be a continuous matrix-valued function in the open domain* $D \subset \bar{\mathbb{C}}$ *and let* $A(\xi)$ *have only a single eigenvalue* $\lambda(\xi)$, $\xi \in D$. *Then there exists a neighborhood* $\mathcal{N}$ *of* ξ *and a constant* $\kappa > 0$ *such that for all* $\zeta \in \mathcal{N}$

$$\|A(\zeta)^n\| \le \kappa(|\lambda(\zeta)|^n + n^{m-1}|\lambda(\zeta)|^{n-m+1}), \qquad n = m, m+1, \ldots .$$

Proof. For every $\zeta \in D$ there exists by Lemma (A.3.2) a unitary matrix $U(\zeta)$ such that the matrix $R(\zeta)$,

$$R(\zeta) = [r_{ij}(\zeta)]_{i,j=1}^{m} = U^H(\zeta)A(\zeta)U(\zeta),$$

is upper triangular. As

$$|r_{ij}(\zeta)| \leq |R(\zeta)| = |A(\zeta)|$$

and $\zeta \to A(\zeta)$ is continuous in ξ there exists a neighborhood $\mathcal{N} \subset \overline{\mathbb{C}}$ of ξ such that

$$r = \sup_{\zeta \in \mathcal{N}} |r_{ij}(\zeta)| \leq \sup_{\zeta \in \mathcal{N}} |A(\zeta)| < \infty.$$

Let now A^* denote the matrix of the absolute values of the elements of A and let

$$Q = [q_{ij}]_{i,j=1}^{m}, \quad q_{ij} = \begin{cases} 1 & i < j \\ 0 & \text{else} \end{cases}.$$

Then the following inequality holds elementwise,

$$R(\zeta)^* \leq |\lambda(\zeta)|I + rQ.$$

Because $Q^m = 0$ we therefore obtain for $n \geq m$ elementwise

$$(R(\zeta)^*)^n \leq |\lambda(\zeta)|^n I + \binom{n}{1}|\lambda(\zeta)|^{n-1} rQ + \ldots + \binom{n}{m-1}|\lambda(\zeta)|^{n-m+1}(rQ)^{m-1}.$$

But

$$|A(\zeta)^n| = |R(\zeta)^n| \leq |(R(\zeta)^n)^*| \leq |(R(\zeta)^*)^n|$$

by the Theorem of Perron and Frobenius (cf. e.g. Varga [62]) hence, writing $\Theta = |rQ|$,

$$|A(\zeta)^n| \leq |\lambda(\zeta)|^n + \binom{n}{1}|\lambda(\zeta)|^{n-1}\Theta + \ldots + \binom{n}{m-1}|\lambda(\zeta)|^{n-m+1}\Theta^{m-1}$$

$$\leq |\lambda(\zeta)|^n + \kappa_{r,m} \max_{0 \leq j \leq m-2}\{\sup_{\zeta \in \mathcal{N}}|\lambda(\zeta)|^j\} n^{m-1} |\lambda(\zeta)|^{n-m+1}$$

which yields the assertion by the continuity of λ in ζ and the norm equivalence theorem.

Notice that $\kappa = \kappa_{r,m}$ for $\zeta \in \mathcal{N} \cap R$ if $\sup_{\zeta \in R}|\lambda(\zeta)| \leq 1$.

(A.3.4) Lemma. *Let* $A: D \to \mathbb{C}^{m\times m}$ *be a continuous matrix-valued function in the open domain* $D \subset \overline{\mathbb{C}}$ *and let* $A(\xi)$ *have* ℓ *different eigenvalues,* $\lambda_1(\xi),\ldots,\lambda_\ell(\xi)$, $1 < \ell \leq m$, $\xi \in D$. *Then there exists a neighborhood* $\mathcal{N}$ *of* ξ *and a continuous matrix-valued function* $T: \mathcal{N} \to \mathbb{C}^{m\times m}$ *such that for all* $\zeta \in \mathcal{N}$

$$H(\zeta)^{-1}A(\zeta)H(\zeta) = \begin{bmatrix} A_1(\zeta) & & \bigcirc \\ & \ddots & \\ \bigcirc & & A_\ell(\zeta) \end{bmatrix}$$

where the matrices $A_i(\zeta)$ *have different eigenvalues and* $A_i(\xi)$ *has the single eigenvalue* $\lambda_i(\xi)$, $i = 1,\ldots,\ell$.

Proof. For $i = 1,\ldots,\ell$, let B_i be an open disk with center $\lambda_i(\xi)$ such that $B_i \cap B_j = \emptyset$ for $i \neq j$. Then there exists a neighborhood $\mathcal{N}_1$ of ξ such that for $\zeta \in \mathcal{N}_1$ every B_i contains exactly m_i eigenvalues of $A(\zeta)$ where m_i denotes the multiplicity of the eigenvalue $\lambda_i(\xi)$. We define the projections

$$P_i(\zeta) = \frac{1}{2\pi i} \int_{\partial B_i} (sI - A(\zeta))^{-1} ds, \qquad \forall\, \zeta \in \mathcal{N}_1,\ i = 1,\ldots,\ell,$$

where ∂B_i denotes the positively oriented boundary of B_i. By Kato [66 , Section 2.5.3] the matrix-valued functions $\zeta \to P_i(\zeta)$ are continuous in ξ. Let further

$$Z_i(\zeta) = \{P_i(\zeta)z,\ z \in \mathbb{C}^m\}$$

be the ranges of $P_i(\zeta)$. Then, for $\zeta \in \mathcal{N}_1$, $Z_i(\zeta)$ is a linear subspace of dimension m_i which is invariant with respect to $A(\zeta)$, and

$$\mathbb{C}^m = Z_1(\zeta) \oplus Z_2(\zeta) \oplus \ldots \oplus Z_\ell(\zeta);$$

cf. Kato [66 , Section 1.5.4]. We choose a basis of $Z_i(\zeta)$,

$$z_1^i(\zeta),\ldots, z_{m_i}^i(\zeta), \qquad i = 1,\ldots,\ell.$$

and write

$$\text{(A.3.5)} \qquad z_k^i(\zeta) = P_i(\zeta)z_k^i(\xi), \qquad k = 1,\ldots,m_i,\ i = 1,\ldots,\ell,$$

then the functions $\zeta \to z_k^i(\zeta)$ are continuous in ξ and hence there exists a neighborhood $\mathcal{N} \subset \mathcal{N}_1$ of ξ such that the vectors (A.3.5) are linear independent in $\mathcal{N}$. Now the matrix with the columns $z_k^i(\zeta)$,

$$H(\zeta) = [z_1^1(\zeta),\ldots,z_{m_1}^1(\zeta),\ \ldots\ ,z_1^\ell(\zeta),\ldots,z_{m_\ell}^\ell(\zeta)]$$

has the desired properties for $\zeta \in \mathcal{N}$.

Proof of the Theorem. By Lemma (A.3.4) there exists for every $\xi \in D$ a neighborhood $\mathcal{N}(\xi)$ and a constant κ such that for all $n \in \mathbb{N}$ with $n \geq m^*(\xi)$

(A.3.6) $|A(\zeta)^n| \le \kappa(\mathrm{spr}(A(\zeta))^n + n^{m^*(\xi)-1}\mathrm{spr}(A(\zeta))^{n-m^*(\xi)+1}) \qquad \forall\, \zeta \in \mathcal{N}(\xi) \cap D.$

But R is closed in $\overline{\mathbb{C}}$ hence there exists a finite number of open sets $\mathcal{N}_j$, j = 1,...,J, which cover R and have the property that (A.3.6) holds for all $\zeta \in \mathcal{N}_j$. Because $\mathrm{spr}(A(\zeta)) \le 1$ for all $\zeta \in R$ we have

$$\sup_{\zeta \in \mathcal{N}_j \cap R} |A(\zeta)^n| \le \kappa_j(1 + n^{m^*(R)-1}) \le \kappa_j^* n^{m^*(R)-1}$$

hence

$$\sup_{\zeta \in R} |A(\zeta)^n| \le \max_{1 \le j \le J} \sup_{\zeta \in \mathcal{N}_j \cap R} |A(\zeta)^n| \le \kappa_R n^{m^*(R)-1}$$

which is the desired result.

A.4. Examples to Chapters I and IV

In this section we give some examples of multistep multiderivative methods for differential systems of first order:

(i) The general consistent single step method with one derivative has the characteristic polynomial

$$\pi(\zeta,\eta) = (1 - \omega\eta)\zeta - (1 + (1 - \omega)\eta),$$

cf. Section 4.2.

(ii) The general single step method with two derivatives has the characteristic polynomial

(A.4.1) $\pi(\zeta,\eta) = (\alpha_{01} + \eta\alpha_{11} + \eta^2\alpha_{21})\zeta + (\alpha_{00} + \eta\alpha_{10} + \eta^2\alpha_{20}).$

a) For the maximum attainable order p = 4 we obtain a uniquely determined method with the polynomial

$$\pi(\zeta,\eta) = (12 - 6\eta + \eta^2)\zeta - (12 + 6\eta + \eta^2).$$

This method has the stability region $S = \{\eta \in \mathbb{C},\ \mathrm{Re}\,\eta \le 0\} \cup \{\infty\}$ and the error constant $\chi_4 = 1/720$ (cf. (1.3.6)).

b) The methods (A.4.1) of order p = 3 have the characteristic polynomial

$$\pi(\zeta,\eta) = (12 - 12(1 + \gamma)\eta + (4 + 6\gamma)\eta^2)\zeta + (-12 + 12\gamma\eta + (2 + 6\gamma)\eta^2).$$

They are A-stable iff $\gamma \geq -1/2$ and have the error constant $\chi_3 = (1 + 2\gamma)/24$. For $\gamma = -2/3$, i.e., for $\alpha_{21} = 0$, the resulting method is not A_0-stable (cf. Definition (1.6.1)). The condition

$$\sigma_2(\eta) = (12 - 12(1 + \gamma)\eta + (4 + 6\gamma)\eta^2) = 12(1 - \delta\eta)^2$$

leads to $\gamma = \pm\sqrt{3}/3$. For $\gamma = \sqrt{3}/3$ the resulting method is A-stable with $\infty \in \mathring{S}$ (Calahan's method, cf. e.g. Fried [79]).

c) The methods (A.4.1) of order p = 2 have the characteristic polynomial

$$\text{(A.4.2)} \quad \pi(\zeta,\eta) = (2 - 2(1+\gamma)\eta + (1+2\gamma-2\delta)\eta^2)\zeta + (-2 + 2\gamma\eta + \delta\eta^2).$$

They are A-stable iff $\gamma \geq -1/2$ and $0 \leq \delta \leq (2\gamma + 1)/4$ and have the error constant $\chi_2 = (1 + 3\gamma - 6\eta)/6$. In particular, we obtain for $\gamma = -1/2$ and $\delta = 0$ the trapezoidal rule and for $\gamma = -1/2$ and $\delta = 1$ a method due to Jeltsch [78a] which is stable on the imaginary axis but not A-stable. The stability region of this method is given in figure (A.4.7).

(ii) Backward differentiation methods are linear multistep methods with the characteristic polynomial

$$\pi(\zeta,\eta) = \sum_{i=0}^{k}\alpha_i\zeta^i - \eta\beta\zeta^k$$

(cf. e.g. Lambert [73 , p. 242]). The coefficients of the k-step methods of order p = k are given for k = 1,...,6 in table (A.4.3). The respective root locus curves and the modified root locus curves are given in figure (A.4.8) and (A.4.12).

<u>(A.4.3) Table of Backward Differentiation Methods.</u>

k	β	α_0	α_1	α_2	α_3	α_4	α_5	α_6
1	1	-1	1					
2	2	1	-4	3				
3	6	-2	9	-18	11			
4	12	3	-16	36	-48	25		
5	60	-12	75	-200	300	-300	137	
6	60	10	-72	225	-400	450	-360	147 .

(iii) The methods of Cryer [73] are linear k-step methods of order p = k with the characteristic polynomial

$$\pi(\zeta,\eta) = \rho(\zeta) - \eta(\zeta + d)^k.$$

For a given d these methods are defined in a unique way by Lemma (1.1.12) because of the prescribed order. For $d = -1 + 2/(1 + 2^{k+1})$ and $k = 1,\dots,7$ the root locus curves and the modified root locus curves are given in figure (A.4.9) and (A.4.13).

(iv) The methods of Enright [74a,b] are nonlinear k-step methods with the characteristic polynomial

$$\pi(\zeta,\eta) = \zeta^k - \zeta^{k-1} - \eta\sum_{i=0}^{k}\beta_i\zeta^i - \eta^2\gamma\zeta^k.$$

The k+2 free parameters can be chosen such that the order is $p = k+2$. The coefficients of the resulting methods are given for $k = 1,\dots,7$ in table (A.4.5) and the associated root locus curves are given in figure (A.4.10). If we choose $\sigma_k(\eta) = (1 - \delta\eta)^2$ as in Calahan's method, i.e.,

$$\text{(A.4.4)} \quad \gamma = -(\beta_k/2)^2,$$

then the remaining k+1 free parameters can be chosen such that the order is k+1. However, in this case the resulting methods are no longer uniquely determined. The methods of order $p = k+1$ with (A.4.4) proposed by Enright are given in table (A.4.6) and the associated root locus curves are given in figure (A.4.11). These curves consist of two segments for $k \geq 4$ but the methods remain $A(\alpha)$-stable at least up to $k = 6$.

(A.4.5) Table of Enright's Methods I.

k	p	γ	β_0	β_1	β_2	β_3	β_4	β_5	β_6
1	3	$\frac{-1}{6}$	$\frac{1}{3}$	$\frac{2}{3}$					
2	4	$\frac{-1}{8}$	$\frac{-1}{48}$	$\frac{5}{12}$	$\frac{29}{48}$				
3	5	$\frac{-19}{180}$	$\frac{7}{1080}$	$\frac{-1}{20}$	$\frac{19}{40}$	$\frac{307}{540}$			
4	6	$\frac{-3}{32}$	$\frac{-17}{5760}$	$\frac{1}{45}$	$\frac{-41}{480}$	$\frac{47}{90}$	$\frac{3133}{5760}$		
5	7	$\frac{-863}{10080}$	$\frac{41}{25200}$	$\frac{-529}{40320}$	$\frac{373}{7560}$	$\frac{-1271}{10080}$	$\frac{2837}{5040}$	$\frac{317731}{604800}$	
6	8	$\frac{-275}{3456}$	$\frac{-731}{725760}$	$\frac{179}{20160}$	$\frac{-5771}{161280}$	$\frac{8131}{90720}$	$\frac{-13823}{80640}$	$\frac{12079}{20160}$	$\frac{247021}{483840}$
7	9	$\frac{-33953}{453600}$	$\frac{8563}{12700800}$	$\frac{-35453}{5443200}$	$\frac{86791}{3024000}$	$\frac{-2797}{36288}$	$\frac{157513}{1088640}$	$\frac{-133643}{604800}$	$\frac{1147051}{1814400}$
7	9	$\beta_7 = \frac{1758023}{3528000}$							

(A.4.6) Table of Enright's Methods II.

k	p	Z	β_0	β_1	β_2	β_3	β_4	β_5	β_6
1	2	$\sqrt{2}$	$-1 + Z$	$2 - Z$					
2	3	$\sqrt{6}$	$\frac{2}{9} - \frac{Z}{9}$	$\frac{-5}{9} + \frac{4Z}{9}$	$\frac{4}{3} - \frac{Z}{3}$				
3	4	$\sqrt{5}$	$\frac{-257}{2904} + \frac{6Z}{121}$	$\frac{137}{363} - \frac{27Z}{121}$	$\frac{-1103}{2904} + \frac{54Z}{121}$	$\frac{12}{11} - \frac{3Z}{11}$			
4	5	$\sqrt{1419}$	$\frac{1057}{22500} + \frac{Z}{625}$	$\frac{-3661}{15000} - \frac{16Z}{1875}$	$\frac{3853}{7500} + \frac{12Z}{625}$	$\frac{-12449}{45000} - \frac{16Z}{625}$	$\frac{24}{25} + \frac{Z}{75}$		
5	6	$\sqrt{5118}$	$\frac{-261979}{9009120} - \frac{10Z}{18769}$	$\frac{2416169}{13513680} + \frac{125Z}{37538}$	$\frac{-2083057}{4504560} - \frac{500Z}{56307}$	$\frac{2889973}{4504560} + \frac{250Z}{18769}$	$\frac{-5534137}{27027360} - \frac{250Z}{18769}$	$\frac{120}{137} + \frac{5Z}{822}$	
6	7	$\sqrt{117573}$	$\frac{1231883}{62233920} + \frac{5Z}{64827}$	$\frac{-20297}{144060} - \frac{4Z}{7203}$	$\frac{124541}{288120} + \frac{25Z}{14406}$	$\frac{-2887799}{3889620} - \frac{200Z}{64827}$	$\frac{587501}{768320} + \frac{25Z}{7203}$	$\frac{-10783}{72030} - \frac{20Z}{7203}$	$\frac{40}{49} + \frac{Z}{822}$

(A.4.7) Figure. Stability region of the method defined by $\pi(\zeta,\eta) = (2 - \eta - 2\eta^2)\zeta - (2 + \eta - 2\eta^2)$.

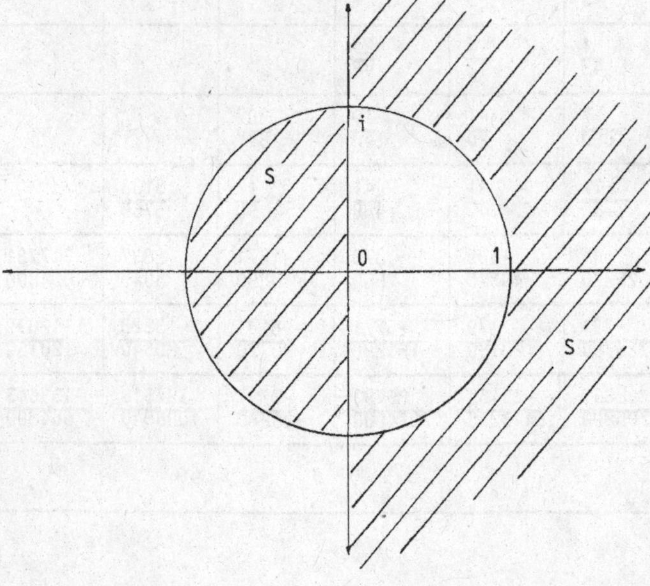

(A.4.8) Figure. Root locus curves of backward differentiation methods.

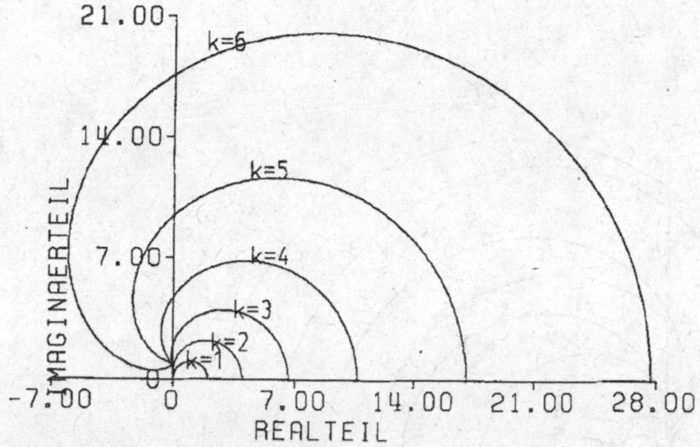

(A.4.9) Figure. Root locus curves of Cryer's methods.

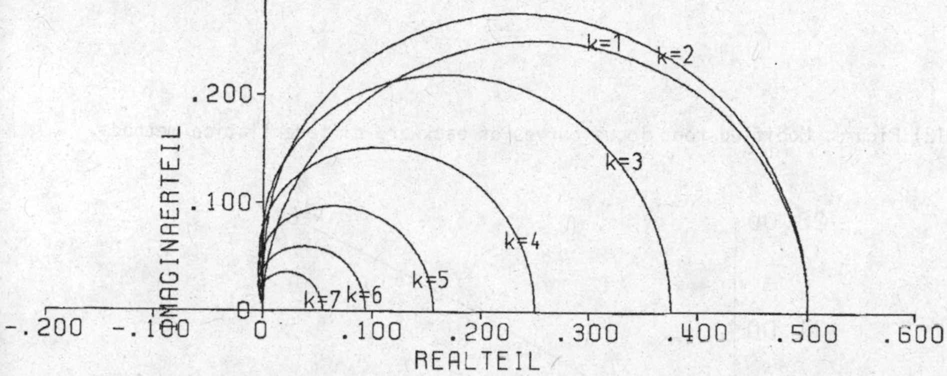

(A.4.10) Figure. Root locus curves of Enright's methods I.

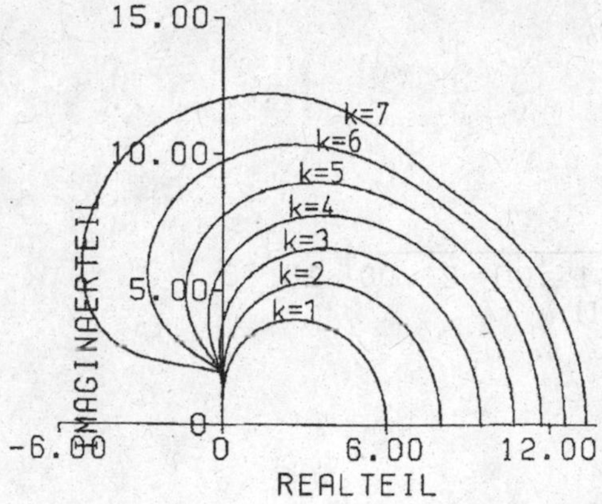

(A.4.11) Figure. Root locus curves of Enright's methods II.

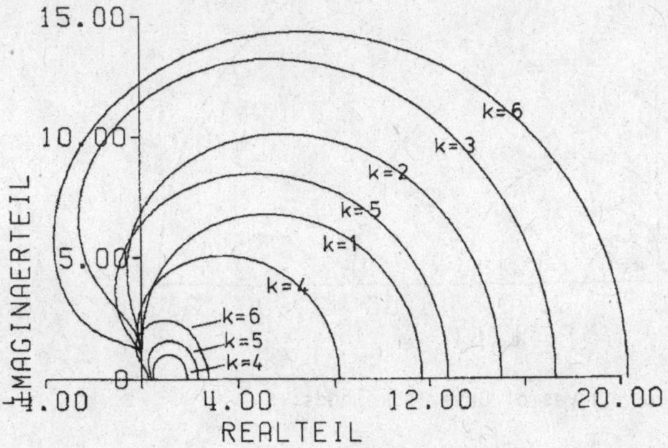

(A.4.12) Figure. Modified root locus curves of backward differentiation methods.

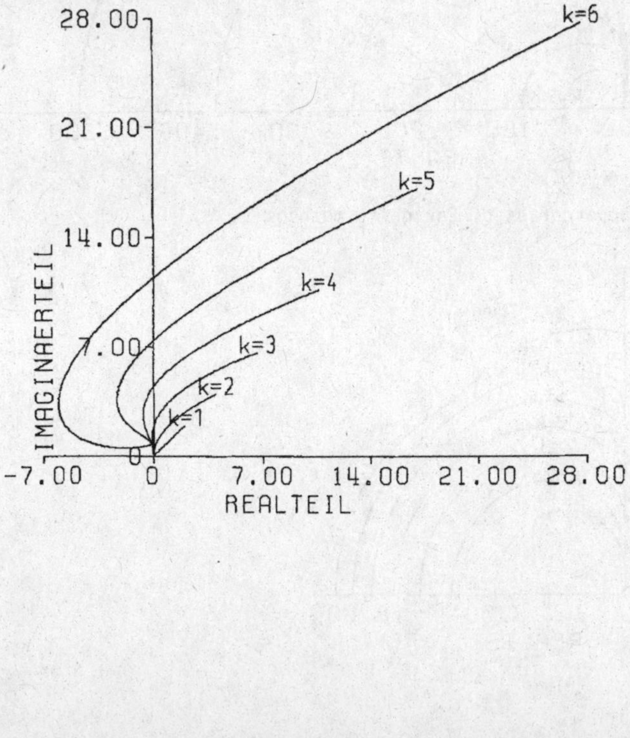

(A.4.13) Figures. Modified root locus curves of Cryer's methods with enlargements.

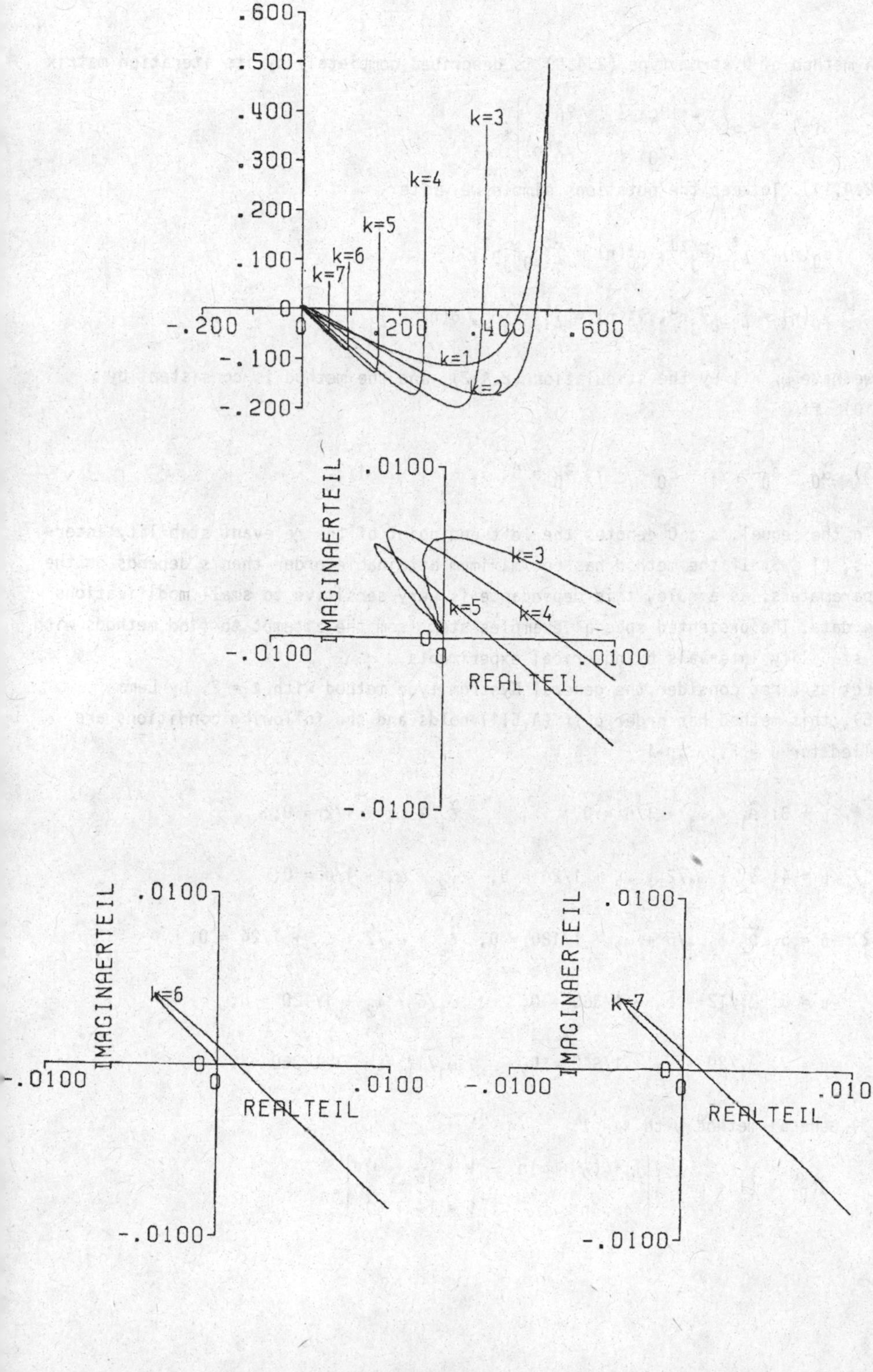

A.5. Examples of Nyström Methods

A method of Nyström type (2.4.9) is described completely by its iteration matrix

$$G(\eta) = -\sigma(\eta)^{-1}\begin{bmatrix}\sigma_0(\eta) & \sigma_0^*(\eta)\\ \chi_0(\eta) & \chi_0^*(\eta)\end{bmatrix},$$

cf. (2.4.11). To keep the notations simple we write

$$\sigma_0(\eta) = \sum_{j=0}^{\ell}\tilde{\alpha}_j\eta^j,\ \sigma_0^*(\eta) = \sum_{j=0}^{\ell}\tilde{\beta}_j\eta^j,$$

$$\chi_0(\eta) = \sum_{j=0}^{\ell}\tilde{\gamma}_j\eta^j,\ \chi_0^*(\eta) = \sum_{j=0}^{\ell}\tilde{\delta}_j\eta^j,\ \sigma(\eta) = \sum_{j=1}^{\ell}\omega_j\eta^j.$$

Then we have $\omega_0 = 1$ by the stipulation (2.4.2), and the method is consistent by (2.4.10) iff

(A.5.1) $\tilde{\alpha}_0 = \tilde{\beta}_0 = \tilde{\gamma}_1 = \tilde{\delta}_0 = -1,\ \tilde{\gamma}_0 = 0,\ \tilde{\alpha}_1 + \omega_1 = -1/2.$

In the sequel, $s \geq 0$ denotes the left end point of the relevant stability interval $[-s, 0] \subseteq S$. If the method has not maximum attainable order then s depends on the free parameters. As a rule, this dependance is very sensitive to small modifications of the data. The presented special examples stem from the attempt to find methods with large stability intervals by numerical experiments.

Let us first consider the general Nyström type method with $\ell = 2$. By Lemma (2.4.5), this method has order p if (A.5.1) holds and the following conditions are fulfilled for $\mu = 3,\dots,p+1$:

$\mu = 3$: $\tilde{\beta}_1 + \omega_1 + 1/6 = 0,\quad \tilde{\delta}_1 + \omega_1 + 1/2 = 0;$

$\mu = 4$: $\tilde{\alpha}_2 + \omega_1/2 + \omega_2 + 1/24 = 0,\quad \tilde{\gamma}_2 + \omega_1 + 1/6 = 0;$

(A.5.2) $\mu = 5$ $\tilde{\beta}_2 + \omega_1/6 + \omega_2 + 1/120 = 0,\quad \tilde{\delta}_2 + \omega_1/2 + \omega_2 + 1/24 = 0;$

$\mu = 6$: $\omega_1/12 + \omega_2 + 1/360 = 0,\quad \omega_1/6 + \omega_2 + 1/120 = 0;$

$\mu = 7$: $\omega_1/20 + \omega_2 + 1/840 = 0,\quad \omega_1/12 + \omega_2 + 1/360 = 0.$

(A.5.3) General method with $\ell = 1$:

$$G(\eta) = (1 + \omega\eta)^{-1}\begin{bmatrix}1 + (\frac{1}{2} + \omega)\eta & 1 + (\frac{1}{6} + \omega)\eta\\ \eta & 1 + (\frac{1}{2} + \omega)\eta\end{bmatrix}.$$

This method has order $p = 2$ and no stability interval $[-s, 0]$ for all $\omega \in \mathbb{R}$.

(A.5.4) (i) Full implicit method with $\ell = 2$ and order $p = 4$: $\omega_1, \omega_2 \in \mathbb{R}$,

$$G(\eta) = (1+\omega_1\eta+\omega_2\eta^2)^{-1}\begin{bmatrix} 1+(\frac{1}{2}+\omega_1)\eta + (\frac{1}{24}+\frac{\omega_1}{2}+\omega_2)\eta^2 & 1+(\frac{1}{6}+\omega_1)\eta + (\frac{1}{120}+\frac{\omega_1}{6}+\omega_2)\eta^2 \\ \eta + (\frac{1}{6}+\omega_1)\eta^2 & 1+(\frac{1}{2}+\omega_1)\eta + (\frac{1}{24}+\frac{\omega_1}{2}+\omega_2)\eta^2 \end{bmatrix}.$$

(ii) Maximum order $p = 5$: $\omega_1 = -1/15$, $\omega_2 = 1/360$; $s = 0$.

(iii) Order $p = 4$, $\omega_2 = 0$: $s = 8$ for $\omega_1 = 0.0416$.

(iv) Order $p = 4$, $\omega_1 = \omega_2 = 0$, i.e. explicit method: $s = 7.06$.

(v) Order $p = 4$, $\omega_1 = \omega$, $\omega_2 = \omega^2/4$, i.e. $\sigma(\eta)$ complete square: $s = 8.18$ for $\omega = 0.0445$.

(vi) Order $p = 3$, $\omega_1 = \omega$, $\omega_2 = \omega^2/4$, $\tilde{\beta}_2 = \tilde{\delta}_2 = 0$: $s = \infty$ for $\omega \geq 1.92$.

In the remaining part of this section we consider explicit Runge-Kutta-Nyström methods for the initial value problem

(A.5.5) $y'' = f(t,y)$, $t > 0$, $y(0) = y_0$, $y'(0) = y_0^*$

with conservative differential equation. As it was emphasized several times, these methods coincide formally with Nyström type methods for the problem (2.4.8),

(A.5.6) $y'' = A^2y + c(t)$, $t > 0$, $y(0) = y_0$, $y'(0) = y_0^*$,

with exception of the treatment of the time-dependent right side $c(t)$.

Recalling the notations for Runge-Kutta methods introduced in Section 5.1,

$$t_{n,i} = (n+\tau_i)\Delta t,\ 0 \leq \tau_i \leq 1,\ v_{n,i} = v(t_{n,i}),\ f_{n,i}(v) = f(t_{n,i},v),\ i = 1,\dots,r,$$

a Runge-Kutta-Nyström method for (A.5.5) reads

(A.5.7) $v_{n,i} = v_n + \tau_i \Delta t w_n + \Delta t^2 \sum_{j=1}^r \alpha_{ij} f_{n,j}(v_{n,j})$, $\quad i = 1,\dots,r$,

(A.5.8) $v_{n+1} = v_n + \Delta t w_n + \Delta t^2 \sum_{j=1}^r \beta_j f_{n,j}(v_{n,j})$

(A.5.9) $\Delta t w_{n+1} = \Delta t w_n + \Delta t^2 \sum_{j=1}^r \gamma_j f_{n,j}(v_{n,j})$, $\quad n = 0,1,\dots$.

We use the same notations as in Section 5.1 but write

$$a = (\tau_1,\dots,\tau_r)^T,\ b = (\beta_1,\dots,\beta_r)^T,\ q = (\gamma_1,\dots,\gamma_r)^T.$$

Then we obtain in complete analogy to (5.1.9) for the linear problem (A.5.6) with time-varying matrix $A(t)^2$ the computational device

$$\begin{aligned} v_{n+1} &= (I + \Delta t^2 b^T \underline{A}_n^2(\underline{I} - \Delta t^2 \underline{PA}_n^2)^{-1} z)v_n \\ &+ (I + \Delta t^2 b^T \underline{A}_n^2(\underline{I} - \Delta t^2 \underline{PA}_n^2)^{-1} a)\Delta t w_n + \Delta t^2 b^T(\underline{I} - \Delta t^2 \underline{PA}_n^2)^{-1}\underline{c}_n \\ \Delta t w_{n+1} &= \Delta t^2 q^T \underline{A}_n^2(\underline{I} - \Delta t^2 \underline{PA}_n^2)^{-1} z v_n \\ &+ (I + \Delta t^2 q^T \underline{A}_n^2(\underline{I} - \Delta t^2 \underline{PA}_n^2)^{-1} a)\Delta t w_n + \Delta t^2 q^T(\underline{I} - \Delta t^2 \underline{PA}_n^2)^{-1}\underline{c}_n,\quad n = 0,1,\dots\ . \end{aligned}$$

In particular, we find for the test equation $y'' = \lambda^2 y$ writing $V_n = (v_n, \Delta t w_n)^T$ and $\eta = \Delta t^2\lambda^2$

$$V_{n+1} = G(\eta)V_n,\quad G(\eta) = \begin{bmatrix} g_1(\eta) & g_2(\eta) \\ g_3(\eta) & g_4(\eta) \end{bmatrix}$$

with the following rational functions in η

$$\text{(A.5.10)}\quad \begin{aligned} &g_1(\eta) = 1 + \eta b^T(I - \eta P)^{-1}z,\ g_2(\eta) = 1 + \eta b^T(I - \eta P)^{-1}a, \\ &g_3(\eta) = \eta q^T(I - \eta P)^{-1}z,\ g_4(\eta) = 1 + \eta q^T(I - \eta P)^{-1}a. \end{aligned}$$

A Runge-Kutta-Nyström method is described by the vectors a, b, and q, and by the matrix P, i.e., by the matrix

$$(q,b,a,P) = \begin{bmatrix} \gamma_1 & \beta_1 & \tau_1 & \alpha_{11} & \cdots & \alpha_{1r} \\ \vdots & \vdots & \vdots & \vdots & & \vdots \\ \gamma_r & \beta_r & \tau_r & \alpha_{r1} & \cdots & \alpha_{rr} \end{bmatrix} \equiv \Xi.$$

The method is *explicit* if P is a lower triangular matrix with zeros in the diagonal. As in Runge-Kutta methods, the order of consistence can have different values for the general nonlinear problem (A.5.5) and the linear problem with constant matrix (A.5.6). In the following examples, p denotes the order with respect to (A.5.5) and with respect to the corresponding Nyström type method, i.e., with respect to (A.5.6). p is said to be maximum - with respect to (A.5.5) and (A.5.6) - if it is the maximum order for all *explicit* methods with the same stage number r.

(A.5.11) $r = \ell = 1$, $p = 1$ (max.), $S = [-4, 0]$, cf. Hairer [77]:

$$\Xi = (\gamma_1, \beta_1, \tau_1, 0) = (1, 1/2, 1/2, 0).$$

(A.5.12) $r = \ell = 2$, $p = 2$ (max.), cf. Hairer [77]; $s = 3.67$ for $\alpha = 0.265$:

$$\Xi = \begin{bmatrix} \frac{1}{4(3\alpha^2-3\alpha+1)} & \frac{1-\alpha}{4(3\alpha^2-3\alpha+1)} & \alpha & 0 & 0 \\ \frac{3(1-2\alpha)^2}{4(3\alpha^2-3\alpha+1)} & \frac{(1-2\alpha)(1-3\alpha)}{4(3\alpha^2-3\alpha+1)} & \frac{2-3\alpha}{3(1-2\alpha)} & \frac{2(3\alpha^2-3\alpha+1)}{9(1-2\alpha)^2} & 0 \end{bmatrix}, \; 0 \leq \alpha \leq 1.$$

(A.5.13) $r = \ell = 3$, $p = 3$ (max. for (A.5.5)), cf. Hairer [77]; $s = 6.69 \; \forall \alpha \in \mathbb{R}$:

$$\Xi = \begin{bmatrix} \frac{1}{2}-\alpha & (\frac{1}{2}-\alpha)(\frac{1}{2}\pm\frac{\sqrt{3}}{6}) & \frac{1}{2}\mp\frac{\sqrt{3}}{6} & 0 & 0 & 0 \\ \frac{1}{2} & \frac{1}{2}(\frac{1}{2}\mp\frac{\sqrt{3}}{6}) & \frac{1}{2}\pm\frac{\sqrt{3}}{6} & \frac{1}{6}\pm\frac{\sqrt{3}}{12} & 0 & 0 \\ \alpha & \alpha(\frac{1}{2}\pm\frac{\sqrt{3}}{6}) & \frac{1}{2}\mp\frac{\sqrt{3}}{6} & 0 & \frac{1}{2\alpha}(\frac{1}{6}\pm\frac{\sqrt{3}}{12}) & 0 \end{bmatrix}, \; \forall \alpha \in \mathbb{R}.$$

(A.5.14) $r = \ell = 3$, $p = 3$ (max. for (A.5.5)), cf. Hairer [77];

$$\Xi = \begin{bmatrix} \frac{1}{6(1-2\alpha)^2} & \frac{1-\alpha}{6(1-2\alpha)^2} & \alpha & 0 & 0 & 0 \\ \frac{4(6\alpha^2-6\alpha+1)}{6(1-2\alpha)^2} & \frac{2(6\alpha^2-6\alpha+1)}{6(1-2\alpha)^2} & \frac{1}{2} & \frac{(1-4\alpha)(1-2\alpha)}{8(6\alpha^2-6\alpha+1)} & 0 & 0 \\ \frac{1}{6(1-2\alpha)^2} & \frac{\alpha}{6(1-2\alpha)^2} & 1-\alpha & 2\alpha(1-2\alpha) & \frac{1}{2}(1-2\alpha)(1-4\alpha) & 0 \end{bmatrix}, \; 0 \leq \alpha \leq 1.$$

This method has order $p = 4$ with respect to (A.5.6) for $\alpha = (5 \pm \sqrt{5})/20$; $s = 6.47$ for $\alpha = (5 + \sqrt{5})/20$ and $s = 15.33$ for $\alpha = (5 - \sqrt{5})/20$.

(A.5.15) $r = \ell = 5$, $p = 5$, cf. Albrecht [55]; $s = 9.24$:

$$\Xi = \begin{bmatrix} 7/90 & 7/90 & 0 & 0 & 0 & 0 & 0 & 0 \\ 32/90 & 24/90 & 1/4 & 1/32 & 0 & 0 & 0 & 0 \\ 12/90 & 6/90 & 1/2 & -1/24 & 1/6 & 0 & 0 & 0 \\ 32/90 & 8/90 & 3/4 & 3/32 & 1/8 & 1/16 & 0 & 0 \\ 7/90 & 0 & 1 & 0 & 3/7 & -1/14 & 1/7 & 0 \end{bmatrix}.$$

(A.5.16) Figure. Root locus curves of method (A.5.4)(iii).

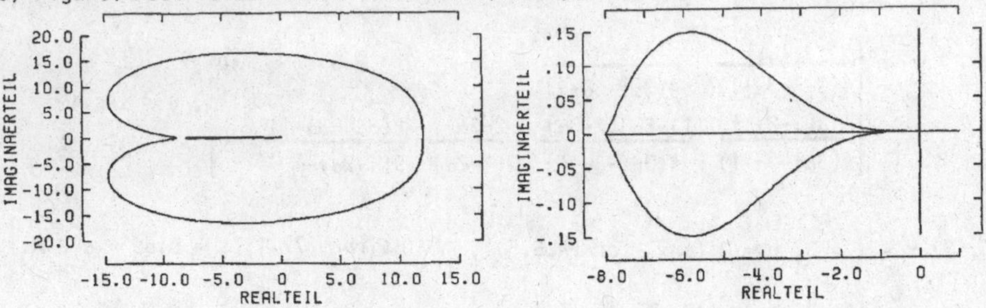

(A.5.17) Figure.
Root locus curves
of method (A.5.4)(iv).

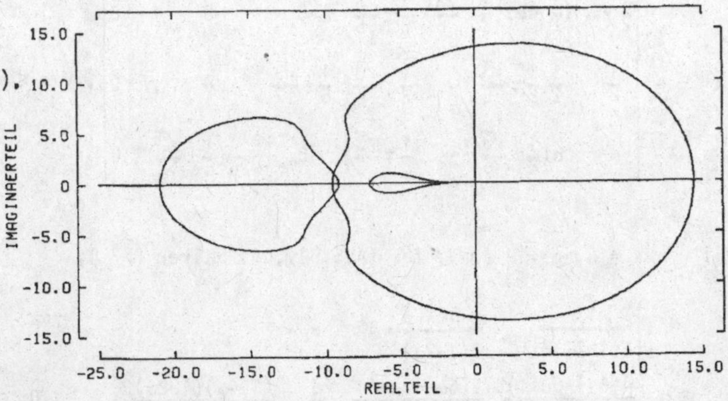

(A.5.18) Figure. Root locus curves of method (A.5.4)(v).

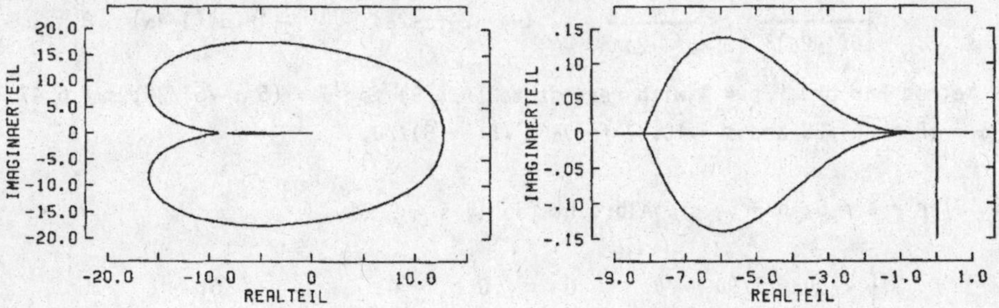

(A.5.19) Figure.
Root locus curves of
method (A.5.4)(vi)
for $\omega = 1.92$.

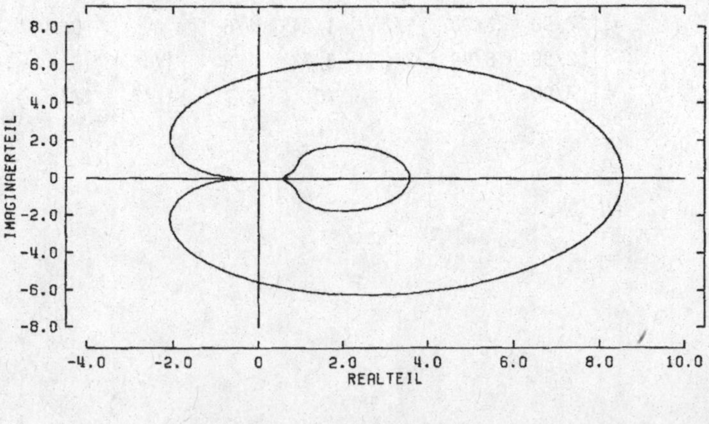

(A.5.20) Figure. Root locus curves of method (A.5.12).

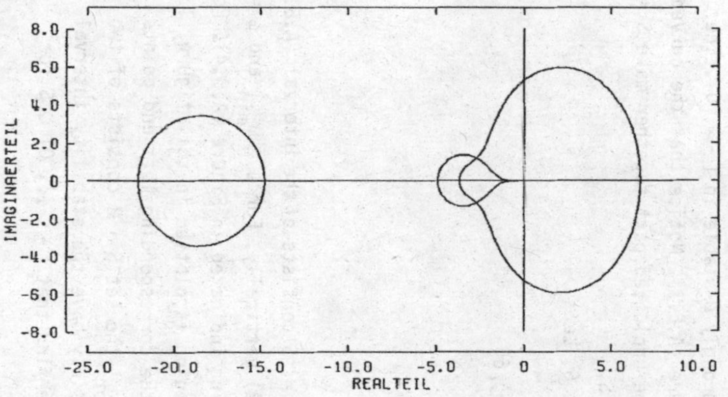

(A.5.21) Figure. Root locus curves of method (A.5.13).

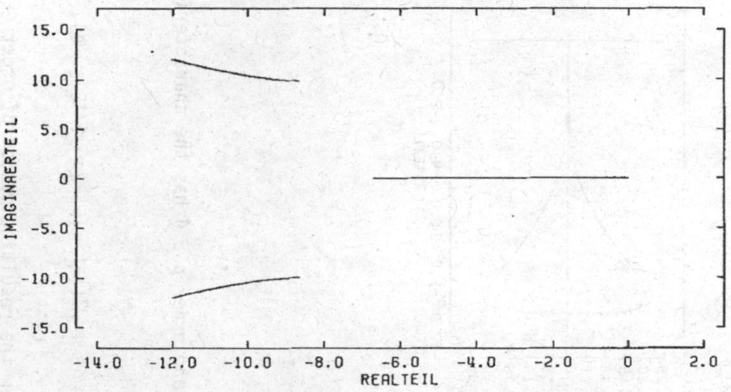

(A.5.22) Figure. Root locus curves of method (A.5.14), $\alpha = (5+\sqrt{5})/20$.

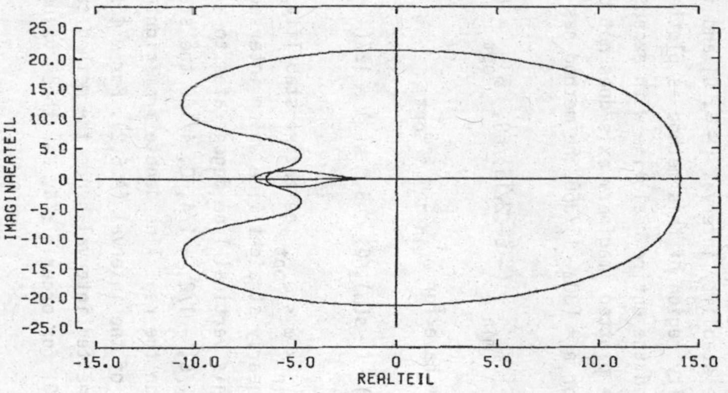

(A.5.23) Figure. Root locus curves of method (A.5.14), $\alpha = (5-\sqrt{5})/20$.

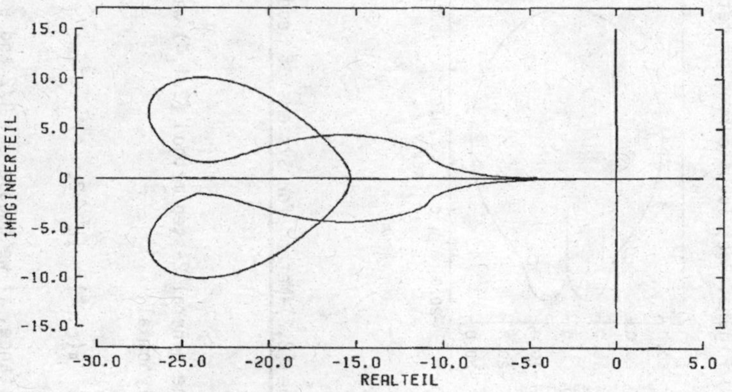

(A.5.24) Figure. Root locus curves of method (A.5.15).

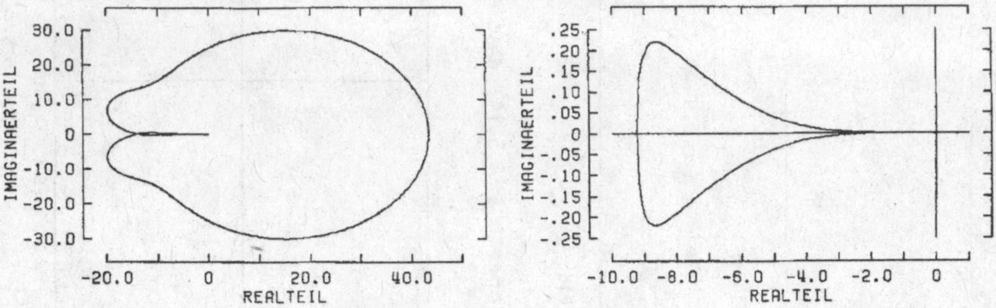

A.6. The (2,2)-Method for Systems of Second Order

The general 2-step method (2.1.4) with $\ell = 2$ and order $p \geq 4$ has the characteristic polynomial

(A.6.1) $$\pi(\zeta,\eta^2) = (\zeta^2 - 2\zeta + 1) - \eta^2(\omega\zeta^2 + (1 - 2\omega)\zeta + \ ) + \eta^4(\theta\zeta^2 + (\omega - 2\theta - 1/12)\zeta + \theta).$$

For instance, if we choose $\omega = 1/2$ and $\theta = 1/4$ then the resulting method of order 4 has the stability interval $[-\infty, 0]$ and is also strongly D-stable in $[-\infty, 0]$. The stability region of this method is plotted in Figure (A.6.3). Notice that the curved line and the entire real line with exception of the unplotted part together make $S = \partial S$. The plotted imaginary axis does not belong to S.

For $\theta = (30\omega - 1)/360$ the method has order $p \geq 6$. Let

$$\omega_{min} = (-5 - 2\sqrt{15})/60, \quad \omega_{opt} = (-5 + 2\sqrt{15})/60$$

then we have for $\omega_{min} \leq \omega \leq \omega_{opt}$

(A.6.2) $$[-s(\omega), 0] \subset S, \quad s(\omega) = 12/(1 - 12\omega).$$

For $\omega_{min} < \omega < \omega_{opt}$ the entire stability region $S = \partial S$ consists of the interval (A.6.2) and a nearly straight line which intersects (A.6.2) vertically. For $\omega = \omega_{min}$ and $\omega = \omega_{opt}$ this vertical line degenerates to a point being indicated in Figure (A.6.4). For $\omega = -1/6, -1/2, -1/4, 0, 4/100$ the stability region is plotted in this figure. The marks on the real line denote from right to left the corresponding left end points $-s(\omega)$ of the interval (A.6.2). For $\omega \notin [\omega_{min}, \omega_{opt}]$ the set $S \cap \mathbb{R}$ consists of two disconnected interval hence the method cannot possibly have the stability interval $(-\infty, 0]$ for order $p \geq 6$. For these values of ω we find that $[-s(\omega), 0] \subset S$ for

$$s(\omega) = 6[15(1-4\omega) - \sqrt{5}(720\omega^2 + 120\omega - 7)^{1/2}]/[13 - 120\omega]$$

hence $s(\omega) \to 0$ for $\omega \to -\infty$ and $s(\omega) \to 6$ for $\omega \to +\infty$.

For $\omega = 11/252 = 0.0436\ldots$ and $\theta = 13/15120$ the method given by (A.6.1) has order $p = 8$. Since $\omega_{min} < 11/252 < \omega_{opt} = 0.0457\ldots$ the stability region has here the same form as in Figure (A.6.4) with $S \cap \mathbb{R} \cong [-25.2, 0]$. In comparison with this the largest stability interval of a method given by (A.6.1) of order 6 was $[-s(\omega_{opt}), 0] \cong [-26.6, 0]$.

(A.6.3) Figure.

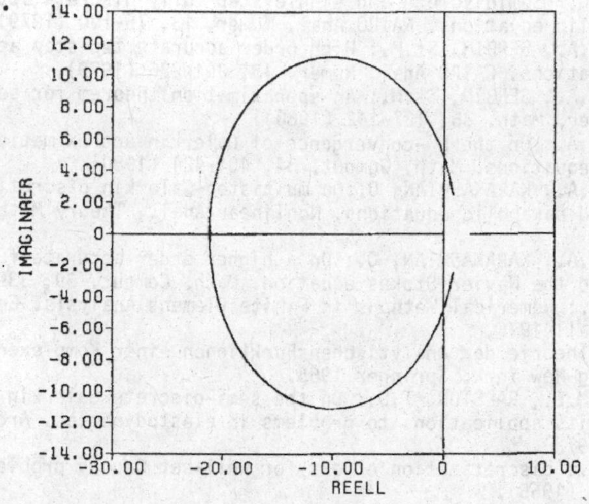

(A.6.4) Figure.

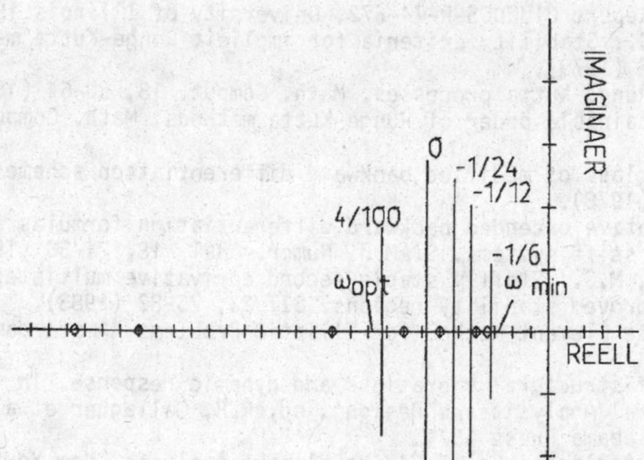

References

AHLFORS, L.K.: Complex Analysis. New York: Mc Graw-Hill 1953.

ALBRECHT, J.: Beiträge zum Runge-Kutta-Verfahren. Z. Angew. Math. Mech. 35, 100-110 (1955).

BAKER, G.A.: A finite element method for first order hyperbolic equations. Math. Comput. 29, 995-1006 (1975).

BAKER, G.A.: Error estimates for finite element methods for second order hyperbolic equations. SIAM J. Numer. Anal. 13, 564-576 (1976).

BAKER, G.A., BRAMBLE, J.H., THOMEE, V.: Single step Galerkin approximations for parabolic problems. Math. Comput. 31, 818-847 (1977).

BAKER, G.A., BRAMBLE, J.H.: Semidiscrete and single-step fully discrete approximations for second order hyperbolic equations. RAIRO Anal. Numer. 13, 75-100 (1979).

BAKER, G.A., DOUGALIS, V.A., SERBIN, St.M.: High order accurate two-step approximations for hyperbolic equations. RAIRO Anal. Numer. 13, 201-226 (1979).

BAKER, G.A., DOUGALIS, V.A., SERBIN, St.M.: An approximation theorem for second-order evolution equations. Numer. Math. 35, 127-142 (1980).

BAKER, G.A., DOUGALIS, V.A.: On the L^∞-convergence of Galerkin approximations for second-order hyperbolic equations. Math. Comput. 34, 401-424 (1980).

BAKER, G.A., DOUGALIS, V.A., KARAKASHIAN, O.:On multistep-Galerkin discretizations of semilinear hyperbolic and parabolic equations. Nonlinear Anal., Theory Methods Appl. 4, 579-597 (1980).

BAKER, G.A., DOUGALIS, V.A., KARAKASHIAN, O.: On a higher order accurate fully discrete Galerkin approximation to the Navier-Stokes equation. Math. Comput. 39, 339-375 (1982).

BATHE, K.J., WILSON, E.L.: Numerical Methods in Finite Element Analysis. Englewood Cliffs, N.J.: Prentice-Hall 1976.

BEHNKE, H., SOMMER, F.: Theorie der Analytischen Funktionen einer Komplexen Veränderlichen. Berlin-Heidelberg-New York: Springer 1965.

BENTHIEN, G.W., GURTIN, M.E., RALSTON, T.D.: On the semi-discrete Galerkin method for hyperbolic problems and its applications to problems in elastodynamics. Arch. Ration. Mech. Anal. 48, 51-63 (1972).

BIRKHOFF, G., VARGA, R.S.: Discretization errors for well-set Cauchy problems I. J. Math. and Phys. 44, 1-23 (1965).

BRAMBLE, J.H., SCHATZ, A.H., THOMEE, V., WAHLBIN, L.B.: Some convergence estimates for semidiscrete Galerkin type approximations for parabolic equations. SIAM J. Numer. Anal. 14, 218-241 (1977).

BROWN, R.L.: Multi-derivative numerical methods for the solution of stiff ordinary differential equations. Report UIUCDCS-R-74-672. University of Illinois 1974.

BURRAGE, K., BUTCHER, J.C.: Stability criteria for implicit Runge-Kutta methods. SIAM J. Numer. Anal. 16, 46-57 (1979).

BUTCHER, J.C.: Implicit Runge-Kutta processes. Math. Comput. 18, 50-64 (1964).

BUTCHER, J.C.: On the attainable order of Runge-Kutta methods. Math. Comput. 19, 408-417 (1965).

CASH, J.R.: A note on a class of modified backward differentiation schemes. J. Inst. Math. Appl. 21, 301-313 (1978).

CASH, J.R.: Second derivative extended backward differentiation formulas for the numerical integration of stiff systems. SIAM J. Numer. Anal. 18, 21-36 (1981).

CHAKRAVARTI, P.C., KAMEL, M.S.: Stiffly stable second derivative multistep methods with higher order and improved stability regions. BIT 23, 75-83 (1983).

CIARLET, Ph.G.: The Finite Element Method for Elliptic Problems. Amsterdam: North-Holland 1979.

CLOUGH, R.W.: Analysis of structural vibrations and dynamic response. In 'Recent Advances in Matrix Structural Analysis and Design', ed. R.H. Gallagher et al., 441-486. Alabama: University of Alabama Press 1971.

COOK, R.D.: Concepts and Applications of Finite Element Analysis. New York-London: Wiley 1974.

CROUZEIX, M.: Sur l'approximation des équations différentielles opérationelles linéaires par des méthodes de Runge-Kutta. Thesis. Université de Paris 1975.

CROUZEIX, M.: Une méthode mulitpas implicite-explicite pour l'approximation des équations d'évolution paraboliques. Numer. Math. 35, 257-276 (1980).

CROUZEIX, M., RAVIART, P.-A.: Approximations des Problèmes d'Evolution. Preprint. Université de Rennes 1980.
CRYER, C.W.: A new class of highly-stable methods: A_0-stable methods. BIT 13, 153-159 (1973).
CURTISS, C.F., HIRSCHFELDER, J.O.: Integration of stiff systems. Proc. Nat. Acad. Sci. USA 38, 235-243 (1952).
DAHLQUIST, G.: Stability and Error Bounds in the Numerical Integration of Ordinary Differential Equations. Transactions Royal Inst. Techn. Stockholm Nr. 130, 1959.
DAHLQUIST, G.: A special stability problem for linear multistep methods. BIT 3, 27-43 (1963).
DAHLQUIST, G.: Error analysis for a class of methods for stiff non-linear initial value problems. In 'Numerical Analysis', ed. G.A. Watson, 60-74. Lecture Notes no. 506. Berlin-Heidelberg-New York: Springer 1975.
DAHLQUIST, G.: On accuracy and unconditional stability of linear multistep methods for second order differential equations. BIT 18, 133-136 (1978a).
DAHLQUIST, G.: G-stability is equivalent to A-stability. BIT 18, 384-401 (1978b).
DANIEL, J.W., MOORE, R.E.: Computation and Theory in Ordinary Differential Equations. San Francisco: Freeman and Co. 1970.
DEKKER, K.: Stability of linear mutistep methods on the imaginary axis. BIT 21, 66-79 (1981).
DENDY, J.E.: An analysis of some Galerkin schemes for the solution of nonlinear time-dependent problems. SIAM J. Numer. ANAL. 12, 541-565 (1975).
DESCLOUX, J.: On the numerical integration of the heat equation. Numer. Math. 15, 371-381 (1970).
DIETRICH, G.: On the efficient and accurate solution of the skew-symmetric eigenvalue problem. ISD Report No. 264. Universität Stuttgart 1981.
DIEUDONNE, J.: Foundations of Modern Analysis. New York-London: Academic Press 1960.
DOUGALIS, V.A.: Multistep-Galerkin methods for hyperbolic equations. Math. Comput. 33, 563-584 (1979).
DOUGALIS, V.A., KARAKASHIAN, O.: On some higher order accurate fully discrete Galerkin methods for the Korteweg-de Vries equation. Math. Comput. 40, 419-433 (1983).
DOUGLAS, J., DUPONT, T.: Galerkin methods for parabolic problems. SIAM J. Numer. Anal. 7, 575-626 (1970).
DUPONT, T.: L^2-estimates for Galerkin methods for second order hyperbolic equations. SIAM J. Numer. Anal. 10, 880-889 (1973).
DUPONT, T., FAIRWEATHER, G., JOHNSEN, J.P.: Three-level Galerkin methods for parabolic equations. SIAM J. Numer. Anal. 11, 392-409 (1974).
ENRIGHT, W.H.: Second derivative multistep methods for stiff ordinary differential equations. SIAM J. Numer. Anal. 11, 321-333 (1974a).
ENRIGHT, W.H.: Optimal second derivative multistep methods for stiff systems. In 'Stiff Differential Systems', ed. R.A. Willoughby, 95-119. New York: Plenum Press 1974b.
FAIRWEATHER, G.: Finite Element Galerkin Methods for Differential Equations. New York: Dekker 1978.
FEHLBERG, E.: Klassische Runge-Kutta-Nyström-Formeln mit Schrittweitenkontrolle für Differentialgleichungen $\ddot{x} = f(t,x)$. Computing 10, 305-315 (1972).
FEHLBERG, E.: Klassische Runge-Kutta-Nyström-Formeln mit Schrittweitenkontrolle für Differentialgleichungen $\ddot{x} = f(t,x,\dot{x})$. Computing 14, 371-385 (1975).
FINLAYSON, B.A.: Convergence of the Galerkin method involving chemical reactions. SIAM J. Numer. Anal. 8, 316-324 (1971).
FRIED, I.: Numerical Solution of Differential Equations. New York-London: Academic Press 1979.
FRIEDLI, A., JELTSCH, R.: An algebraic test for A_0-stability. BIT 18, 402-414 (1978).
GAUTSCHI, W.: On inverses of Vandermonde and confluent Vandermonde matrices. Numer. Math. 4, 117-123 (1962).
GEAR, C.W.: The automatic integration of stiff ordinary differential equations. In 'Information Processing', ed. A.J. Morell, 187-193. Amsterdam: North-Holland 1969.
GEAR, C.W.: The stability of numerical methods for second order differential equations. SIAM J. Numer. Anal. 15, 188-197 (1978).
GEKELER, E.: Linear multistep methods and Galerkin procedures for initial boundary value problems. SIAM J. Numer. Anal. 13, 536-548 (1976).
GEKELER, E.: Galerkin-Runge-Kutta methods and hyperbolic initial boundary value problems. Computing 18, 79-88 (1977).

GEKELER, E., JOHNSEN, Th.L.: Galerkin-Obrechkoff methods and hyperbolic initial boundary value problems with damping. Comput. Methods Appl. Mech. Eng. 10, 359-370 (1977).
GEKELER, E.: A-priori error estimates for Galerkin backward differentiation methods in time-inhomogeneous parabolic problems. Numer. Math. 30, 369-383 (1978).
GEKELER, E.: Uniform stability of linear multistep methods in Galerkin procedures for parabolic problems. J. Math. Math. Sciences 2, 651-667 (1979).
GEKELER, E.: Stability of Galerkin mulitstep procedures in time-inhomogeneous hyperbolic problems. Computing 24, 315-324 (1980).
GEKELER, E.: On the pointwise matrix product and the mean value theorem. Linear Algebra Appl. 35, 183-191 (1981).
GEKELER, E.: On the stability of backward differentiation methods. Numer. Math. 38, 467-471 (1982a).
GEKELER, E.: Linear multistep methods for stable differential equations $\ddot{y} = Ay + B\dot{y} + c(t)$. Math. Comput. 39, 481-490 (1982b).
GERADIN, M.: A classification and discussion of integration operators for transient structural response. AIAA paper no. 74-105 (1974).
GODLEWSKI, E., PUECH-RAOULT, A.: Equations d'évolution linéaires du second ordre et méthodes multipas. RAIRO Anal. Numer. 13, 329-353 (1979).
Grigorieff, R.D.: Numerik Gewöhnlicher Differentialgleichungen I, II. Stuttgart: Teubner 1972, 1977.
GUNZBURGER, M.D.: On the stability of Galerkin methods for initial-boundary value problems for hyperbolic systems. Math. Comput. 31, 661-675 (1977).
HACKMACK, U.: Fehlerschranken für linear Mehrschrittverfahren bei stabilen Differentialgleichungen mit variablen Koeffizienten. Dissertation. Universität Stuttgart 1981.
HAIRER, E., WANNER, G.: Multistep-multistage-multiderivative methods for ordinary differential equations. Computing 11, 287-303 (1973).
HAIRER, E., WANNER, G.: A theory of Nyström methods. Numer. Math. 25, 383-400 (1976).
HAIRER, E.: Méthodes de Nyström pour l'équation différentielle $y'' = f(x,y)$. Numer. Math. 27, 283-300 (1977).
HAIRER, E.: Unconditionally stable methods for second order differential equations. Numer. Math. 32, 373-379 (1979).
HENRICI, P.: Discrete Variable Methods for Ordinary Differential Equations. New York-London: Wiley 1962.
HENSEL, K., LANDSBERG, G.: Theorie der algebraischen Funktionen einer Variablen. Leipzig: Teubner 1902.
HILLE, E.: Analytic Function Theory II. Boston: Grim and Co. 1962.
HOUSHOLDER, A.S.: The Theory of Matrices in Numerical Analysis. New York: Dover Publ. 1964.
HUMMEL, P.M., SEEBECK, C.L.: A generalization of Taylor's expansion. Am. Math. Mon. 56, 243-247 (1949).
HYMAN, J.M.: The method of lines solution of partial differential equations. NYU Report COO-3077-139.
ISERLES, A.: On the A-stability of implicit Runge-Kutta processes. BIT 18, 157-169 (1978).
JELTSCH, R.: Multistep Multiderivative Methods for the Solution of Initial Value Problems for Ordinary Differential Equations. Seminar Notes. Lexington: University of Kentucky 1976a.
JELTSCH, R.: Stiff stability and its relation to A_0- and A(0)-stability. SIAM J. Numer. Anal. 13, 8-17 (1976b).
JELTSCH, R.: Stiff stability of multistep multiderivative methods. SIAM J. Numer. Anal. 4, 760-772 (1977).
JELTSCH, R.: Corrigendum: Stiff stability of multistep multiderivative methods. SIAM J. Numer. Anal. 16, 339-345 (1979).
JELTSCH, R.: Stability on the imaginary axis and A-stability of linear multistep methods. BIT 18, 170-174 (1978a).
JELTSCH, R.: Complete characterization of multistep methods with an interval of periodicity for solving $y'' = f(x,y)$. Math. Comput. 32, 1108-1114 (1978b).
JELTSCH, R.: On the stability regions of multistep multiderivative methods. In 'Numerical Treatment of Differential Equations', ed. R. Bulirsch et al., 63-80. Lecture Notes no. 631. Berlin-Heidelberg-New York: Springer 1978c.
JELTSCH, R., NEVANLINNA, O.: Largest disk of stability of explicit Runge-Kutta methods. BIT 18, 500-502 (1978d).

JELTSCH, R., NEVANLINNA, O.: Stability of explicit time-discretizations for solving initial value problems. Numer. Math. 37, 61-91 (1981).
JELTSCH, R., NEVANLINNA, O.: Stability and accuracy of time discretizations for initial value problems. Numer. Math. 40, 245-296 (1982a).
JELTSCH, R., NEVANLINNA, O.: Stability of semidiscretizations of hyperbolic problems. Preprint. To appear in SIAM J. Numer. Anal. (1982b).
JENSEN, P.S.: Stability analysis of structures by stiffly stable methods. Comput. Struct. 4, 615-626 (1974).
JENSEN, P.S.: Stiffly stable methods for undamped second order equations of motion. SIAM J. Numer. Anal. 13, 549-563 (1976).
KATO, T.: Perturbation Theory for Linear Operaotrs. Berlin-Heidelberg-New York: Springer 1966.
KREISS, H.O.: Über die Stabilitätsdefinition für Differenzengleichungen, die partielle Differentialgleichungen approximieren. BIT 2, 153-181 (1962).
KRETH, H.: Die numerische Lösung nichtlinearer Evolutionsgleichungen durch implizite Ein- und Zweischrittverfahren. Habilitationsschrift. Universität Hamburg 1981.
LAMBERT, J.D.: Computational Methods in Ordinary Differential Equations. New York-London: Wiley 1973.
LAMBERT, J.D., WATSON, I.A.: Symmetric multistep methods for periodic initial value problems. J. Inst. Math. Appl. 18, 189-202 (1976).
LEROUX, M.-N.: Semi-discrétisation en temps pour les équations d'évolution paraboliques lorsque l'opérateur dépend du temps. RAIRO Anal. Numer. 13, 119-137 (1979a).
LEROUX, M.-N.: Semidiscretization in time for parabolic problems. Math. Comput. 33, 919-931 (1979b).
LEROUX, M.-N.: Méthodes multipas pour des équations paraboliques non linéaires. Numer. Math. 35, 143-162 (1980).
LEROUX, M.-N.: Variable step size multistep methods for parabolic problems. SIAM J. Numer. Anal. 19, 724-741 (1982).
MARDEN, M.: The Geometry of Polynomials. Providence, R.I.: Amer. Math. Soc. 1966.
MOORE, P.: Finite element multistep multiderivative schemes for parabolic equations. J. Inst. Math. Appl. 21, 331-344 (1978).
MUIR, Th.: The Theory of Determinants, I. New York: Dover Publ. 1960.
NASSIF, N.R., DESCLOUX, J.: Stability study for time-dependent linear parabolic equations and its application to hermitean methods. In 'Topics in Numerical Analysis III', ed. J.H. Miller, 293-316. New York-London: Academic Press 1977.
NEVANLINNA, O.: On the numerical integration of nonlinear initial value problems by linear multistep methods. BIT 17, 58-71 (1979).
NEVANLINNA, O., ODEH, F.: Multiplier techniques for linear multistep methods. Numer. Funct. Anal. Optimization 3, 377-423 (1981).
NORSETT, S.P.: A criterion for A(α)-stability of linear multistep methods. BIT 9, 259-263 (1969).
ORTEGA, J.M., RHEINBOLDT, W.C.: Iterative Solution of Nonlinear Equations in Several Variables. New York-London: Academic Press 1970.
PRZEMIENIECKI, J.S.: Theory of Matrix Structural Analysis. New York: Mc Graw-Hill 1968.
RIESZ, F., NAGY, B.: Lecons d'Analyse Fonctionelle. Budapest: Akademiai Kiado 1952.
SERBIN, St.M.: On a fourth order unconditionally stable scheme for damped second-order systems. Comput. Methods Appl. Mech. Eng. 23, 333-340 (1980).
STETTER, H.J.: Analysis of Discretization Methods for Ordinary Differential Equations. Berlin-Heidelberg-New York: Springer 1973.
STOER, J., BULIRSCH, R.: Introduction to Numerical Analysis. Berlin-Heidelberg-New York: Springer 1980.
STRANG, G., FIX, G.J.: An Analysis of the Finite Element Method. Englewood Cliffs, N.J.: Prentice-Hall 1973.
STREHMEL; K., WEINER, R.: Adaptive Nyström-Runge-Kutta Methoden für gewöhnliche Differentialgleichungen. Computing 30, 35-47 (1983).
Van der Houwen, P.J.: Construction of Integration Formulas for Initial Value Problems. Amsterdam: North-Holland 1977.
Van der Houwen, P.J.: Stabilized Runge-Kutta methods for second order differential equations without first derivatives. SIAM J. Numer. Anal. 16, 523-537 (1979).
VARGA, R.S.: Matrix Iterative Analysis. Englewood Cliffs, N.J.: Prentice-Hall 1962.
WAHLBIN, L.: A modified Galerkin procedure with Hermite cubics for hyperbolic problems. Math. Comput. 29, 978-984 (1975).

WANNER, G., HAIRER, E., NORSETT, S.P.: Order stars and stability theorems. BIT 18, 475-489 (1978a).
WANNER, G., HAIRER, E., NORSETT, S.P.: When I-stability implies A-stability. BIT 18, 503 (1978b).
WHEELER, M.F.: L_∞ estimates of optimal orders for Galerkin methods for one-dimensional second order parabolic and hyperbolic equations. SIAM J. Numer. Anal. 10, 908-913 (1973).
WIDLUND, O.B.: A note on unconditionally stable linear multistep methods. BIT 7, 65-70 (1967).
Zlamal, M.: Unconditionally stable finite element schemes for parabolic equations. In 'Topics in Numerical Analysis II', ed. J.H. Miller, 253-261. New York-Londondon: Academic Press 1974.
Zlamal, M.: Finite element multistep methods for parabolic equations. In 'Finite elemente und Differenzenverfahren', ed. L. Collatz und J. Albrecht, 177-186. ISNM Bd. 28. Basel: Birkhäuser-Verlag 1975.
Zlamal, M.: Finite element multistep discretizations for parabolic boundary value problems. Math. Comput. 29, 350-359 (1975).

Glossary of Symbols

iff := if and only if;

$\mathbb{N}$ set of positive integers, $\mathbb{R}^m$ set of real m-tuples, $\mathbb{C}^m$ set of complex m-tuples, $\mathbb{N}_0 = \mathbb{N} \cup \{0\}$, $\mathbb{R}^+ = \{x \in \mathbb{R},\ x > 0\}$, $\overline{\mathbb{C}} = \mathbb{C} \cup \{\infty\}$;

$(a, b] = \{x \in \mathbb{R},\ a < x \leq b\}$,

$X \oplus Y = \{x + y,\ x \in X,\ y \in Y$ when $X \cap Y = \emptyset\}$;

$\mathring{S}$ interior of the set $S \subset \mathbb{C}$, $\overline{S}$ closed hull of S, ∂S boundary of S, $\mathbb{C} \setminus S$ complement of S in $\mathbb{C}$;

$\Theta = \partial/\partial t$, τ: $y(t) \to y(t + \Delta t)$ shift operator;

$\mathrm{Re}\,\eta$ (or $\mathrm{Re}(\eta)$) real part of $\eta \in \mathbb{C}$, $\mathrm{Im}\,\eta$ imaginary part of $\eta \in \mathbb{C}$, $\arg(re^{i\phi}) = \phi$, $r > 0$, $0 \leq \phi < 2\pi$, $[\eta]$ largest integer not greater than $\eta \in \mathbb{R}$, $\mathrm{sgn}(\eta)$ sign of $\eta \in \mathbb{R}$;

$\deg(\sigma(\eta))$ degree of the polynomial $\sigma(\eta)$;

$\|x\|$ arbitrary vector norm, $\|x\|_p = (\sum_{\mu=1}^m |x_\mu|^p)^{1/p}$, $1 \leq p \leq \infty$, $|x| = \|x\|_2$, $x \in \mathbb{C}^m$;

$\|A\| = \max_{\|x\|=1} \|Ax\|$, Sp(A) set of the eigenvalues of A, spr(A) spectral radius of A, det(A) determinant of A, $\mathrm{Re}(A) = (A + A^H)/2$, A (m,m)-matrix;

the superscript T stands for 'transpose', the superscript H stands for 'conjugate transpose';

$P \geq Q \Leftrightarrow x^H(P - Q)x \geq 0\ \forall\ x \in \mathbb{C}^m \Leftrightarrow \mathrm{Re}(P - Q)$ positiv semidefinit, P,Q (m,m)-matrices;

$C^p(\mathbb{R};\mathbb{R}^m)$ set of p-times continuously differentiable functions $f: \mathbb{R} \to \mathbb{R}^m$,

$|||f|||_n = \max_{0 \leq t \leq n\Delta t} |f(t)|$, $f: \mathbb{R} \to \mathbb{R}^m$, $n \in \mathbb{N}$, $\Delta t > 0$;

For symbols used only in Chapter VI see Section 6.1.

Subject Index